Springer Protocols Handbooks

More information about this series at http://www.springer.com/series/8623

Arun K. Shukla
Editor

Chemical and Synthetic Approaches in Membrane Biology

Humana Press

Editor
Arun K. Shukla
Indian Institute of Technology
Kanpur, India

ISSN 1949-2448 ISSN 1949-2456 (electronic)
Springer Protocols Handbooks
ISBN 978-1-4939-6835-0 ISBN 978-1-4939-6836-7 (eBook)
DOI 10.1007/978-1-4939-6836-7

Library of Congress Control Number: 2017942745

Printed on acid-free paper

This Humana Press imprint is published by Springer Nature
The registered company is Springer Science+Business Media LLC
The registered company address is: 233 Spring Street, New York, NY 10013, U.S.A.

Preface

Cells in our body are surrounded by lipid membranes that act as selectivity barriers for the flow of signals and messages from outside to cell interior and vice-versa. Biological membranes have been a focal point of intense research over the years in terms of understanding their composition and contribution to specific cellular functions. Over the last decade or so, many novel tools and techniques have been developed to directly probe the physico-chemical properties, overall architecture, and network of interactions that determine the basic features of biological membranes. Especially, the interaction of proteins embedded in the membranes with their surrounding lipid environment and how these interactions shape the membrane structure and function is at the center stage of membrane biology research. Development and optimization of streamlined protocols to assemble and reconstitute synthetic membrane-like platforms in vitro for direct investigation of lipid-protein interactions in a defined setting has provided many meaningful insights in the area of membrane research. Moreover, directly probing the structure, function, and dynamics of proteins embedded in the biological membranes such as receptor, ion channels, and transporters has emerged as one of the most productive research areas in modern biology. Considering that integral membrane proteins represent the largest class of drug targets, research in this area has direct implications for novel therapeutics for many different human diseases. In this volume of Springer Protocols entitled *Chemical and Synthetic Approaches in Membrane Biology*, we bring together a broad range of topics related to membrane biology research with particular emphasis on novel approaches, technology platforms, and emerging tools in this area.

The first three chapters in this book pertain to artificial or designer membrane mimetics that can be utilized for in vitro studies of membrane-lipid interactions. Reed and coworkers describe a protocol for preparing and characterizing lipid-coated gold nanoparticles and how they can be used to generate hybrid lipid membranes for investigating protein-membrane interactions. Roemer and coauthors elegantly present the assembly protocol of giant unilamellar vesicles as an approach to study novel paradigms of endocytosis in a minimal biomimetic membrane environment. Surface patterning with supported lipid bilayers has been gaining momentum lately where photolithographic and soft lithographic methods have made it possible to pattern different types of biomolecules for wide ranging applications. Carrer and Sanchez present a streamlined protocol for generating lipid bilayer-based micropattern using one-photon lithography and microcontact printing.

The next nine chapters in the book cover the approaches and methodologies to directly investigate membrane protein structure, localization, and dynamics. Mahalakshmi and colleagues provide a comprehensive discussion of the commonly utilized approaches to extract and purify transmembrane β-barrel proteins and also touch upon the methods for efficient refolding and biophysical studies of these proteins. Carla and coauthors describe a method using total internal reflection fluorescence microscopy to investigate the trafficking of a glutamate transporter, an approach that can be potentially extended to other integral membrane proteins. Janesh Kumar and colleagues present an overview of a fluorescence-based technique that is being widely utilized these days for rapid, efficient, and relatively inexpensive screening of membrane proteins in terms of construct engineering and membrane extraction.

High-resolution imaging of membrane proteins in their native environment is a very important aspect in membrane protein research. Francisco Barrantes provides a detailed protocol for imaging synaptic proteins using single-molecule super-resolution microscopy. The steps described here for fixing, handling, and imaging primary culture of hippocampal neuronal cells as a case example should be transferrable to other neuronal cultures.Quantum dots have emerged as a wonderful tool not only for probing molecular dynamics and imaging but also for profiling at single-molecule level and in the context of large cellular population. Rosenthal and coworkers present a layout for using the quantum dot tool box to investigate localization and trafficking patterns of neurotransmitter transporters and discuss specific advantages associated with the use of quantum dots over other methods.

Post-translational modifications and ligand-induced trafficking of membrane proteins are critical regulatory steps in their functional targeting and downregulation, respectively. Hang and Yuan introduce various types of protein fatty-acylation, their roles in membrane interaction, activity, and trafficking followed by a step-by-step protocol for monitoring protein fatty-acylation via metabolic labeling protocol using chemical reporters. Ikeda and Kumagai present a Halo Tag-based pulse-chase labeling platform for simple and versatile assessment of GPCR endocytosis. They also discuss the potential advantages of this approach over conventional methods to monitor agonist-induced GPCR internalization.

High-quality antibodies are one of the most coveted tools in the area of membrane protein research, and phage display-based methodologies are emerging as a robust platform for antibody generation. Uysal and Kossiakoff describe a strategy that has been successfully used for generating synthetic antibody fragments against membrane protein targets. This technology shows great promise in yielding synthetic antibodies that are wonderful tools not only for structural purposes but also for molecular, cell biological, and therapeutic applications. Despite recent advances in the area of membrane protein production and isolation, NMR-based studies on membrane protein targets still remain a technically challenging task. In the final chapter of this book, Brown and coauthors describe a strategy that they have developed and successfully implemented to prepare sensory rhodopsin in native membrane environment and subsequently use it for solid-state NMR analysis.

I take this opportunity to sincerely thank Springer's production team who did a wonderful job of getting the chapters ready for publication. I hope that you enjoy the topics covered in this volume and find it useful in your own research endeavors. Please feel free to share your feedback with me.

Department of Biological Sciences and Bioengineering
Indian Institute of Technology, Kanpur 208016 India

Arun K. Shukla

Contents

Contents

Contributors

Danielle M. Bailey Department of Chemistry, Vanderbilt University, Nashville, TN, USA

Department of Pharmacology, Vanderbilt University, Nashville, TN, USA

Department of Interdisciplinary Materials Science, Vanderbilt University, Nashville, TN, USA

Francisco J. Barrantes Laboratory of Molecular Neurobiology, Institute of Biomedical Research (BIOMED), UCA-CONICET, Buenos Aires, Argentina

Leonid S. Brown Department of Physics and Biophysics Interdepartmental Group, University of Guelph, Guelph, ON, Canada

Yuheng Cai Department of Chemistry, University of Colorado Denver, Denver, CO, USA

Dolores C. Carrer Instituto de Investigación Médica Mercedes y Martín Ferreyra (INIMEC), CONICET-Universidad Nacional de Córdoba, Córdoba, Argentina

Eliana S. Di Cairano Laboratory of Molecular and Cellular Physiology, Department of Pharmacological and Biomolecular Sciences, Università degli Studi di Milano, Milan, Italy

Surbhi Dhingra National Centre for Cell Science, Pune, Maharashtra, India

M. Florencia Sánchez Instituto de Investigación Médica Mercedes y Martín Ferreyra (INIMEC), CONICET-Universidad Nacional de Córdoba, Córdoba, Argentina

Desmond J. Hamilton Department of Chemistry, University of Colorado Denver, Denver, CO, USA

Howard C. Hang Laboratory of Chemical Biology and Microbial Pathogenesis, The Rockefeller University, New York, NY, USA

Yuichi Ikeda Department of Cardiovascular Medicine, Graduate School of Medicine, The University of Tokyo, Tokyo, Japan

Rupinder Kaur Department of Chemistry, University of Colorado Denver, Denver, CO, USA

Anthony Kossiakoff Faculty of Medicine, Bezmialem Vakif University, Istanbul, Turkey

Oleg Kovtun Department of Chemistry, Vanderbilt University, Nashville, TN, USA

Hidetoshi Kumagai Department of Cardiovascular Medicine, Graduate School of Medicine, The University of Tokyo, Tokyo, Japan

Department of Advanced Clinical Science and Therapeutics, The University of Tokyo, Tokyo, Japan

Janesh Kumar Wellcome Trust/DBT India Alliance, Hyderabad, India

National Centre for Cell Science, Pune, Maharashtra, India

Jyoti Kumari National Centre for Cell Science, Pune, Maharashtra, India

Vladimir Ladizhansky Department of Physics and Biophysics Interdepartmental Group, University of Guelph, Guelph, ON, Canada

Marilyn R. Mackiewicz Department of Chemistry, Portland State University, Portland, OR, USA

Josef Madl Faculty of Biology, Albert-Ludwigs-University Freiburg, Freiburg, Germany

Centre for Biological Signalling Studies (BIOSS), Albert-Ludwigs-University Freiburg, Freiburg, Germany

Freiburg Center for Interactive Materials and Bioinspired Technology (FIT), Albert-Ludwigs-University Freiburg, Freiburg, Germany

R. Mahalakshmi Wellcome Trust/DBT India Alliance, Hyderabad, India

Molecular Biophysics Laboratory, Department of Biological Sciences, Indian Institute of Science Education and Research, Govindpura, Bhopal, India

Paola Marciani Laboratory of Molecular and Cellular Physiology, Department of Pharmacological and Biomolecular Sciences, Università degli Studi di Milano, Milan, Italy

Grant W. Marquart Department of Chemistry, Portland State University, Portland, OR, USA

Stefania Moretti Laboratory of Molecular and Cellular Physiology, Department of Pharmacological and Biomolecular Sciences, Università degli Studi di Milano, Milan, Italy

Carla Perego Laboratory of Molecular and Cellular Physiology, Department of Pharmacological and Biomolecular Sciences, Università degli Studi di Milano, Milan, Italy

Scott M. Reed Department of Chemistry, University of Colorado Denver, Denver, CO, USA

Winfried Römer Faculty of Biology, Albert-Ludwigs-University Freiburg, Freiburg, Germany

Centre for Biological Signalling Studies (BIOSS), Albert-Ludwigs-University Freiburg, Freiburg, Germany

Freiburg Center for Interactive Materials and Bioinspired Technology (FIT), Albert-Ludwigs-University Freiburg, Freiburg, Germany

Sandra J. Rosenthal Department of Chemistry, Vanderbilt University, Nashville, TN, USA

Department of Pharmacology, Vanderbilt University, Nashville, TN, USA

Department of Interdisciplinary Materials Science, Vanderbilt University, Nashville, TN, USA

Department of Chemical and Biomolecular Engineering, Vanderbilt University, Nashville, TN, USA

Department of Physics and Astronomy, Vanderbilt University, Nashville, TN, USA

Vanderbilt Institute of Nanoscale Science and Engineering, Vanderbilt University, Nashville, TN, USA

Materials Science and Technology Division, Oak Ridge National Laboratory, Oak Ridge, TN, USA

Lucas B. Thal Department of Chemistry, Vanderbilt University, Nashville, TN, USA

Serdar Uysal Faculty of Medicine, Bezmialem Vakif University, Istanbul, Turkey

Sarah Villringer Faculty of Biology, Albert-Ludwigs-University Freiburg, Freiburg, Germany

Centre for Biological Signalling Studies (BIOSS), Albert-Ludwigs-University Freiburg, Freiburg, Germany

Meaghan E. Ward Department of Physics and Biophysics Interdepartmental Group, University of Guelph, Guelph, ON, Canada

Xiaoqiu Yuan Laboratory of Chemical Biology and Microbial Pathogenesis, The Rockefeller University, New York, NY, USA

Lipid-Coated Gold Nanoparticles as Probes for Membrane Binding

Desmond J. Hamilton*, Yuheng Cai*, Rupinder Kaur, Grant W. Marquart, Marilyn R. Mackiewicz, and Scott M. Reed

Abstract

A synthetic protocol is described for preparing lipid-coated gold nanoparticles (AuNPs) of varying shape, size, and membrane composition. Using ultraviolet-visible spectroscopy (UV-Vis), the localized surface plasmon resonance (LSPR) of the AuNPs is used to examine nanoparticle stability. Highly spherical AuNPs are obtained by selectively etching octahedral nanoparticles with cyanide. Alkanethiols are used to anchor the membranes to the gold surface, creating hybrid lipid membranes. Cyanide etch studies are described that test membrane stability and coverage and are performed by monitoring changes to the LSPR. A strategy is provided for minimizing the amounts of excess lipids and thiols present while using minimal purification steps. The design strategy can be used to produce a library of stable, hybrid lipid-coated AuNPs for applications including drug delivery, imaging, and sensing. Methods for using these AuNPs to study protein-membrane interactions using UV-Vis spectroscopy and dynamic light-scattering (DLS) measurements are also described.

Keywords Beta amyloid (Aβ) peptides, C-reactive protein, Gold nanoparticles, Hybrid membranes, Protein-membrane interactions, Sensors

1 Introduction

Lipid membranes supported on two-dimensional surfaces, such as glass [1] or gold [2], as well as three-dimensional surfaces, such as silica [3] or gold [4–6] nanoparticles, have many applications. For example, membrane-coated nanoparticles have been developed as drug delivery/imaging agents [7–9]. Membranes on solid planar supports have been used as biomimetic platforms to study and understand complex protein-membrane interactions in biology [2, 7, 10–13]. On three-dimensional supports, the membrane conforms to the shape of the hard nanoparticle surface. Consequently, high

*These authors contributed equally to the work.

Arun K. Shukla (ed.), *Chemical and Synthetic Approaches in Membrane Biology*, Springer Protocols Handbooks, (2017) 1–16, DOI 10.1007/8623_2016_8, © Springer Science+Business Media LLC 2016, Published online: 25 October 2016 degrees of positive or negative curvature are accessible that mimic localized regions of membrane curvature found in biological cell membranes [14]. In addition, metal nanoparticles have unique optical and electronic properties that allow detection of protein-membrane interactions in solution [10, 11]. The LSPR of metal nanoparticles is sensitive to changes in the refractive index of its surroundings. Therefore, they can report on protein-membrane interactions without the use of labels on either the lipids or the protein.

Many methods describing the preparation of lipid-coated nanoparticles with a variety of size, shape, and membrane compositions have been reported and proceed either by the reduction of gold chloride in the presence of the lipids [4, 8, 14–18] or lipid coating of pre-synthesized AuNPs [8, 19]. In many of these examples, there is an excess of lipids not bound to the nanoparticle surface. This is problematic as free lipids can undergo lipid exchange with lipid-coated nanoparticles and can complicate binding studies involving the use of nanoparticles. Consequently, these nanoparticles must be purified, which can be challenging.

One approach to addressing membrane instability on surfaces is through the use of hydrophobic thiols to anchor membranes to the surface. We, and others, have demonstrated that hydrophobic thiols form hybrid bilayer membranes on both planar and nanoparticle-based surfaces that are more stable [4, 19–23].

Here we report a protocol for the preparation of hybrid lipid-coated AuNPs that can be applied to any size and shape AuNP. The design strategy controls the amount of lipids and other components present during the coating process. Multiple approaches for the preparation of hybrid lipid-coated AuNPs are included. In one example, uniform spherical AuNPs up to 100 nm diameter are obtained from octahedral AuNPs. In another, citrate-capped AuNPs are used directly as a simpler route to smaller spherical hybrid lipid-coated AuNPs. In order to increase the affinity of lipids to the AuNPs, a method for firmly attaching the membrane to the gold surface using hydrophobic alkanethiols as anchoring groups is described. A protocol that optimizes the amount of lipids, alkanethiols, and other components during the coating process is also described. Transmission electron microscopy (TEM) is used to characterize the size and shape of the AuNPs, while dynamic light scattering (DLS) and UV-Vis are used to probe protein binding to the lipid-coated AuNPs as well as examine their stability. This protocol allows researchers to produce a library of hybrid lipid-coated AuNPs.

2 Materials

2.1 Synthesis of AuNPs

Ethylene glycol (Oakwood Chemical, Estill, SC), poly(diallyldimethylammonium chloride) (Sigma Aldrich, St. Louis MO), phosphoric acid (J.T. Baker, Center Valley, PA), hydrogen tetrachloroaurate (III) (STREM Chemicals, Newburyport, MA), sodium cyanide (Alfa Aesar, Ward Hill, MA), FCF300-Cu formvar carbon film on 300 mesh copper grids (Electron Microscopy Sciences, Hatfield, PA).

2.2 Lipids and Other Agents for Assembly of Coated AuNPs

Sodium oleate (TCI AMERICA, Portland, OR), L-α-phosphatidylcholine (PC) 30% (Sigma Aldrich, St. Louis MO), L-α-phosphatidylcholine 95%, 1,2-dilauroyl-*sn*-glycero-3-phosphocholine (DLPC), and 1,2-dioleoyl-*sn-glycero*-3-[phospho-L-serine] (Avanti Polar Lipids, Alabaster, AL), Chloroform (BDH, VWR, West Chester, PA), activated alumina (EM science, Gibbstown, NJ), propanethiol (Acros Organics, Morris Plains, NJ), hexanethiol (Alfa Aesar, Ward Hill, MA), decanethiol (Alfa Aesar, Ward Hill, MA), cholesterol (Sigma Aldrich), hydrogenated phosphatidylcholine (Avanti Polar Lipids, Alabaster, AL). Phosphate buffered saline (PBS) (Amresco, Solon, OH): sodium phosphate dibasic anhydrous, potassium phosphate monobasic, potassium chloride, sodium chloride. PIPES buffer: piperazine-*N,N'*-bis(2-ethanesulfonic acid), calcium chloride, sodium chloride. HEPES buffer: 4-(2-hydroxyethyl)-1-piperazineethanesulfonic acid.

2.3 Protein-Binding Studies with Lipid-Coated AuNPs

Human C-reactive protein (Academy Biomedical Company, Houston, TX), Hexafluoroisopropanol (HFIP) (Sigma Aldrich). $A\beta_{1-42}$ (lyophilized powder stored at −20°C in HFIP) (Anaspec, Fremont, CA).

3 Methods

AuNPs are synthesized by reducing Au^{III} ions to Au^0 with a reducing agent such as sodium citrate or sodium borohydride in the presence of an appropriate capping agent [4, 5, 24, 25]. Furthermore, the size and shape of the nanoparticles can be tuned (and through that the optical and electronic properties altered) by changing the synthetic reaction conditions [26], such as the gold source (chloroauric acid or bromoauric acid), reducing agents [25], temperature, solvents [4, 5], and additives used [18, 27]. Here we describe a method for preparing ultrasmooth spherical AuNPs where lipid membranes coat the surface and allow for the study of protein-membrane interactions.

3.1 Route 1: Synthesis of Ultrasmooth, Spherical AuNPs from Octahedral AuNPs

3.1.1 Synthesis of Octahedral Gold Nanoparticles

Uniform octahedral AuNPs are synthesized using modifications to published methods where the size of the octahedral AuNPs can be manipulated by using varied concentrations of acid [28, 29]. Charge a 250 mL glass bottle with 200 mL of ethylene glycol, 4.00 g poly(diallyldimethylammonium chloride) and 8.00 mL of 1 M H_3PO_4 and stir for 15 min. Add 200 μL of 0.5 M $HAuCl_4$ in H_2O and stir for 15 min. Transfer 10 mL aliquots into 20 mL scintillation vials, add a stir bar to each, and seal with electrical tape.

Place sealed vials into a 197°C oil bath with a stir bar and stir the solution rapidly for 30 min. Remove the samples from the hot oil bath and allow them to cool while stirring. Uniform heating of the solutions is imperative to minimize the formation of polydisperse nanoparticles.

3.1.2 Conversion of Octahedral to Spherical Gold Nanoparticles by Etching

Using a modified etch procedure from the literature [28], convert octahedral AuNPs to spherical AuNPs with the addition of NaCN. The size of the spherical AuNPs produced is directly related to the edge length of the starting octahedral AuNPs. (1) Calculate the total number of Au^{III} atoms per mL of octahedral AuNPs used in the initial synthesis. (2) Experimentally determine the moles of CN^- required to etch octahedral AuNPs to spheres. Start by using the moles of CN- calculated using Eq. (1). Add that amount of CN^- to the unpurified octahedral AuNP solution and stir rapidly for 20 h. Increase the amount of cyanide sequentially until spheres are obtained. Start by using the moles of CN^- calculated using Eq. (1) (*see* **Note 1**). (3) After 20 h of stirring, purify the spherical AuNPs using four rounds of centrifugation and sonication. Centrifuge the AuNPs until the supernatant is clear, remove the supernatant, resuspend the pellet in 1 mL Milli-Q H_2O, and sonicate in a bath sonicator for 10 min. (4) Obtain UV-Vis spectra of the AuNPs before and after etching and of the purified AuNPs (Fig. 1). Examine the UV-Vis spectra and note if a shift of the LSPR band for the octahedral AuNPs from longer (Fig. 1, blue trace) to shorter wavelengths (Fig. 1, red trace) is observed after CN^- etching and purification (Fig. 1, green trace). (5) Perform TEM (FEI Tecnai G2 TEM with a Gatan Ultrascan 1000 digital camera) with AuNPs before and after CN^- etching of octahedral-shaped AuNPs to confirm shape conversion. Count the number of particles in the TEM images and obtain a ratio of octahedral to spherical AuNPs. Octahedral AuNPs are expected to have distinguishable vertices and edges (Fig. 2a), while AuNPs after CN^- etching will have a

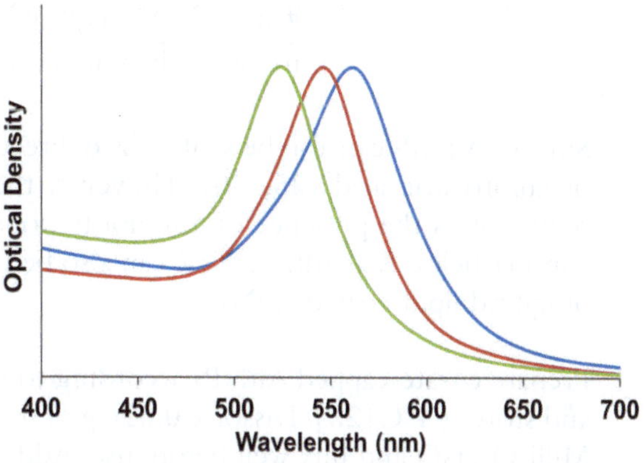

Fig. 1 Representative UV-Vis spectra of octahedral AuNPs converted to spherical AuNPs. Traces are of unpurified octahedral AuNPs (*blue*), purified x4 octahedral AuNPs (*red*), and purified x4 spherical AuNPs (*green*)

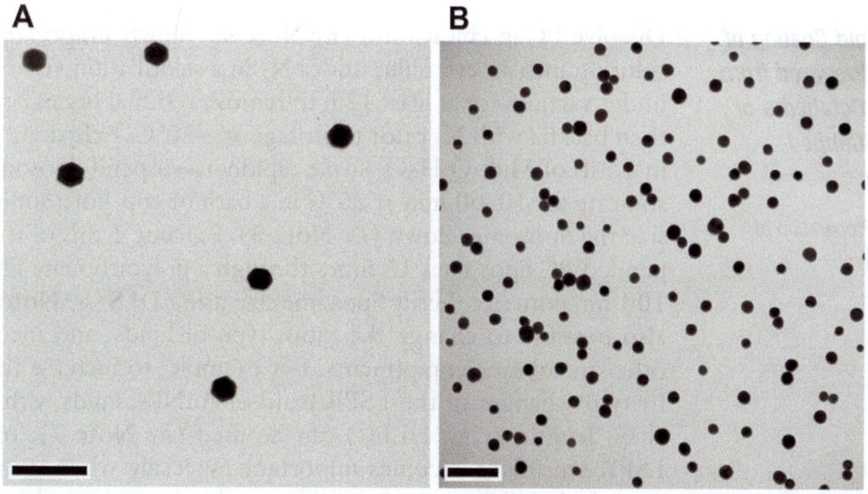

Fig. 2 Representative TEM images of spherical AuNPs before (**a**) and after (**b**) etching octahedron AuNPs. Scale bar = 200

rounded morphology (Fig. 2b). If the TEM images show greater than 5% of octahedral-shaped AuNPs compared to spherical AuNPs, more CN^- is required.

Equation 1: *Moles of* CN^- *to add when etching octahedron AuNPs to spherical AuNPs.*

$$\frac{\left[\left(\frac{\#\text{ of Au ions used in synthesis per mL}}{\frac{\frac{\sqrt{2}}{3}((a\text{ nm})^3)}{\frac{(0.408)^3}{4}}}\right)\left(\frac{\left(\frac{\sqrt{2}}{3}((a\text{ nm})^3)\right)-\left(\frac{4\pi a^3}{3(\sqrt{6})^3}\right)}{\left(\frac{(0.408)^3}{4}\right)}\right)\times(2)\right]}{6.022\times10^{23}}$$

$$=$$

$$\frac{\#\,\mathrm{mol\,of\,CN^-\,required}}{1\,\mathrm{mL\,of\,octahedral\,solution}}$$

3.2 Route 2: Preparation of Citrate-Capped Spherical AuNPs

Smaller AuNPs can also be synthesized directly by citrate reduction of chloroauric acid [25, 30]. However, the drawback to citrate reduction is the presence of a minority population of asymmetric nanoparticles. Regardless, these can also be used for the formation of hybrid lipid-coated AuNPs.

3.2.1 Sodium Oleate-Capped AuNPs

Prepare citrate-capped AuNPs according to the Turkevich method and store at 4°C [25]. Dissolve 0.016 g of sodium oleate in 5 mL of Milli-Q H_2O and mix well before use. Add 5 μL of sodium oleate (10 mM) in H_2O to 1 mL of AuNPs (Optical Density (OD) = 0.7) and stir for 30 min to produce sodium oleate-coated AuNPs prior to liposome encapsulation (*see* **Note 2**). Protect the oleate solution from light by wrapping the tube with aluminum foil (*see* **Note 3**).

3.3 Lipid Coating of AuNPs Prepared from Etched Octahedra or Citrate-Capped Spheres

3.3.1 Preparation of Liposomes

Dissolve PC in chloroform (*see* **Note 4**). Slowly evaporate the PC solution into an even film under N_2 in a scintillation vial and place under vacuum for at least 12 h to remove residual organic solvents, then backfill with N_2 prior to storage at −20°C. Rehydrate the film in 1 mL of Milli-Q H_2O, shake rapidly to suspend liposomes, and sonicate for 10–60 min at 25°C in a bath or cup-horn sonicator to size the liposomes down (*see* **Note 5**). Extrude 1 mL of the resuspended PC liposomes 15 times through a polycarbonate filter with 100-nm pore size. Verify liposome size using DLS (*see* **Note 6**). It is also possible to change the ratio, type of lipids, and incorporate other membrane components. For example, to increase the sensitivity to changes of the LSPR band of AuNPs, lipids with shorter chain lengths (e.g., DLPC) can be used (*see* **Note 7**). Increased LSPR sensitivity becomes important especially when using lipid-coated AuNPs as nanosensors to detect and study protein-membrane interactions (Sect. 3.7). Since the refractive index change of AuNPs increase near the surface of AuNPs, thinner lipid coating is expected to increase the sensitivity of the nanosensor system. It is also possible to modify the membrane compositions to include cholesterol, PS, or hydrogenated PC (*see* **Note 8**).

3.3.2 Preparation of Alkanethiol Solution

Prepare a 1 mM alkanethiol solution. Shorter alkanethiols such as propanethiol can be prepared by dissolving 1 μL of neat propanethiol in 10 mL of Milli-Q H_2O that is degassed with N_2 to provide a 1 mM stock. Longer-chain thiols that are more hydrophobic, such as hexanethiol or decanethiol, are prepared in ethanol degassed with N_2 (*see* **Note 9**).

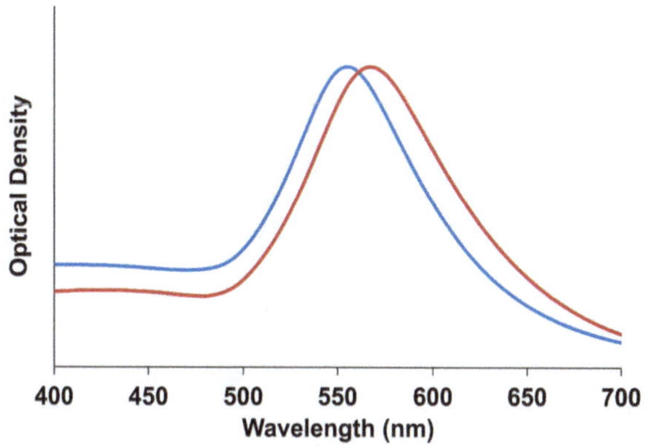

Fig. 3 Representative UV-Vis spectra of purified spherical AuNPs without hybrid lipid membranes (*blue*) and spherical AuNPs with hybrid lipid membranes (*red*)

3.3.3 Coating of Hybrid Lipid Membranes on AuNPs

Add the optimized concentration of PC liposomes (*see* Sect. 3.4.2 for optimization method) to the spherical AuNPs (Fig. 3, blue trace) (obtained from either Sects. 3.1.2 or 3.2.1), stir for 30 min, and monitor the LSPR (*see* **Note 10**). Add the optimal amount of alkanethiol (*see* Sect. 3.4.1 for optimization method) to this solution, stir for 1 h, and monitor LSPR band (Fig. 3, red trace). If the LSPR band stops shifting within 30 min of alkanethiol addition, the coating process is finished. However, if CN^- etch tests show a blue shift in the LSPR band or decrease in OD, then too little thiol has been added.

3.4 Optimization of PC and Alkanethiol Concentrations for Complete Encapsulation of AuNPs by Liposomes

The optimal concentration of liposomes and thiols is determined using cyanide stability assays. The optimal concentration of alkanethiols is determined before optimizing the lipids, as this provides a more efficient route to assessing the concentration of membrane components that lead to complete AuNP coverage and stability. This approach can be used to optimize coating of any size or shape nanoparticle.

3.4.1 Determine the Optimal Concentration of Alkanethiols

Purify the spherical AuNPs as described in Sect. 3.1.2 with four rounds of centrifugation and sonication. Next, for the purpose of alkanethiol optimization, add an excess amount of lipids to coat 1 mL of 40 nm spherical AuNPs (OD = 0.5) based on a calculation of the surface area and the assumption that a single lipid occupies 0.72 nm^2 [31, 32], stir, and monitor the LSPR band by UV-Vis spectroscopy. To this AuNP solution, add the alkanethiol stock solution (1 mM) in serial additions (start with 1–5 μL each addition). Incubate the lipid-coated AuNP solution with NaCN (26 μL of 0.1 M in Milli-Q H_2O) for 12 h and take a UV-Vis spectrum.

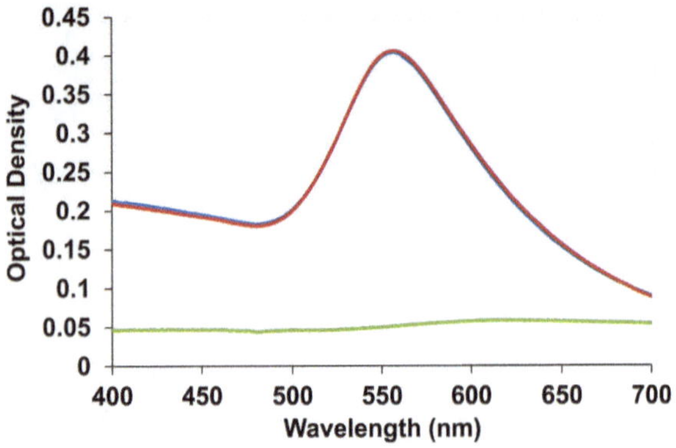

Fig. 4 Representative UV-Vis spectra of lipid-coated spherical AuNPs before addition of NaCN (*blue*), 12 h after addition of NaCN (*red*), and spherical AuNPs without lipid coating 12 h after addition of NaCN (*green*)

Examine the LSPR band before and after CN^- addition. If no LSPR change is observed, the surface of the AuNPs are coated with enough alkanethiol and lipids to form a complete hybrid membrane that is stable to cyanide etch (*see* **Note 11**). This cyanide etch test is important as it confirms that there are no bare patches of gold surface available for proteins to bind to instead of the membrane surrounding the gold core.

3.4.2 Optimal Concentration of Liposomes

As a starting point, add various amounts of PC liposomes (0, 0.10, 0.20, 0.30,. . .1.0 μmoles of 10 mM) to separate 1 mL aliquots of purified spherical AuNPs (OD between 0.5 and 1.0) and stir for 30 min. Add the experimentally determined amount of alkanethiol (Sect. 3.4.1), stir for 1 h, and take a UV-Vis spectrum. To each solution add 26 μL of NaCN (0.1 M in H_2O), stir for at least 12 h, and take another UV-Vis spectrum. Overlay and compare the two spectra, and determine if there is a change in OD or shift in LSPR band. No change in the OD or shift in LSPR demonstrates CN^- impermeability and success in fully encapsulating the AuNP with the membrane (Fig. 4, blue and red traces). If a decrease in the OD and change in color from red to colorless are observed, these changes indicate the presence of bare patches of gold surface or membrane permeability where CN^- ions are able to etch the gold surface transforming Au^0 to Au^I (Fig. 4, green trace).

3.5 Nanoparticle Purification

As prepared, nanoparticles are often heterogeneous in size and shape. Furthermore, synthesis and coating procedures often result in unbound ligands. For these reasons, purification is often required. For example, lipid-coated nanoparticles can be separated from free liposomes, nanoprisms [27] can be separated from spheres, and octahedra can be separated from spheres. Therefore, to prepare monodisperse nanoparticle samples of one size or shape free of excess liposomes, use the sucrose gradient method to purify the AuNPs.

3.5.1 Sucrose Gradient to Purify Nanoparticles

a. Add 2 mL each of 80% sucrose (w/w), 70% sucrose, 60% sucrose, 50% sucrose, and 40% sucrose to a 15 mL centrifuge tube to make a gradient with the highest concentration on the bottom of the tube. Add 5 mL of crude AuNPs (OD = 0.9) on top of the sucrose gradient slowly to minimize disturbance of the boundary.

b. Centrifuge the sample at 2000*g* for 2 h at 4°C and 1 h at 20°C. Upon centrifugation, the crude nanoparticles will spread across the sucrose gradient as pink or purple bands. The larger AuNPs settle at the bottom of the centrifuge tube, while most of the nanospheres with smaller sizes (pink) remain in top sucrose fractions between 40% and 70%. Retrieve the desired AuNPs using a pipette (*see* **Note 12**).

c. Transfer the collected pellets to a 1.5 mL Eppendorf tube. Dilute the pellets to 1.5 mL with Milli-Q H_2O, and centrifuge the solution at 5223*g* for 10 min. Remove the clear supernatant from the pellet, and repeat the centrifugation five times to reduce the sucrose concentration to less than 0.3% (*see* **Note 13**). Resuspend the nanoparticles in 1 mL of Milli-Q H_2O (*see* **Note 14**).

3.6 Nanoparticle Characterization

3.6.1 Nanoparticle Characterization by UV-Vis Spectroscopy

UV-Vis spectroscopy is a convenient method to examine the optical and electronic properties of AuNPs that are governed by the size and shape of the nanomaterials. In addition, it is a good technique for evaluating if there is complete membrane coverage of the hybrid lipid-coated AuNPs [14, 20, 33]. Use UV-Vis spectroscopy to determine if octahedral particles have been etched as shown in Fig. 1, to monitor the construction of the hybrid lipid membrane as shown in Fig. 3, and to determine if the hybrid lipid membrane forms an ion-impermeable membrane as shown in Fig. 4 (*see* **Note 15**). UV-Vis can also be used to determine the size and shape of the AuNPs [34].

3.6.2 Nanoparticle Characterization by TEM

TEM is an efficient method for evaluating the size and morphology of nanomaterials. OD of samples used for TEM deposition is between 0.5 and 1.0.

a. *Method 1*: Drop 3 μL of AuNP solution onto a piece of parafilm. Place the grid with the carbon-coated (darker) side upside-down on the AuNP droplet using fine-tipped tweezers. Incubate for 3 min, and wash the grid with H_2O droplet to remove excess AuNPs. Blot the additional H_2O off the grid using the edge of filter paper and dry overnight before TEM data collection (*see* **Note 16**).

b. *Method 2*: Drop 5 μL of the AuNP solution onto the carbon-coated copper grid, let sample sit for 5 min, then "wick off" excess material with the edge of a piece of filter paper, or allow the H_2O to evaporate (*see* **Note 17**). Repeat twice before the samples are allowed to dry in open air overnight before TEM data collection.

3.6.3 Nanoparticle Characterization by DLS

DLS is an efficient method to determine the hydrodynamic diameter of the AuNPs upon coating, and changes in size can be used to detect protein binding. This method is most reliable for spherical nanoparticles than for nanoparticles that are anisotropic. To measure the hydrodynamic diameter of membrane-coated nanomaterials, warm up the DLS instrument 1 h before use. Prepare solutions of AuNPs with an OD between 0.2 and 1.0. The OD of the AuNP solutions will affect the DLS data quality. That is, if the OD of the AuNPs is too low or high, this will result in a decrease or increase in the amount of scattered light by the nanoparticles as recorded by DLS and an inaccurate assessment of the actual hydrodynamic diameter in solution.

3.7 Protein-Binding Studies with Hybrid Membrane-Coated AuNPs

Here we describe the use of hybrid lipid-coated AuNPs to detect and study protein-membrane interactions. A few studies are described in detail below and include interactions with C-reactive protein (CRP) and in Beta amyloid (Aβ) peptides. CRP is an inflammatory cardiovascular biomarker that is linked to stroke, death, and myocardial infarction [35]. It is known to bind to lipids through a calcium-bridging mechanism [36, 37] and to low-density lipoprotein particles, enhancing their uptake by macrophages and the formation of foam cells [38]. Aβ is a peptide found in high concentrations as oligomers or fibrils in the brain and is thought to play a significant role in Alzheimer's disease [39–42]. While the physiochemical properties of Aβ oligomers are not well understood, they are known to permeate and disrupt membranes [41, 43], induce oxidative lipid damage, and cause lipid release in neurons [44, 45]. In vitro studies have demonstrated that Aβ binds to membranes with high cholesterol and glycosphingolipids content and causes sporadic cytosolic-calcium signaling in both astrocytes and neurons [46, 47]. Consequently, CRP and Aβ are excellent substrates for evaluating their interactions with biomimetic membrane platforms as described here.

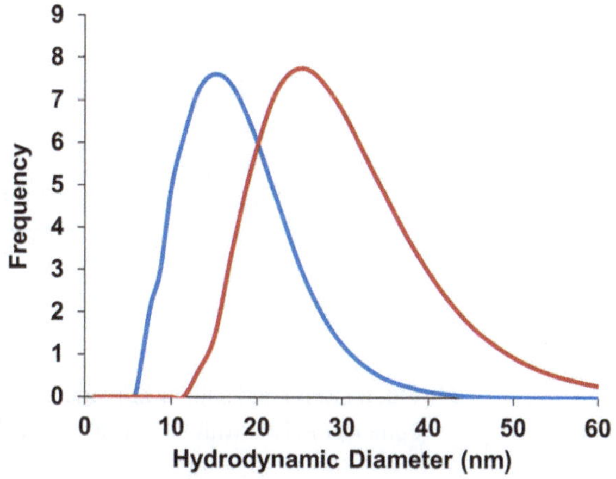

Fig. 5 Representative DLS distributions of the average hydrodynamic diameter AuNPs coated in PC and hexanethiol before (*blue*) and 15 min after the addition of 2-μM Aβ_{1-42} monomers (*red*) in 10-mM sodium phosphate buffer at pH 8.0

3.7.1 Aβ Peptides

Prepare Aβ oligomers by adding Aβ_{1-42} peptides (10 μL of 222 μM in HFIP) to 90 μL of 10 mM HEPES for pH 6.5 or 10 mM sodium phosphate pH 8.0. Incubate the Aβ_{1-42} peptide solutions at 4°C for 24 h to allow for oligomerization. Buffer the coated AuNPs with 10 μL of 1 M HEPES buffer at pH 6.5 or 10 μL of 1 M sodium phosphate buffer at pH 8.0. Add Aβ monomers or oligomers to the hybrid lipid-coated AuNP solutions to yield final concentrations of 2 μM for monomers or 2 μM for oligomers. Take DLS measurement in a quartz cuvette before and 15 min after the introduction of Aβ. A typical DLS result is shown in Fig. 5. If an increase in the hydrodynamic diameter is observed in the presence of lipid-coated AuNPs (Fig. 5, blue trace) and Aβ, this indicates the Aβ species is binding to the membranes encapsulating the gold core (Fig. 5, red trace).

3.7.2 C-Reactive Protein

Prepare 1 mL of hybrid lipid-coated AuNPs (OD = 0.7). Add 0.5 mL of PIPES buffer with Ca^{2+} (pH 6.4, 30 mM PIPES, 2 mM $CaCl_2$, 140 mM NaCl) to AuNPs and stir for 30 min. Add 5 μL of CRP (1 mg/mL) in the solution and stir for 30 min. For a control, use 0.5 mL of PIPES buffer without Ca^{2+} (pH 6.4, 30 mM PIPES, 140 mM NaCl) to buffer the solution. With Ca^{2+} in the solution (Fig. 6, red trace), addition of CRP causes a 3 nm red-shift of the LSPR band. In the control group (Fig. 6, blue trace),

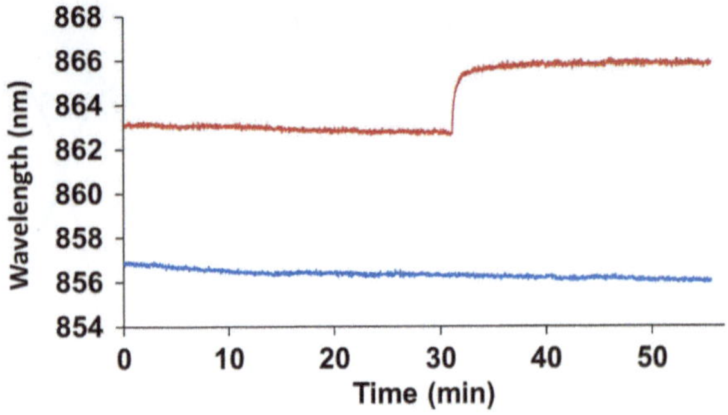

Fig. 6 LSPR band shift of CRP addition (at 30 min) to PC and hexanethiol-coated spherical AuNPs with Ca^{2+} (*red*) and without Ca^{2+} (*blue*)

without Ca^{2+} in the solution, addition of CRP does not show a significant shift of the LSPR band (Fig. 6).

4 Notes

1. Cyanide assists in the selective oxidation of Au^0 atoms at the vertices of octahedral AuNPs, and when an optimized concentration of NaCN is used, spherical AuNPs are formed. Experimentally, between 2 and 2.5 times, the calculated amount of CN^-, with Eq. (1), is required to produce mostly spherical AuNPs. Too little CN^- results in truncated octahedral AuNPs, while too much CN^- leads to misshaped spherical AuNPs. Equation (1) was developed by assuming a 100% yield of octahedral AuNPs from the initial Au^{III} used and the amount of Au^0 atoms that are required to be removed from an octahedral AuNP to form spherical AuNPs.
2. Sodium oleate addition is not required when using etched octahedral AuNPs.
3. The sodium oleate can be mixed by sonicator or vortex mixer to reduce bubbles formed during mixing and to ensure all the solid is dissolved. Sodium oleate solutions are stored under N_2 gas.
4. Chloroform is filtered through basic alumina to remove HCl impurities that promote decomposition of the lipids.
5. The liposome solution will appear cloudy indicating large multilamellar liposomes prior to sonication and will become transparent and clear after sizing to smaller liposomes.
6. Prepare the sample for DLS instrument (Zetasizer Nano S90) by adding 10 μL of 10 mM extruded liposomes to 1 mL of Milli-Q H_2O and mix well. Use liposome stocks within 5 days

of hydration to minimize oxidation present in liposome formulations.

7. Prepare DLPC film by dissolving 124 mg of DLPC in 20 mL of cyclohexane, and remove the solvent by slow evaporation using N_2 gas, dry the films under vacuum for 24 h, and store at −20°C under N_2 gas. Resuspend the lipid film in 1 mL of 10 mM HEPES buffer pH 6.5, sonicate the liposome stock solution for 1 h, and extrude 1 mL of liposome stock solution 15 times through a polycarbonate filter with 100-nm pore size. Verify the size of the resulting liposomes by DLS.
8. Membranes comprised of HyPC only are prepared by dissolving HyPC in $CHCl_3$ and using 42 μL of 21.6 mM to prepare the film. For mixed membrane compositions such as PC-cholesterol, these are prepared by dissolving PC and cholesterol in $CHCl_3$, which are then combined to form a thin film where the PC-cholesterol ratio is 1:1 ratio (25 μL of 21.6 mM PC: 25 μL of 21.6 mM cholesterol). The ratio, lipid type, and composition can all be varied. Other compositions can be a 1:1:1 ratio PC-cholesterol-sphingomyelin (50 μL of 21.6 mM PC: 50 μL of 21.6 mM cholesterol: 100 μL of 10.8 mM sphingomyelin). Once films are prepared, desiccate the lipid films under vacuum for 12 h to remove residual organic solvents that hinder liposome formation. Reconstitute all films with 2 mL 10 mM PBS buffer (~ pH 7.0), and sonicate lipid solution for 90 min at 25°C.
9. Alkanethiol stock solutions should be used within 24 h, as thiols oxidize readily.
10. Liposomes must be added first as the particles will aggregate if alkanethiols are added first.
11. An excess of alkanethiol, usually 10–20% of the experimentally determined optimal concentration, is used as this yields more consistent results when determining complete membrane encapsulation of AuNPs in order to achieve ion impermeability prior to protein binding studies.
12. Collect as little sucrose solution as possible to reduce the difficulty of purifying AuNPs from sucrose solution. Collect only the most AuNP concentrated band to limit the volume of sucrose solution. Regardless, there will be some loss of AuNPs during separation.
13. The concentration is calculated based on an assumption that the 80% sucrose solution collected is less than 0.5 mL. Then, every centrifugation after this will reduce 66% of the sucrose. Normally, five rounds of centrifugation are enough to bring the sucrose concentration down to below 0.3%.

14. The speed and time of centrifugation can be varied. An increase in speed can increase the recovery of gold nanoparticles but increase nanoparticle aggregation. To prevent aggregation of gold nanoparticles, decreasing centrifugation speed or time for each repeated centrifugation is a recommended method since the solution becomes less viscous as the sucrose concentration decreases.
15. When octahedral AuNPs are converted to spheres, a blue shift in LSPR band is observed; and when the hybrid lipid membrane encapsulates AuNPs, a red-shift in LSPR band is expected, as the refractive index at the surface of the AuNPs has increased. During CN^- stability assays, if a blue shift or loss in OD is observed, the membrane likely does not completely encapsulate the AuNPs.
16. To optimize the sample attachment to the grid, glow discharging can be applied to the grid before incubation. Also, the incubation can be longer than 3 min if the sample is dilute. To visualize the lipid membrane on the surface of the AuNPs, the lipid-coated AuNPs can be stained by uranyl acetate or ammonium molybdate, which should be prepared fresh before use.
17. The drop of a sample should cover the grid, and if static electricity is experienced upon drop casting then try the alternative method described in Sect. 3.6.2, *Method 1.*

Acknowledgments

This work was supported by a grant from the NIH (2R15GM088960-02) (to SMR) and the Collins Medical Trust (to MMR). We thank the Electron Microscopy Center in the University of Colorado Anschutz Medical Campus School of Medicine and Dorothy Dill for assistance in running the TEM. Support from The Wenner-Gren Foundation for SMR is acknowledged. We also appreciate input from many group members that assisted in the development of these protocols over many years including Min Wang, Heather Hodges, Reid Messersmith, and Benjamin Ayres.

References

1. Tanaka M, Sackmann E (2005) Polymer-supported membranes as models of the cell surface. Nature 437(7059):656–663. doi:10.1038/nature04164
2. Plant AL (1999) Supported hybrid bilayer membranes as rugged cell membrane mimics. Langmuir 15(15):5128–5135. doi:10.1021/la981662t
3. Baksh MM, Jaros M, Groves JT (2004) Detection of molecular interactions at membrane surfaces through colloid phase transitions. Nature 427(6970):139–141. doi:10.1038/nature02209
4. Mackiewicz MR, Ayres BR, Reed SM (2008) Reversible, reagentless solubility changes in phosphatidylcholine-stabilized gold

nanoparticles. Nanotechnology 19:115607. doi: 10.1088/0957-4484/19/11/115607

5. Brust M, Walker M, Bethell D, Schiffrin DJ, Whyman R (1994) Synthesis of thiol-derivatised gold nanoparticles in a two-phase liquid-liquid system. J Chem Soc Chem Commun 7:801–802. doi:10.1039/c39940000801
6. Hong K, Friend DS, Glabe CG, Papahadjopoulos D (1983) Liposomes containing colloidal gold are a useful probe of liposome-cell interactions. Biochim Biophys Acta 732(1):320–323. doi:10.1016/0005-2736(83)90220-1
7. Chhour P, Naha PC, O'Neill SM, Litt HI, Reilly MP, Ferrari VA, Cormode DP (2016) Labeling monocytes with gold nanoparticles to track their recruitment in atherosclerosis with computed tomography. Biomaterials 87:93–103. doi: 10.1016/j.biomaterials.2016.02.009
8. Kang JH, Ko YT (2015) Lipid-coated gold nanocomposites for enhanced cancer therapy. Int J Nanomed 10 Spec Iss:33–45. doi: 10.2147/IJN.S88307
9. Boisselier E, Astruc D (2009) Gold nanoparticles in nanomedicine: preparations, imaging, diagnostics, therapies and toxicity. Chem Soc Rev 38(6):1759–1782. doi: 10.1039/b806051g
10. Sackmann E (1996) Supported membranes: scientific and practical applications. Science 271(5245):43–48. doi: 10.1126/science.271.5245.43
11. Groves JT, Boxer SG (2002) Micropattern formation in supported lipid membranes. Acc Chem Res 35(3):149–157. doi:10.1021/ar950039m
12. Lahiri J, Kalal P, Frutos AG, Jonas SJ, Schaeffler R (2000) Method for fabricating supported bilayer lipid membranes on gold. Langmuir 16(20):7805–7810. doi:10.1021/la000468f
13. Love JC, Estroff LA, Kriebel JK, Nuzzo RG, Whitesides GM (2005) Self-assembled monolayers of thiolates on metals as a form of nanotechnology. Chem Rev 105(4):1103–1170. doi:10.1021/cr0300789
14. Wang MS, Messersmith RE, Reed SM (2012) Membrane curvature recognition by C-reactive protein using lipoprotein mimics. Soft Matter 8(30):7909–7918. doi:10.1039/c2sm25779c
15. He P, Urban MW (2005) Phospholipid-stabilized Au–nanoparticles. Biomacromolecules 6(3):1224–1225. doi:10.1021/bm0501961
16. Robertson D, Tiersch B, Kosmella S, Koetz J (2007) Preparation of crystalline gold nanoparticles at the surface of mixed phosphatidylcholine-ionic surfactant vesicles. J Colloid Interface Sci 305(2):345–351. doi:10.1016/j.jcis.2006.09.079
17. Marie-Edith M, Olivier L, Bernard D, Chrystel F (2006) Synthesis of stable, gold-particle-containing onion-type multilamellar vesicles. Influence of particle size on the onions' internal structure. Nanotechnology 17(5):1193. doi:10.1088/0957-4484/17/5/005
18. He P, Zhu X (2007) Phospholipid-assisted synthesis of size-controlled gold nanoparticles. Mater Res Bull 42(7):1310–1315. doi:10.1016/j.materresbull.2006.10.014
19. Levin CS, Kundu J, Janesko BG, Scuseria GE, Raphael RM, Halas NJ (2008) Interactions of ibuprofen with hybrid lipid bilayers probed by complementary surface-enhanced vibrational spectroscopies. J Phys Chem B 112(45):14168–14175. doi:10.1021/jp804374e
20. Mackiewicz MR, Hodges HL, Reed SM (2010) C-Reactive protein induced rearrangement of phosphatidylcholine on nanoparticle mimics of lipoprotein particles. J Phys Chem B 114(16):5556–5562. doi:10.1021/jp911617q
21. Knoll W, Köper I, Naumann R, Sinner E-K (2008) Tethered bimolecular lipid membranes—a novel model membrane platform. Electrochim Acta 53(23):6680–6689. doi:10.1016/j.electacta.2008.02.121
22. Yang JA, Murphy CJ (2012) Evidence for patchy lipid layers on gold nanoparticle surfaces. Langmuir 28(12):5404–5416. doi:10.1021/la300325p
23. Sitaula S, Mackiewicz MR, Reed SM (2008) Gold nanoparticles become stable to cyanide etch when coated with hybrid lipid bilayers. Chem Commun 26:3013–3015. doi: 10.1039/B801525B
24. Brown LO, Hutchison JE (2001) Formation and electron diffraction studies of ordered 2-D and 3-D superlattices of amine-stabilized gold nanocrystals. J Phys Chem B 105(37):8911–8916. doi:10.1021/jp011231a
25. Turkevich J, Stevenson PC, Hillier J (1951) A study of the nucleation and growth processes in the synthesis of colloidal gold. Discuss Faraday Soc 11:55–75. doi:10.1039/df9511100055
26. Chen H, Kou X, Yang Z, Ni W, Wang J (2008) Shape- and size-dependent refractive index sensitivity of gold nanoparticles. Langmuir 24(10):5233–5237. doi:10.1021/la800305j
27. Ayres BR, Reed SM (2014) A minor lipid component of soy lecithin causes growth of triangular prismatic gold nanoparticles. Environ Sci Nano 1(1):37–44. doi:10.1039/c3en00015j

28. Lee Y-J, Schade NB, Sun L, Fan JA, Bae DR, Mariscal MM, Lee G, Capasso F, Sacanna S, Manoharan VN, Yi G-R (2013) Ultrasmooth, highly spherical monocrystalline gold particles for precision plasmonics. ACS Nano 7 (12):11064–11070. doi:10.1021/nn404765w
29. Li C, Shuford KL, Chen M, Lee EJ, Cho SO (2008) A facile polyol route to uniform gold octahedra with tailorable size and their optical properties. ACS Nano 2(9):1760–1769. doi:10.1021/nn800264q
30. Ji X, Song X, Li J, Bai Y, Yang W, Peng X (2007) Size control of gold nanocrystals in citrate reduction: the third role of citrate. J Am Chem Soc 129(45):13939–13948. doi:10.1021/ja074447k
31. Thurmond RL, Dodd SW, Brown MF (1991) Molecular areas of phospholipids as determined by 2H NMR spectroscopy. Comparison of phosphatidylethanolamines and phosphatidylcholines. Biophys J 59(1):108–113. doi:10.1016/S0006-3495(91)82203-2
32. Nagle JF, Tristram-Nagle S (2000) Lipid bilayer structure. Curr Opin Struct Biol 10 (4):474–480
33. Messersmith RE, Nusz GJ, Reed SM (2013) Using the localized surface plasmon resonance of gold nanoparticles to monitor lipid membrane assembly and protein binding. J Phys Chem C 117(50):26725–26733. doi:10.1021/jp406013q
34. Haiss W, Thanh NTK, Aveyard J, Fernig DG (2007) Determination of size and concentration of gold nanoparticles from UV–Vis spectra. Anal Chem 79(11):4215–4221. doi:10.1021/ac0702084
35. Ridker PM, Hennekens CH, Buring JE, Rifai N (2000) C-reactive protein and other markers of inflammation in the prediction of cardiovascular disease in women. N Engl J Med 342 (12):836–843. doi: 10.1056/NEJM200003233421202
36. Shrive AK, Cheetham GM, Holden D, Myles DA, Turnell WG, Volanakis JE, Pepys MB, Bloomer AC, Greenhough TJ (1996) Three dimensional structure of human C-reactive protein. Nat Struct Biol 3(4):346–354. doi:10.1038/nsb0496-346
37. Pepys MB, Hirschfield GM (2003) C-reactive protein: a critical update. J Clin Invest 111 (12):1805–1812. doi:10.1172/JCI18921
38. Fu T, Borensztajn J (2002) Macrophage uptake of low-density lipoprotein bound to aggregated C-reactive protein: possible mechanism of foam-cell formation in atherosclerotic lesions. Biochem J 366(Pt 1):195–201. doi:10.1042/BJ20020045
39. Cleary JP, Walsh DM, Hofmeister JJ, Shankar GM, Kuskowski MA, Selkoe DJ, Ashe KH (2005) Natural oligomers of the amyloid-beta protein specifically disrupt cognitive function. Nat Neurosci 8(1):79–84 doi:10.1038/nn1372
40. Gong Y, Chang L, Viola KL, Lacor PN, Lambert MP, Finch CE, Krafft GA, Klein WL (2003) Alzheimer's disease-affected brain: presence of oligomeric Aβ ligands (ADDLs) suggests a molecular basis for reversible memory loss. Proc Natl Acad Sci 100 (18):10417–10422. doi:10.1073/pnas.1834302100
41. Lambert MP, Barlow AK, Chromy BA, Edwards C, Freed R, Liosatos M, Morgan TE, Rozovsky I, Trommer B, Viola KL, Wals P, Zhang C, Finch CE, Krafft GA, Klein WL (1998) Diffusible, nonfibrillar ligands derived from Aβ1–42 are potent central nervous system neurotoxins. Proc Natl Acad Sci 95 (11):6448–6453
42. Walsh DM, Klyubin I, Fadeeva JV, Rowan MJ, Selkoe DJ (2002) Amyloid-beta oligomers: their production, toxicity and therapeutic inhibition. Biochem Soc Trans 30(4):552–557. doi: 10.1042/bst0300552
43. Wong PT, Schauerte JA, Wisser KC, Ding H, Lee EL, Steel DG, Gafni A (2009) Amyloid-β membrane binding and permeabilization are distinct processes influenced separately by membrane charge and fluidity. J Mol Biol 386 (1):81–96. doi:10.1016/j.jmb.2008.11.060
44. Jang H, Zheng J, Lal R, Nussinov R (2008) New structures help the modeling of toxic amyloidbeta ion channels. Trends Biochem Sci 33(2):91–100. doi:10.1016/j.tibs.2007.10.007
45. Berthelot K, Cullin C, Lecomte S (2012) What does make an amyloid toxic: morphology, structure or interaction with membrane? Biochimie 20:20. doi: 10.1016/j.biochi.2012.07.011
46. Williams TL, Serpell LC (2011) Membrane and surface interactions of Alzheimer's Abeta peptide—insights into the mechanism of cytotoxicity. FEBS J 278(20):3905–3917. doi: 10.1111/j.1742-4658.2011.08228.x
47. Abramov AY, Ionov M, Pavlov E, Duchen MR (2011) Membrane cholesterol content plays a key role in the neurotoxicity of β-amyloid: implications for Alzheimer's disease. Aging Cell 10(4):595–603. doi:10.1111/j.1474-9726.2011.00685.x

Delving into Lipid-Driven Endocytic Mechanisms Using Biomimetic Membranes

Josef Madl*, Sarah Villringer*, and Winfried Römer

Abstract

Endocytosis is an essential mechanism in cell biology. Besides clathrin-mediated or caveolin-dependent processes, also routes independent of these players have been identified. One example is the formation of endocytic structures which is driven by lectin-induced clustering of glycosphingolipids. For example, the binding of the B-subunit of Shiga toxin to its lipid receptor Gb3 in the plasma membrane induces the formation of tubular invaginations and finally results in the cellular uptake of the lipid-binding lectins. Studying such processes in cells is challenging due to the heterogeneous membrane composition. Synthetic membrane systems such as giant unilamellar vesicles (GUV) allow reconstituting of such processes in a minimal biomimetic membrane environment. Thereby, the molecular mechanism can be studied and understood in more detail. We present the basic principles for the preparation of GUVs and how they can be employed for investigating lipid-driven endocytic processes.

Keywords Electroformation, Endocytosis, Giant unilamellar vesicle, Glycolipid, Shiga toxin, Synthetic membrane system

1 Introduction

Biological cells, the principal building blocks of living organisms, employ protein-loaded lipid bilayer membranes to enclose their cytoplasm and thereby maintain defined concentrations of a plethora of biomolecules, such as energy carriers, signaling molecules, structural elements for shaping the cell, or nucleic acids for storing genetic information. The cellular plasma membrane is tight and practically impassable for large molecules such as proteins or nucleic acid strands. Especially charged or polar molecules cannot readily permeate through the lipid barrier, even if they are small. Nevertheless, the exchange of substances is absolutely required for maintenance of metabolic processes, to allow cell growth and

*These authors contributed equally to the work.

Arun K. Shukla (ed.), *Chemical and Synthetic Approaches in Membrane Biology*, Springer Protocols Handbooks, (2017) 17–36, DOI 10.1007/8623_2016_7, © Springer Science+Business Media LLC 2016, Published online: 28 September 2016

proliferation, as well as for receiving or sending of signals. Accordingly, in order to enable such exchange, controllable cellular import/export routes are crucial. Endocytosis is vital for substance uptake in every living cell. There are different types of endocytosis, e.g., clathrin-mediated endocytosis, caveolin-dependent endocytosis, clathrin- and caveolin-independent endocytosis processes, and mechanisms such as phagocytosis or macropinocytosis [1]. Any endocytic process requires that either a membrane invagination is formed or that the cargo is engulfed via cell protrusions followed by cellular uptake. Membrane curvature can be induced by various mechanisms, for example, by a change in the lipid composition, by applying a force on the membrane, or by direct membrane scaffolding [2]. In any case, energy is required for a significant membrane deformation. The best studied endocytosis process is clathrin-mediated endocytosis [1]. Here, clathrin coat proteins are required to stabilize an initial negative membrane curvature. By subsequent action of a complex protein machinery, a membrane bud is formed which grows until finally a vesicle can be pinched off from the plasma membrane by the GTP-consuming action of dynamin. Also actin is essential for many endocytosis mechanisms, where the driving force of membrane deformation is provided by actin polymerization or by myosin movement along actin filaments.

Nonetheless, it was shown that some cargo proteins can induce formation of endocytic structures for cell entry by themselves even without relying on an active endocytosis machinery [3, 4]. Such proteins, e.g., the B-subunit of Shiga toxin (StxB), bind to glycosphingolipids in the plasma membrane. StxB has 15 binding sites for its host cell receptor, the glycosphingolipid globotriaosylceramide (also called Gb3, CD77, or P^k blood group antigen). Upon binding, the protein/glycolipid complexes interact to form nanoclusters. Thereby, due to protein-assisted membrane demixing and local enrichment of Gb3 molecules, a negative curvature is imposed onto the membrane [3, 5]. Ultimately, this process results in the formation of polymorphic tubular structures that are connected to the cell surface. In the case of StxB, actin polymerization was found to be necessary for completing the endocytosis, i.e., for scission of such tubular structures from the plasma membrane, at least in a physiological context at 37°C [5]. On a side note, by significantly lowering the temperature, such tubular structures were found to undergo scission spontaneously, without further energy consumption, at least when cholesterol was present. The latter requirement underlines the importance of the lipid membrane composition for endocytic processes [5]. While the completion of endocytosis requires cell-supplied energy, the initial formation of endocytic structures can also occur without additional energy, e.g., in the form of ATP or GTP consumption, but only with the binding energy that is provided from the binding of the ligand to its receptor in the plasma membrane. Here, the released binding energy is sufficient to overcome the bending rigidity of the membrane and to form small endocytic structures. Aside from StxB, also other pathogens or pathogenic factors, like *Simian virus 40*, *polyomavirus*, or the B-subunit of cholera toxin (CtxB), are able to exploit host cell plasma membrane glycosphingolipids to enter and eventually infect cells [4, 6]. After binding to the respective glycosphingolipids, these molecules or particles enter the cells via endocytic routes that do not involve the elaborate network of clathrin-mediated endocytosis. As StxB, these bacterial toxins and viruses can induce tubular plasma membrane invaginations without the help of cytosolic protein machinery through the dynamic assembly of protein/lipid nanodomains whose intrinsic properties lead to membrane bending and ultimately invagination of the plasma membrane. In the given examples, controlled depletion of the cells' energy or inhibition of endocytic key players as dynamin or actin allows to uncouple the membrane invagination from the subsequent scission step suggesting that cellular factors are not required for the formation of the invaginations, but for their processing [4, 7].

The highly complex structural and functional organization of cellular membranes is mainly governed by specific interactions between various membrane proteins and lipids. Usually top-down strategies are applied in order to gain understanding about cell biological principles such as membrane reorganization or reshaping. One approach is to visualize the molecules of interest using fluorescence microscopy after labeling them chemically or by fusion to fluorescent proteins. Furthermore, the activity of key players can be blocked by specific inhibitors, and their expression levels can be reduced by genetic knockdown. Of course, these are just a few of the potential strategies for investigating cellular processes. However, despite these versatile methods and toolboxes, studying mechanistic details of molecular interactions between distinct proteins or lipids directly within the membrane of living cells is still challenging due to the fact that the membrane composition is very heterogeneous and because cellular processes are highly interlaced. Here, bottom-up approaches are very promising [8]. The use of minimal artificial membrane systems allows obtaining a more detailed understanding of the role of individual key players in such processes. For instance, giant unilamellar vesicles (GUVs) have been utilized for the successful reconstitution of complex biological processes such as the above-described induction of endocytic structures by bacterial or viral carbohydrate-binding protein complexes [3, 5, 6, 9]. Especially, the Gb3-driven membrane tubulation was successfully reconstituted in a minimal membrane system composed of a simple phospholipid species, i.e., DOPC, cholesterol, and the Gb3 glycosphingolipid with a fraction up to 5 mol%. Upon addition of StxB, e.g., 200 nM, to these minimal protocells, the same processes as in native cells occurred, starting from binding of the ligand to the membrane receptors, emergence of nanoclusters, and ultimately the formation

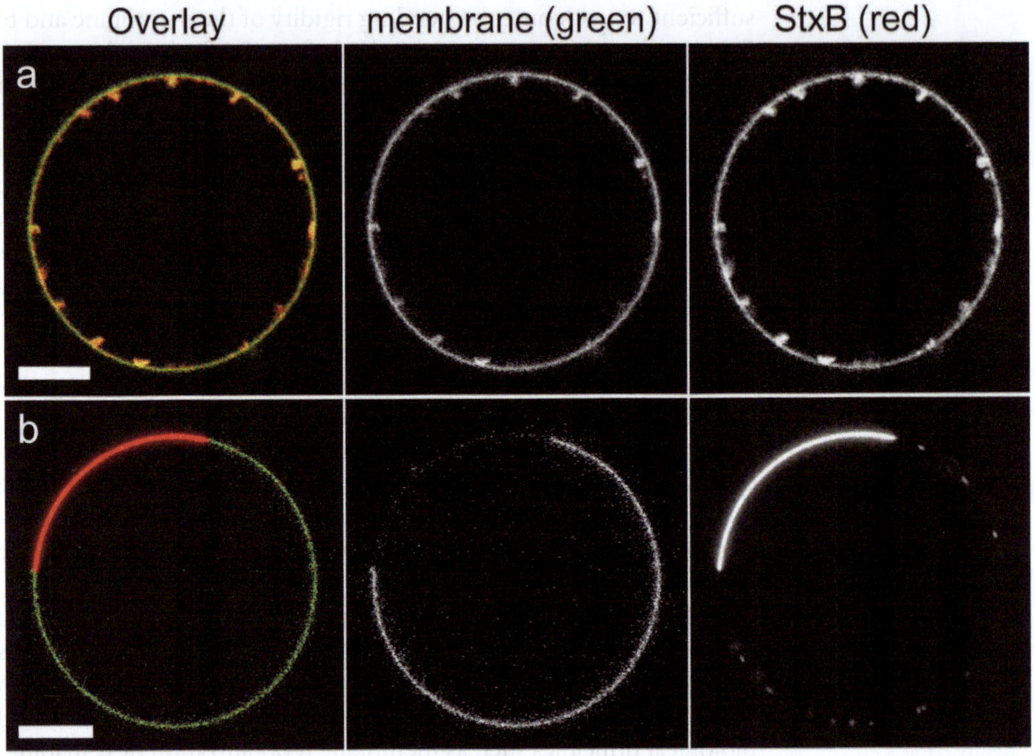

Fig. 1 StxB binding to Gb3-containing GUVs. (**a**) 200 nM Cy3-labeled StxB (*red*) was added to GUVs composed of 64 mol% DOPC, 30 mol% cholesterol, 5 mol% Gb3, and 1 mol% of BODIPY-C_5-HPC as membrane marker (*green*). StxB binds to the GUV membrane and induces tubule formation. (**b**) GUVs were electroformed at 53°C using a lipid mixture of 3:1:3 DOPC–cholesterol–sphingomyelin with 5 mol% Gb3. After cooling to room temperature, phase separation occurs, which leads to formation of different domains with a high and a low degree of lipid ordering. The membrane marker (*green*) is excluded from the highly ordered domains. Gb3 is localized in the ordered domains indicated by StxB-Cy3 binding to these domains (*red*, 200 nM). Here, due to the higher membrane tension, no membrane tubulation is induced. Scale bars 10 μm

of membrane tubules with a topology closely resembling those from native plasma membranes (*see* Fig. 1a). The functional reconstitution of this process in the minimal biomimetic system was direct proof of a proposed mechanism in living cells, which suggested that cargo complexes themselves can manipulate a cellular membrane for their endocytosis. In addition, biomimetic membranes have the advantage that mechanistic details can be studied at a level of controllability, which cannot be reached in cells. For example, the role of the fatty acyl chains of Gb3 was investigated upon engineering the fatty acid backbone and introducing them into GUVs [3]. Only cone-shaped lipids (Gb3 analogue with a single unsaturated acyl chain) allowed the formation of inward-pointing membrane tubules (negative curvature). Binding of StxB to more cylindrically shaped lipids (Gb3 analogue with a saturated acyl chain) yielded less or no membrane reshaping and tubulation [3]. Furthermore, GUVs also provide the possibility to analyze the role of the carbohydrate-binding molecules with high precision. For instance, a study on a trimeric lectin from *Ralstonia solanacearum* (RSL) and reengineered RSL neo-lectins showed that the valency, i.e., the number of binding pockets, is important for membrane bending but that especially the intramolecular distances between the binding pockets are crucial for tubule formation [4, 10].

Beyond that, GUVs offer even more advantages since factors such as geometrical or mechanical parameters can be varied or directly determined. This allows developing simple theoretical models based on experimental data and to confirm or reject a hypothesis on molecular mechanisms [11]. For instance, the lipid zipper, a process where a pathogenic bacterium, i.e., *Pseudomonas aeruginosa*, is triggering its cellular invasion by binding to Gb3 in the plasma membrane, could be reproduced in GUVs simply by introducing this glycosphingolipid into the GUV membrane and "infecting" the liposomes with living bacteria [11]. Here, theoretical modeling confirmed that the binding energy of the bacterial lectin LecA, which is localized on the surface of *P. aeruginosa*, to Gb3 in the host membrane is sufficient to induce membrane engulfment of the bacterium. Due to the simple geometry, GUVs can be utilized to investigate other endocytic processes such as phagocytosis by performing force measurements using optical tweezers. By this means, Meinel et al. were able to develop models, which yield the force and energy landscapes that particles encounter during induced phagocytic uptake into a biomimetic liposome [12]. Furthermore, by applying assays such as micropipette aspiration on GUVs, mechanical parameters like membrane tension can be measured directly or altered during the experiment [13]. Even without sophisticated technical setups, it is possible to modify the mechanical GUV characteristics, e.g., by adapting the osmolarity of the surrounding buffer solution to induce an osmotic pressure for altering the membrane tension or by utilizing photoinduced membrane softening [3]. The possibility to fully control or determine various factors and parameters allows testing biophysical models of membrane-bending mechanisms with high fidelity. The abovementioned examples represent only a small fraction of potential applications, but they highlight the advantages of using minimal biomimetic membrane systems such as GUVs for studying cell biological principles.

Besides GUVs, supported lipid bilayer (SLB) membranes can be used as biomimetic surfaces to study the binding of ligands to their respective membrane receptors (*see* Fig. 2) [14]. SLBs are usually formed on a rigid support, e.g., on glass or on mica (a mineral which can be easily prepared as an ultra-flat substrate). Since the supports are flat, they are useful for techniques such as conventional widefield epifluorescence microscopy or total internal reflection fluorescence (TIRF) microscopy [15] and also for non-optical high-resolution techniques such as atomic force microscopy

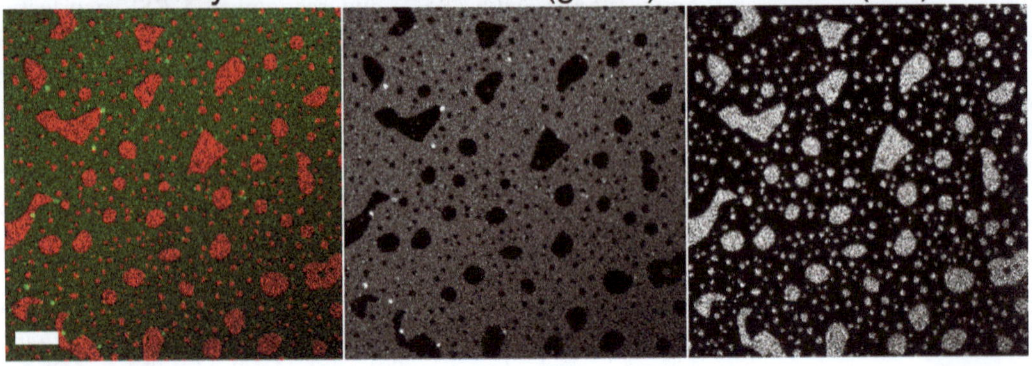

Fig. 2 STxB binds to high-ordered domains on phase-separated SLBs. An SLB was formed on an ultraclean glass coverslip at 55°C using a lipid mixture of 3:1:3 DOPC–cholesterol–sphingomyelin with 5 mol% Gb3. Phase separation occurred after a slow cooldown to room temperature. Comparable to GUVs, domains with a high and a low order are formed. The membrane marker (*green*) is excluded from the highly ordered domains. Cy3-labeled StxB (*red*, 200 nM) binds to Gb3 and is almost exclusively found in the high-order domains. SLBs are suited for studying clustering of StxB and StxB-induced membrane reorganization with techniques such as TIRF or AFM. Scale bar 10 μm

(AFM). AFM is a probe-based microscopy approach that allows imaging the sample topography with a resolution down to a few nanometers under physiological conditions [16]. In addition, AFM can be utilized, e.g., to measure mechanical sample characteristics or interaction forces between ligand molecules on the AFM probe and receptor molecules on the sample surface with a force resolution down to a few piconewton [16]. In the context of endocytosis, supported membrane systems are valuable for characterizing clustering events that occur upon ligand binding and directly precede formation of endocytic structures. Furthermore, AFM allows investigating the StxB-induced reorganization of the lipid membrane [14]. A similar approach showed drastic effects of different natural and synthetic hydroxy enantiomers of Gb3 on StxB-dependent membrane remodeling [17]. However, supported lipid bilayers are not optimal to study the entire process of endocytosis, since the proximity to the support impairs the formation of a significant negative membrane curvature (assuming that the accessible volume represents the cell exterior). Alternatives are polymer-cushioned bilayers or tethered bilayers [18], but even these biomimetic systems offer too little flexibility along the *z*-dimension for observing a full endocytosis event. A possible alternative to supported or cushioned bilayers is the use of black lipid membranes [19], since here the bilayer is freestanding. A limitation of such systems is that they are rather fragile and sometimes exhibit significant membrane fluctuations that might hamper even optical imaging at the highest resolution, while AFM imaging seems completely impossible. An interesting compromise between supported and freestanding lipid bilayers is the use of pore-suspending membranes [20, 21]. Here, a continuous lipid bilayer spans nano- or micrometer-sized pores on an otherwise solid support. Such systems are useful for high-resolution optical microscopy but also for performing AFM experiments [22]. Recently, it has been shown that even highly curved membrane buds could be induced over such pores [23]. Such strongly bended membrane structures might become extremely valuable for studying the role of lipids or soluble proteins in membrane deformation.

In conclusion, biomimetic lipid membranes offer the opportunity to reconstitute distinct stages of endocytosis in a fully controllable environment. Furthermore, the highly adjustable experimental setups provide many advantages for specific investigative purposes. In particular, these artificial membranes represent highly simplified "protocellular" systems which allow overcoming some of the limitations and technical challenges that accompany experiments performed directly within the membrane of living cells.

1.1 Preparation of GUVs

For preparing GUVs with arbitrary concentrations of the desired components, the lipids have to be mixed, which is usually performed in organic solvents such as chloroform, methanol, or chloroform/methanol mixtures [24]. Evaporation of the solvent yields a lipid film on a support, from which unilamellar vesicles can be prepared. One of the first methods that was used to do so is gentle hydration [25]. To this end the thin lipid films on a glass support are slowly hydrated with aqueous solutions. Due to hydrophobic/hydrophilic interactions of the amphiphilic lipids with the polar water molecules, bilayer stacks form by self-assembly, which ultimately swell into larger liposomes upon addition of larger amounts of aqueous solution. This technique was significantly improved by using electric fields to assist the formation of self-assembled bilayer structures and their growth to giant vesicles [26, 27]. First, a lipid film is formed on an electrically conductive surface, such as indium tin oxide (ITO)-coated glass or on platinum wires. Then an aqueous solution is added, and an alternating electric field is applied over some hours, which accelerates and optimizes the process of liposome formation. The electroformation method is more effective than gentle hydration and yields very large unilamellar vesicles, typically between 5 and 100 μm in diameter. Many labs use electroformation as a standard method to produce GUVs. It is also our method of choice, and it will be described in detail in the step-by-step protocol below the introduction section. A drawback of classical electroformation is that it is usually limited to aqueous solutions with low salinity [28]. However, many biomolecular interactions require the presence of various monovalent or divalent ions at high concentrations. By adapting the lipid-to-support deposition protocol and optimizing the voltages and frequencies, successful electroformation was achieved under physiologically relevant conditions [29]. Potential alternatives to electroformation are methods such as gel-assisted GUV swelling [30]. To this end a lipid film is applied on top of a layer of a dry hydrogel. Upon addition of an arbitrary buffer solution, the hydrogel starts to swell, which results in a very efficient formation of liposomes. This technique is rather fast and yields GUVs within a few minutes, while electroformation typically takes more than 1 h [30].

The discussed methods so far have the limitation that it is usually impossible to introduce some kind of polarization or asymmetry into the lipid bilayer. However, such a lipid asymmetry is important in many cellular processes. For example, phosphatidylserine is exclusively located in the inner plasma membrane leaflet of a healthy cell, whereas its presence on the outer leaflet is a general feature of apoptosis [31]. An intriguing liposome preparation technique that enables assembly of GUVs with asymmetric lipid compositions is the inverse emulsion technique [32]. Furthermore, this technique allows enclosing biomolecules such as key players of the actin polymerization machinery inside the GUVs, which of course is very practical for synthetically reconstituting structures such as cytoskeletal elements [33]. Of course, there are many more techniques and combinations available that enable building even more complex biomimetic systems. For instance, GUVs have been decorated with a cholesterol-anchored artificial proto-glycocalyx [34]. Here fully synthetic, lipidated glycopeptides were designed and integrated into the lipid membrane with the aim to create a modular toolbox for studying biological processes that involve glyco-decorated lipid membranes. Furthermore, it was shown that even transmembrane proteins such as ion channels can be incorporated into GUV membranes by, e.g., detergent-mediated incorporation [35]. Furthermore, by using lipid mixtures that roughly resemble the native lipid content of a plasma membrane, phenomena such as the formation of distinct microdomains can be studied. A simple approach to induce such phase-separated domains in biomimetic systems is to use ternary mixtures of, e.g., cholesterol, an unsaturated phospholipid such as DOPC, and a saturated high-melting lipid such as sphingomyelin. Below the phase transition temperature, these lipid mixtures demix and form separated domains with a differing degree in their ordering and also include or exclude distinct lipids, especially glycosphingolipids like GM1 [6] or Gb3 [14] (*see* Fig. 1b). These systems can be used as a crude approximation of signaling platforms in cells, called lipid rafts [36]. Such separated domains are also interesting for studying the role of lipids in membrane curvature that is linked to endocytosis: between the different domains in phase-separated GUVs, a line tension occurs which can influence the mean curvature of these domains [37]. Interestingly, despite the high local concentration of Gb3 or GM1 in such domains, it is not possible to induce tubule formation with StxB or CtxB within domains with a high order due to higher membrane tension within such domains (*see* Fig. 1b).

In the following the use of GUVs for reconstituting lipid-driven endocytic mechanisms of bacterial lectins will be described. A step-by-step protocol is presented that is routinely used in our lab for preparation of GUVs by electroformation. However, it should be noted that using different lipid mixtures or buffer solutions might require optimization of the presented protocols in order to achieve reproducible and artifact-free results. An enlightening review on avoiding potential pitfalls in GUV preparation was written by Morales-Penningston et al. [24].

2 Materials

1. Lipids solved either in pure chloroform or in appropriate mixtures of chloroform/methanol according to manufacturer's instructions. Especially when using glycosphingolipids, it is necessary to add methanol to achieve a good solubility. Recommended lipid stock concentrations are between 1 and 10 mg/mL (**Note 1**). We routinely use phospholipids and cholesterol (Avanti Polar Lipids, Inc.) as main components, Gb3 or GM1 (Matreya, LLC) as glycosphingolipids, and, for visualizing the GUV membranes fluorescently, labeled lipid probes such as BODIPY-FL-C_5-HPC or Texas Red-DHPE (Thermo Fisher Scientific Inc.).
2. Chloroform *p.a.* and methanol *p.a.*
3. Amber glass vials (2 mL) with screw caps and PTFE-coated inlay.
4. Several gastight Hamilton glass syringes for accurate pipetting of chloroform-suspended lipids. Various volumes from 5 μL up to 500 μL are recommended (**Note 2**).
5. Indium tin oxide (ITO)-coated glass slides (25 mm × 75 mm × 1.1 mm, Präzisions Glas & Optik GmbH).
6. Polytetrafluorethylene (PTFE) foil (or similar) with single-side adhesive layer to assemble a spacer between ITO slides.
7. Sigillum wax plates (microhematocrit tube sealant, Vitrex Medical A/S) or similar malleable and physiologically inert material.
8. Vapor pressure osmometer (Vapro 5600, Wescor Inc.).
9. Imaging buffer solution, e.g., Dulbecco's phosphate-buffered saline (PBS).
10. Electroformation solution with osmolarity adjusted to the imaging buffer, e.g., approx 0.3 M sucrose solution in ultrapure water (**Note 3**).
11. Function generator and cables with crocodile clips and multimeter. Optional: ramp generator.

12. Glass capillaries, e.g., Drummond Microcaps 30 μL capillaries.
13. Blocking solution, e.g., 1 mg/mL β-casein solution in 0.1 M Tris, pH 7.5.
14. Microscope glass slides (76 mm × 26 mm × 1 mm) and coverslips #1.5 (22 mm × 22 mm and ø 30 mm).
15. Cell culture cloning cylinders (glass, 8 × 8 mm).
16. UV-curable adhesive, e.g., Norlands Optical Adhesive 63.
17. Holder for coverslips, e.g., interchangeable coverslip dish for round 30 mm coverslips (Bioptechs Inc.).
18. Glycosphingolipid-binding lectins or other proteins of interest (unlabeled or fluorescently labeled, recommended stock concentration 0.5–2 mg/mL, **Note 4**). We use StxB from Sigma-Aldrich Chemie GmbH and CtxB from Thermo Fisher Scientific Inc. The final lectin concentrations in the measurement chambers should be between 10 nM and 1 μM.

3 Methods

3.1 Preparing GUVs by Electroformation

1. Choose a lipid mixture with desired molar percentages of the lipids (e.g., 64 mol% DOPC, 30 mol% cholesterol, 5 mol% Gb3, and 1 mol% BODIPY-FL-C_5-HPC, *see* Fig. 1a; or a 3:1:3 mixture of DOPC–cholesterol–sphingomyelin with 5 mol% Gb3 and 1 mol% BODIPY-FL-C_5-HPC, *see* Fig. 1b) and calculate the required amounts from the lipid stock solutions based on the respective molecular weights and stock concentrations (**Note 5**).
2. Mix the different lipids in amber glass vials using Hamilton gastight syringes (**Note 6**). Add pure chloroform (or chloroform/methanol mixture) to achieve a final concentration of, e.g., 0.5 mg/mL. This ensures that always the same total amount of lipids is used for electroformation, even for different mixtures. Store the working solutions in amber glass vials at −20°C up to several weeks. Cautiously flushing the vials with argon gas is recommended before storage (**Note 7**).
3. Clean the conducting faces of two ITO-coated glass slides successively with 70% ethanol *p.a.* in ultrapure water, then isopropanol *p.a.* and finally chloroform *p.a.* using lint-free tissues such as Kimwipes. In case of doubt, use a multimeter to identify the ITO-coated side.
4. Attach PTFE spacers on one end of each ITO slide to yield a distance between the slides of ~1.1 mm (*see* Fig. 3 and **Note 8**).
5. Stack ITO slides with conductive sides facing each other. Shift the slides by approx 1 cm to leave space for attaching the electrical clips (*see* Fig. 3). Draw congruent circles with

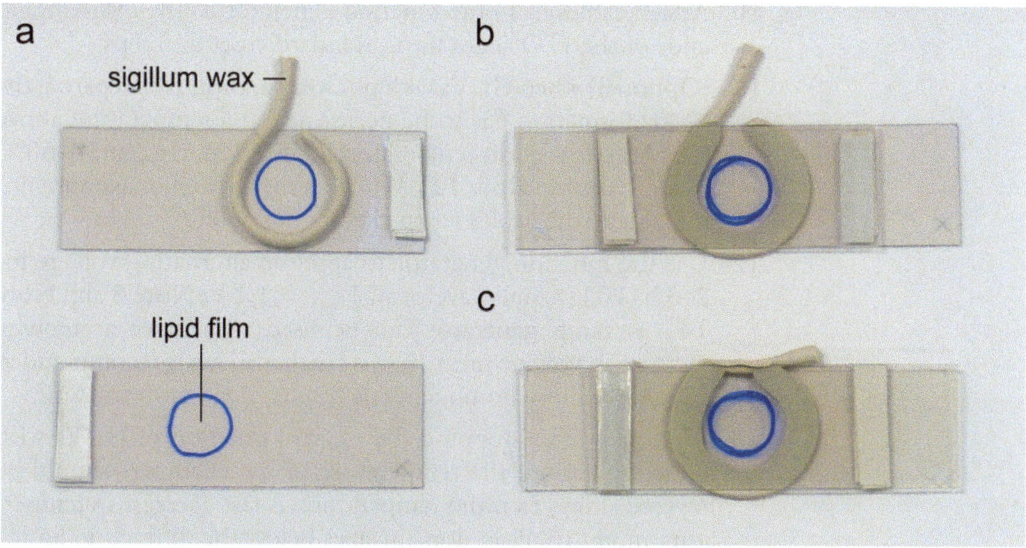

Fig. 3 Preparing electroformation chambers. (**a**) A thin lipid film is spread in central circular areas on the conductive sides of two ITO-coated glass slides. An almost closed ring of sigillum wax is put around the lipid film on one of the slides. (**b**) The chamber is assembled by pressing the second slide on top of the first slide (the conductive sides must face each other, and the two lipid films have to be aligned). Please note that on both sides there is an overlap of the glass slides which is required for attaching the electrical connections to the function generator. (**c**) After filling the chamber with the electroformation solution, the chamber is closed with the remaining sigillum wax strand

~1.5 cm diameter on the backsides of the slides using a non-permanent marker.

6. Pipette 7.5 μL of the lipid mixture onto the conductive side of the slides. Use the side part of the Hamilton pipette needle to spread the lipid droplet carefully across the marked area until all solvent is evaporated. Perform this step twice on each slide, i.e., use 30 μL lipid mixture per electroformation chamber. The result should be a thin and uniform lipid film (**Note 9**).
7. To remove any residual solvent, put the slides under vacuum (10–15 mbar) for at least 30 min up to overnight evaporation (**Note 10**).
8. Prepare a thin strand of sigillum wax by rolling it on a clean surface. Lay a not fully closed ring of sigillum wax (**Note 11**) around the lipid film on one slide (*see* Fig. 3a). Assemble the electroformation chamber by carefully pressing the second slide on top (*see* Fig. 3b). The two lipid films should be aligned, and the two Teflon spacers have to be on opposing sides of the slides.
9. Fill the chamber with electroformation solution (e.g., 0.3 M sucrose), and carefully close the sigillum wax ring without trapping any air bubbles (*see* Fig. 3c and **Note 12**).

10. Attach cables from the function generator to the overhanging ends of the ITO slides using standard crocodile clips.
11. (Optional) when GUVs with phase separation are prepared, the electroformation has to be performed at a temperature above the phase transition temperature (**Note 13**), e.g., at 50–60°C inside an oven (**Note 11**). When using an oven, make sure not to damage the cables when closing the door.
12. Use the function generator to apply an alternating voltage for 2–3 h (10 Hz, sine waveform, $V_{max} = 1.1$ V; **Note 8** and **Note 14**). A ramp generator can be used to achieve a stepwise increase in voltage until V_{max} is reached. This is recommended to achieve more homogeneous results of electroformation.
13. (Optional) after stopping the electroformation of GUVs with phase separation, the temperature of the chambers should be lowered slowly to room temperature. A fast decrease in temperature might result in domain sizes below the diffraction limit.
14. After electroformation, detach growth chamber from cables. Cautiously open the sigillum wax so that a microcapillary can be inserted and used to harvest the liposome-containing sucrose solution (**Note 15**).

3.2 Viewing GUVs by Fluorescence Microscopy

Appropriate sample chambers are required for performing fluorescence microscopy on GUVs. Standard glass-bottom Petri dishes or chambered coverslips will work fine. Nevertheless, we recommend using home-built chambers. When assembling many chambers at the same time, the time effort per chamber is reasonable, while the costs are minimal. Furthermore, the sample volumes can be kept small (<200 μL), which allows using less material such as lectins for the experiments. Also cleaning of the surfaces is more straightforward than on most ready-to-use dishes or wells.

3.2.1 "Closed" Microscopy Chamber System

When assembled correctly these closed chambers have only little evaporation and a large accessible observation area despite a small sample volume. However, it is not possible to add substances once the chambers are closed. Also, after adding the GUVs into the chamber, the binding of lectins will occur immediately. Accordingly, the time for recording the early events is limited.

1. Clean microscope slides and 22 mm × 22 mm coverslips with 70% ethanol *p.a.* using Kimwipes. Carefully press two stacks of four Parafilm stripes (1 cm × 2.5 cm) on a microscope slide. The distance between the stacks should be about 1 cm (*see* Fig. 4a). Cautiously place a coverslip on top of the stacks.
2. Put the microscope slides onto a hot heating plate (~100–150°C), so that the Parafilm stacks are on the plate but a small part of the slides is overhanging for handling. Wait until the Parafilm starts to melt and gets transparent.

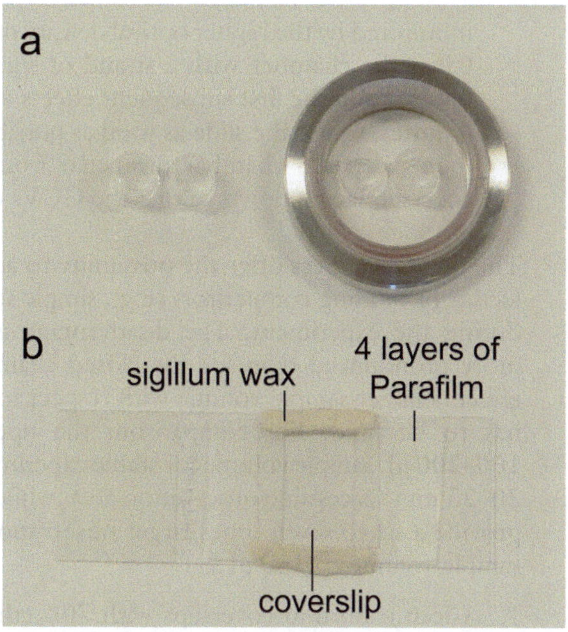

Fig. 4 "Open" and "closed" microscopy chamber systems. (**a**) Two 8 × 8 mm cloning cylinders are glued to an ø 30 mm coverslip, which can be placed into a coverslip holder dish for imaging. The two chambers in the cloning cylinders are accessible, and different agents can be added sequentially, even during the experiment. (**b**) Two stacks of four stripes of Parafilm are placed onto a microscopy slide and heated until they start to melt in order to uniformly attach a 22 mm × 22 mm coverslip on top of it. The chamber has to be closed using sigillum wax and is therefore no longer accessible for adding further substances. However, no evaporation can occur when the chamber is tightly closed. These chambers have a large observation area which makes observation of individual GUVs more convenient

Immediately remove the slide from the heating plate, and apply a soft pressure to the coverslip using the side of a pipette tip to firmly attach the coverslip to the slide while the glass slide cools and the Parafilm hardens (*see* Fig. 4 and **Note 16**).

3. Coat the chamber with β-casein solution for approximately 30 min (**Note 17**). Wash the chamber twice with ultrapure water and once with PBS (drain the liquid by gently touching the chamber gaps with a soft lint-free tissue). Close one of the open side gaps of the chamber with a strand of sigillum wax. Fill the chamber with the desired imaging buffer (e.g., PBS or lectin at desired concentration in PBS) to approximately 90% (**Note 18**). Appropriate lectin concentrations are in the range of 10 nM to 1 μM.
4. Hold the buffer-filled chamber in one hand with the coverslip facing downward, and slowly add ~10 μL of the liposome-containing sucrose directly to the open side gap (**Note 19**). Slowly tilt the imaging chamber from left to right in order to evenly distribute the GUVs in the chamber. The GUVs can be

monitored by the higher optical density of the sucrose solution. Close the chamber with a strand of sigillum wax. The lectin binding and the first subsequent effects will occur within a few minutes, so use the slide as soon as possible for imaging. Until then, keep the chamber protected from light and with the coverslip facing down so that the GUVs can settle.

3.2.2 "Open" Microscopy Chamber Systems

The open chambers offer the possibility to add substances such as lectins or binding competitors (e.g., simple soluble carbohydrates) during the experiment. The disadvantage is that evaporation is more pronounced than for the closed chambers. To reduce this effect, a larger sample volume with respect to the observation area has to be used. For comparison, the open chambers require 100–200 μL sample volume for stable experiments and yield approx 20–25 mm^2 accessible observation area, while the closed chambers provide a six to seven times larger observation area at comparable sample volumes (~150 μL).

1. Clean ø 30 mm coverslips with 70% ethanol *p.a.* using Kimwipes. The 8 × 8 mm cloning cylinders can be cleaned by soaking them in ethanol *p.a.* followed by ultrapure water. Apply a thin layer of UV-curable adhesive to the rim of one side of the 8 × 8 mm cloning cylinders. Place two cloning cylinders into the center of one coverslip. Cure the glue for 10 min under UV light (*see* Fig. 4b and **Note 20**).
2. Completely fill the chambers with β-casein solution, and incubate for 30 min for passivation of the surfaces (**Note 17**). Wash three times with PBS, and fill the chamber with 80–100 μL of the appropriate buffer (e.g., PBS, **Note 18**).
3. Finally, add an appropriate amount of the liposome-containing sucrose solution (~1–10 μL, **Note 19**). The advantage of the open system is that liposomes can be imaged in the native state followed by, e.g., the addition of lectin and further imaging. Thereby the binding, clustering, and tubule formation can be monitored live. Moreover, different lectins or other compounds can be added sequentially. As above, also here appropriate lectin concentrations are in the range of 10 nM to 1 μM. To ensure faster distribution of the lectins in the chamber, they can be pre-diluted in about 50–100 μL buffer before addition. One disadvantage of a completely open system is evaporation, which will change the osmolarity of the buffers in the chamber over time (**Note 21**). Depending on the temperature (e.g., 37°C) and the relative humidity in the microscopy room, evaporation might occur and affect the GUVs, even within 15–30 min. Evaporation can be further reduced by covering the coverslip holder dish with a glass lid (**Note 21**).

3.2.3 Fluorescence Microscopy

For imaging GUVs a confocal fluorescence microscope system is recommended. Due to the optical sectioning, it is possible to obtain high-contrast images from the central parts of GUVs (*see* Fig. 1). Especially for phase-separated GUVs, the recording of *z*-stacks to visualize the 3D structure represents an interesting option. A prerequisite for this is that the GUVs are practically immobile, i.e., loosely adhering to the glass. Depending on the membrane dynamics, a standard laser scanning confocal with conventional high-resolution Galvano scanning might be too slow to capture sharp images. Here, the use of resonant scanning units or faster confocal systems such as spinning disk confocal microscopes might be of advantage. Also other imaging systems such as widefield fluorescence microscopy with fast camera readout can be used for recording lectin-induced GUV dynamics. Thereby, the speed will be adequate for most samples, but in comparison to optical sectioning microscopy techniques, the contrast will be decreased due to out-of-focus fluorescence. In general, the sample chambers should be mounted stably on the microscope sampler holders, yet for time observations longer than 5 min, the drift of the sample holder should be considered. In addition, for fast dynamics it is recommended to record short sequences. In any case care should be taken regarding evaporation, especially when experiments are performed at elevated temperatures such as 37°C.

4 Notes

1. The lipid stock concentrations can be adjusted to values that allow pipetting of convenient amounts (i.e., >1 μL) from the lipid stocks to prepare the electroformation mixtures. If the lipids are purchased in a solvent-suspended form, a diluted stock can be prepared in amber glass vials. Stock and working solutions are stored at −20°C under argon (**Note** 7).
2. Various sizes of Hamilton syringes are recommended: 500 μL for pipetting pure chloroform, 100 μL for aliquoting stocks, and 5–25 μL for preparing the lipid working solutions. It is advisable to assign distinct Hamilton syringes for pipetting nonfluorescent lipids and lipids with spectrally different fluorescent labels to avoid a potential cross contamination with different fluorescent lipids.
3. First, the osmolarity of the imaging buffer is determined. Then, the sucrose solution is prepared by weighing an appropriate amount into ultrapure water to yield an osmolarity slightly higher than the imaging buffer. The fine adjustment is performed by repeatedly adding small amounts of water to the sucrose and measuring the osmolarity until the difference to the imaging buffer is smaller than 5 mOsm/L or less. We prefer

using sucrose solution because due to the higher density compared to PBS, the GUVs sink to the glass support and are less mobile, which is beneficial for imaging.

4. Ideally the proteins should be solved in the same buffer as the imaging buffer. The proteins of interest can be labeled using commercially available fluorophores with functional groups such as NHS, e.g., Alexa-488-NHS (Thermo Fisher Scientific Inc.), according to the manufacturer's instructions.
5. The recommended total lipid amount used for the preparation for one working aliquot is between 100 and 300 μg. This amount is sufficient for about 7–20 individual electroformation runs.
6. To remove any lipid residues from the previous pipetting steps, the Hamilton gastight syringes are rinsed five times successively by filling the syringe with clean chloroform (or 2:1 chloroform/methanol mixture) from a glass vial (2 mL) and swiftly ejecting the solvent into a waste beaker. This procedure is repeated once using a second glass vial with clean solvent.
7. Argon is used to remove air from the vials and to provide an oxygen-free atmosphere to avoid oxidation of unsaturated lipids for long-term storage. Polyunsaturated lipids are more susceptible for oxidation and should be handled in low-oxygen atmosphere at all times when possible.
8. Pieces with a size of 5 mm × 25 mm are cut from the one-sided adhesive foil and stacked, if necessary. The recommended spacer thickness is about 1 mm. The spacer determines the distance between the conductive surfaces and thereby the voltage that must be applied to achieve the desired electric field strength. We normally use a voltage of 1.1 V with a 1.1 mm spacer, i.e., a field strength of 1 V/mm. Depending on the lipid mixture and the electroformation buffer, other settings might result in a better electroformation efficiency [38].
9. Other techniques for spreading the lipids such as spin coating can yield more homogeneous lipid films on the ITO slides [39]. However, spin coating usually requires larger amounts of lipids.
10. After removing the solvent, the slides should be used for electroformation as soon as possible. If necessary, the lipid-coated ITO slides can be stored for 1–2 days at −20°C in a 50 mL Falcon tube filled with argon to prevent oxidation. In this case, allow the slides to reach room temperature before opening the Falcon tubes to avoid condensation on the lipid films.
11. Vitrex sigillum wax is normally used as hematocrit tube sealant, which means that it is physiologically inert and accordingly it should not interfere with GUVs. We use it since it can be easily

shaped into thin strands for closing the electroformation chambers. Alternatives are vacuum grease or soft elastic O-rings with an opening for pipetting. Please note: when the electroformation is performed at higher temperatures, the sigillum wax softens. In order to stabilize the assembled ITO slides even at higher temperatures, we wrap Parafilm and tape around them.

12. Air contact is destructive for liposomes; thus, any air bubbles should be avoided. However, gas bubbles can be also generated during electroformation due to electrolysis [24].

13. The phase transition temperature depends on the used lipids and the exact ratio between the lipids. For many mixtures these values can be found in literature.

14. Electroformation is reasonable fast, but the electroformation time can be varied and optimized for different lipid compositions and electroformation temperatures (from 1 up to over 24 h). In our hands, electroformation for ~3 h generally yields satisfying results both for standard lipid mixtures at room temperature and phase-separating lipid mixtures at 50–60°C. In our protocol, the voltage is increased in 3 min intervals through the following levels, 20 mV–60 mV–100 mV–200 mV–300 mV–500 mV–700 mV–900 mV, and then stably at 1.1 V until the end of electroformation.

15. Microcapillaries can be used with micropipettes. We prefer to cut a few millimeters away from normal plastic tips (size 20–200 μL) and insert a 4 cm-long Microcaps capillary which is fixed with Parafilm. This allows using a standard 20 μL pipette for harvesting the GUVs. Since vesicles often stay attached to the ITO-coated glass slides after swelling, it is important to slowly harvest the GUVs by moving the pipette tip from left to right across the lipid area while aspirating slowly. In general, attention should be paid to avoid a high shear stress by, e.g., pipetting too fast as this can result in disruption of the GUVs.

16. It is important that after attaching the coverslip is oriented exactly parallel to the microscope slide. Thereby, when the microscope slide is mounted on the microscope, also the coverslip will be oriented in the optimal manner. Any angle between slide and coverslip can affect the imaging performance. Thus, attention should be paid to apply an equal amount of pressure when attaching the coverslip to the melted Parafilm.

17. To prevent unspecific adhesion and rupture of the GUVs after deposition to the glass chambers, it is advisable to coat the chambers with casein or albumin.

18. As the volumes of different chambers can vary slightly, the volume of buffer needs to be adjusted. Furthermore, the

chamber does not have to be filled completely, which can be used to save, e.g., lectin.

19. The amount of GUVs can vary between different electroformation experiments even if the same lipid mixture is used. Thus, the amount of liposome-containing sucrose solution needs to be adjusted individually depending on the desired amount of vesicles inside the chamber.
20. The cloning cylinders can be reused several times. After the experiments, aspirate the buffer and immerse the open chambers into acetone for 20–30 min, until the glue has softened. Before reuse, remove remaining glue using tweezers, and soak the cylinders in ethanol *p.a.* followed by ultrapure water.
21. Differences in osmolarity between the GUVs and the surrounding medium can lead to bursting of the GUVs (surrounding medium is hypoosmotic) or to wobbly GUVs (surrounding medium is hyperosmotic). When using the open chambers, we recommend adding a small amount of pure water into the coverslip holder dish next to the GUV-containing glass cylinders and covering the whole dish with glass. This reduces evaporation from the sample solution. Yet, osmolarity differences can also be exploited to change the membrane tension, e.g., to promote or inhibit lectin-induced membrane invaginations.

Acknowledgments

WR acknowledges the support by the Excellence Initiative of the German Research Foundation (EXC 294), the German Federal Ministry of Education and Research (BMBF) in the framework of the EU ERASynBio project SynGlycTis, and a starting grant from the European Research Council (Program "Ideas" ERC-2011-StG 282105).

References

1. Doherty GJ, McMahon HT (2009) Mechanisms of endocytosis. Annu Rev Biochem 78:857–902. doi:10.1146/annurev.biochem.78.081307.110540
2. McMahon HT, Gallop JL (2005) Membrane curvature and mechanisms of dynamic cell membrane remodelling. Nature 438(7068):590–596
3. Römer W, Berland L, Chambon V, Gaus K, Windschiegl B, Tenza D, Aly MRE, Fraisier V, Florent J-C, Perrais D, Lamaze C, Raposo G, Steinem C, Sens P, Bassereau P, Johannes L (2007) Shiga toxin induces tubular membrane invaginations for its uptake into cells. Nature 450(7170):670–675. doi:10.1038/nature05996
4. Schubert T, Römer W (2015) How synthetic membrane systems contribute to the understanding of lipid-driven endocytosis. Biochim Biophys Acta 1853(11, Part B):2992–3005. doi:10.1016/j.bbamcr.2015.07.014
5. Römer W, Pontani L-L, Sorre B, Rentero C, Berland L, Chambon V, Lamaze C, Bassereau P, Sykes C, Gaus K, Johannes L (2010) Actin

dynamics drive membrane reorganization and scission in clathrin-independent endocytosis. Cell 140(4):540–553. doi:10.1016/j.cell.2010.01.010

6. Ewers H, Römer W, Smith AE, Bacia K, Dmitrieff S, Chai W, Mancini R, Kartenbeck J, Chambon V, Berland L, Oppenheim A, Schwarzmann G, Feizi T, Schwille P, Sens P, Helenius A, Johannes L (2010) GM1 structure determines SV40-induced membrane invagination and infection. Nat Cell Biol 12 (1):11–18. doi:10.1038/ncb1999, sup pp 11–12
7. Johannes L, Parton RG, Bassereau P, Mayor S (2015) Building endocytic pits without clathrin. Nat Rev Mol Cell Biol 16(5):311–321. doi:10.1038/nrm3968
8. Arumugam S, Chwastek G, Schwille P (2011) Protein–membrane interactions: the virtue of minimal systems in systems biology. Wiley Interdiscip Rev Syst Biol Med 3(3):269–280. doi:10.1002/wsbm.119
9. Rydell GE, Svensson L, Larson G, Johannes L, Römer W (2013) Human GII.4 norovirus VLP induces membrane invaginations on giant unilamellar vesicles containing secretor gene dependent α1,2-fucosylated glycosphingolipids. Biochim Biophys Acta 1828 (8):1840–1845. doi:10.1016/j.bbamem.2013.03.016
10. Arnaud J, Tröndle K, Claudinon J, Audfray A, Varrot A, Römer W, Imberty A (2014) Membrane deformation by neolectins with engineered glycolipid binding sites. Angew Chem Int Ed 53(35):9267–9270. doi:10.1002/anie.201404568
11. Eierhoff T, Bastian B, Thuenauer R, Madl J, Audfray A, Aigal S, Juillot S, Rydell GE, Muller S, de Bentzmann S, Imberty A, Fleck C, Römer W (2014) A lipid zipper triggers bacterial invasion. Proc Natl Acad Sci U S A 111 (35):12895–12900. doi:10.1073/pnas.1402637111
12. Meinel A, Trankle B, Römer W, Rohrbach A (2014) Induced phagocytic particle uptake into a giant unilamellar vesicle. Soft Matter 10 (20):3667–3678. doi:10.1039/c3sm52964a
13. Kwok R, Evans E (1981) Thermoelasticity of large lecithin bilayer vesicles. Biophys J 35 (3):637–652. doi:10.1016/S0006-3495(81) 84817-5
14. Windschiegl B, Orth A, Römer W, Berland L, Stechmann B, Bassereau P, Johannes L, Steinem C (2009) Lipid reorganization induced by Shiga toxin clustering on planar membranes. PLoS One 4(7), e6238. doi:10.1371/journal.pone.0006238
15. Plochberger B, Stockner T, Chiantia S, Brameshuber M, Weghuber J, Hermetter A, Schwille P, Schütz GJ (2010) Cholesterol slows down the lateral mobility of an oxidized phospholipid in a supported lipid bilayer. Langmuir 26 (22):17322–17329. doi:10.1021/la1026202
16. Hinterdorfer P, Dufrene YF (2006) Detection and localization of single molecular recognition events using atomic force microscopy. Nat Methods 3(5):347–355
17. Schütte Ole M, Patalag Lukas J, Weber Lucas MC, Ries A, Römer W, Werz Daniel B, Steinem C (2015) 2-Hydroxy fatty acid enantiomers of Gb3 impact Shiga toxin binding and membrane organization. Biophys J 108(12):2775–2778. doi:10.1016/j.bpj.2015.05.009
18. Rebaud S, Maniti O, Girard-Egrot AP (2014) Tethered bilayer lipid membranes (tBLMs): interest and applications for biological membrane investigations. Biochimie 107, Part A:135–142. doi:10.1016/j.biochi.2014.06.021
19. Bartsch P, Walter C, Selenschik P, Honigmann A, Wagner R (2012) Horizontal bilayer for electrical and optical recordings. Materials 5 (12):2705
20. Janshoff A, Steinem C (2015) Mechanics of lipid bilayers: what do we learn from pore-spanning membranes? Biochim Biophys Acta 1853(11, Part B):2977–2983. doi:10.1016/j.bbamcr.2015.05.029
21. Römer W, Steinem C (2004) Impedance analysis and single-channel recordings on nano-black lipid membranes based on porous alumina. Biophys J 86(2):955–965. doi:10.1016/S0006-3495(04)74171-5
22. Orth A, Johannes L, Römer W, Steinem C (2012) Creating and modulating microdomains in pore-spanning membranes. Chemphyschem 13(1):108–114. doi:10.1002/cphc.201100644
23. Gleisner M, Mey I, Barbot M, Dreker C, Meinecke M, Steinem C (2014) Driving a planar model system into the 3rd dimension: generation and control of curved pore-spanning membrane arrays. Soft Matter 10 (33):6228–6236. doi:10.1039/c4sm00702f
24. Morales-Penningston NF, Wu J, Farkas ER, Goh SL, Konyakhina TM, Zheng JY, Webb WW, Feigenson GW (2010) GUV preparation and imaging: minimizing artifacts. Biochim Biophys Acta 1798(7):1324–1332. doi:10.1016/j.bbamem.2010.03.011
25. Reeves JP, Dowben RM (1969) Formation and properties of thin-walled phospholipid vesicles. J Cell Physiol 73(1):49–60. doi:10.1002/jcp.1040730108

26. Angelova MI, Dimitrov DS (1986) Liposome electroformation. Faraday Discuss Chem Soc 81:303–311. doi:10.1039/dc9868100303
27. Angelova MI, Soléau S, Méléard P, Faucon F, Bothorel P (1992) Preparation of giant vesicles by external AC electric fields. Kinetics and applications. In: Helm C, Lösche M, Möhwald H (eds) Trends in colloid and interface science VI. Steinkopff, Darmstadt, pp 127–131. doi:10.1007/BFb0116295
28. Rumiana D, Said A, Natalya B, Vesselin N, Karin AR, Reinhard L (2006) A practical guide to giant vesicles. Probing the membrane nanoregime via optical microscopy. J Phys Condens Matter 18(28):S1151
29. Pott T, Bouvrais H, Méléard P (2008) Giant unilamellar vesicle formation under physiologically relevant conditions. Chem Phys Lipids 154(2):115–119. doi:10.1016/j.chemphyslip.2008.03.008
30. Weinberger A, Tsai F-C, Koenderink Gijsje H, Schmidt Thais F, Itri R, Meier W, Schmatko T, Schröder A, Marques C (2013) Gel-assisted formation of giant unilamellar vesicles. Biophys J 105(1):154–164. doi:10.1016/j.bpj.2013.05.024
31. Martin SJ, Reutelingsperger CP, McGahon AJ, Rader JA, van Schie RC, LaFace DM, Green DR (1995) Early redistribution of plasma membrane phosphatidylserine is a general feature of apoptosis regardless of the initiating stimulus: inhibition by overexpression of Bcl-2 and Abl. J Exp Med 182(5):1545–1556. doi:10.1084/jem.182.5.1545
32. Pautot S, Frisken BJ, Weitz DA (2003) Production of unilamellar vesicles using an inverted emulsion. Langmuir 19 (7):2870–2879. doi:10.1021/la026100v
33. Pontani L-L, van der Gucht J, Salbreux G, Heuvingh J, Joanny J-F, Sykes C (2009) Reconstitution of an actin cortex inside a liposome. Biophys J 96(1):192–198. doi:10.1016/j.bpj.2008.09.029
34. Stuhr-Hansen N, Madl J, Villringer S, Aili U, Römer W, Blixt O (2016) Synthesis of cholesterol-substituted glycopeptides for a tailor-made glycocalyxification of artificial membrane systems. ChemBioChem 17 (15):1403-1406. doi:10.1002/cbic.201600258
35. Dezi M, Di Cicco A, Bassereau P, Levy D (2013) Detergent-mediated incorporation of transmembrane proteins in giant unilamellar vesicles with controlled physiological contents. Proc Natl Acad Sci U S A 110(18):7276–7281. doi:10.1073/pnas.1303857110
36. Dietrich C, Bagatolli LA, Volovyk ZN, Thompson NL, Levi M, Jacobson K, Gratton E (2001) Lipid rafts reconstituted in model membranes. Biophys J 80(3):1417–1428. doi:10.1016/S0006-3495(01)76114-0
37. Baumgart T, Hess ST, Webb WW (2003) Imaging coexisting fluid domains in biomembrane models coupling curvature and line tension. Nature 425(6960):821–824, http://www.nature.com/nature/journal/v425/n6960/suppinfo/nature02013_S1.html
38. Politano TJ, Froude VE, Jing B, Zhu Y (2010) AC-electric field dependent electroformation of giant lipid vesicles. Colloids Surf B Biointerfaces 79(1):75–82. doi:10.1016/j.colsurfb.2010.03.032
39. Estes DJ, Mayer M (2005) Electroformation of giant liposomes from spin-coated films of lipids. Colloids Surf B Biointerfaces 42 (2):115–123. doi:10.1016/j.colsurfb.2005.01.016

Lipid Bilayer Patterns Fabrication by One-Photon Lithography

M. Florencia Sánchez and Dolores C. Carrer

Abstract

In the last decade, there has been an increasing interest in patterning surfaces with supported lipid bilayers. The ability of this system to control molecular composition and mobility of this system at micro- and nanoscales have inspired several new routes of biological and biotechnological investigation. Successes in this field include the development of photolithographic and soft lithography techniques which allow the printing of a variety of biomolecules on different substrates. Here, we describe a simple, flexible, and inexpensive protocol for the generation of lipid bilayer micropatterns based on one-photon lithography and microcontact printing (μCP).

Keywords Lipid bilayer, Lithography, Microcontact printing, Patterned surfaces, Photopolymerization, Soft lithography

1 Introduction

Supported lipid bilayers provide a powerful model of natural cell membranes and have been widely used to study a variety of biological process in which membranes are involved [1–3]. These bilayers can be prepared over several solid supports [4, 5], and once prepared, a thin layer of water separates the bilayer from the underlying substrate, allowing lateral mobility of membrane molecules [6]. This model system has often been used in the study of juxtacrine signaling, in which a cell–cell contact is required. In this situation, one cell can be replaced by an artificial lipid bilayer to understand how the other cell responds to specific aspects of the extracellular environment [7]. In particular, different proteins can be tethered to these bilayers, allowing their spatial reorganization under the influence of a target cell. This model has been successful in understanding the immune and cytotoxic synapses, in which communication focuses on a ligand–receptor interaction between a lymphocyte and a target or antigen-presenting cell [1, 8–11].

Arun K. Shukla (ed.), *Chemical and Synthetic Approaches in Membrane Biology*, Springer Protocols Handbooks, (2017) 37–48, DOI 10.1007/8623_2016_6, © Springer Science+Business Media LLC 2016, Published online: 25 September 2016

Usually, the main approach used to form supported lipid bilayers – fusion of lipid vesicles to a surface – creates homogeneous membranes of uniform composition across the substrate [12]. However, most of the physiological cell processes of interest involve local membrane order, domain formation, and diffusional constraints, i.e., the presence of diffusional barriers important to epithelial cell polarity or membrane microdomains which regulate receptor signaling [13–15]. Therefore, the possibility of creating patterns of supported lipid bilayer on a single surface greatly increases the utility of this system in biology and biotechnology applications.

In the past two decades, lithographic methods allowed the generation of micro- and nano-size patterns of a variety of biomolecules over several surfaces [16–21]. The basic photolithographic method generates patterns on a surface (typically a silicon wafer) coated with a photoresist by selectively irradiating the surface with light through a predesigned mask. The mask contains the features of desired shapes and sizes in an opaque material (usually chromium) on a transparent background. UV light shining through the transparent regions of the mask changes the chemical composition of the photoresist, making it soluble (positive) or insoluble (negative) to particular solvents. After exposure, unprotected sites can be either chemically or dry etched, and subsequent processing steps yield surfaces called "masters" that have two (or more) patterned regions for distinctive chemical modification or construction of topographical features [21].

One of the first micropatterns of supported lipid bilayers was based on the generation of diffusional barriers over a substrate using photolithography. The fusion of lipid vesicles occurs in the substrate (glass, silicon) but not in the barriers (plastic, aluminum oxide, or metals) [22]. With the objective of modifying surfaces with biomolecules of interest distributed according to a design pattern, Whitesides and collaborators developed the technique called "soft lithography" [23, 24]. The term refers to a set of fabrication techniques that involve molding, imprinting, or embossing with a template made of polymers, such as the elastomer poly-dimethylsiloxane (PDMS). Two soft lithography approaches are particularly useful for investigating cell interactions: microcontact printing (μCP) and polymer liftoff [25].

In μCP a master, typically prepared by conventional photolithography, is used to create an elastomeric stamp with the desired features. The stamp is coated with the desired biomaterial and then pressed against a surface to deposit the material and passivate the surface. This approach has been successfully applied to obtain lipid bilayer patterns of different compositions which have been used to study different biological processes [26–32]. However, the generation of masters through lithographic technology has disadvantages which include the need of access to special equipment, the high costs, and the requirement of qualified staff to develop the technique.

An alternative developed by Costantino et al. [33] generates structures on standard glass coverslips by moving a diffraction-limited laser spot along a glass surface covered with a commercially available UV-curable resin, following a designed trajectory in which the laser is set to switch on and off when desired. As a consequence of two-photon nonlinear photopolymerization, the material is only solidified in the direct vicinity of the focal spot with micrometer resolution. Some time later, Kunik et al. [34] obtained micropatterned surfaces using a standard one-photon excitation laser: a similar photopolymerization reaction is produced, and microstructures can be obtained. The polymerization efficiency depends on both the laser power and the scanning speed. Among the advantages of this technique are the simplicity of the protocol and the fact that it can be performed in any standard confocal microscope.

In the next sections, we will describe in detail a simple and easy-to-follow protocol that combines one-photon lithography and microcontact printing techniques to fabricate patterned lipid bilayers of micrometric size [35].

2 Materials

2.1 Instrumentation

The one-photon lithography technique [34] can be performed either in commercial confocal microscopes or in a special instrument called lithographer (TOLKET SRL, Buenos Aires, Argentina). In our case, we performed some of the experiments in the confocal microscope FV1000 (Olympus, Tokyo, Japan) using a 405 nm diode laser (200 μW) focused with a 10× NA 0.4 microscope objective. However, the confocal setup we used did not have a motorized stage; thus, the size of the patterned area was limited to the scanning area of the microscope. Therefore, we used the lithographer in those cases in which we required bigger patterned areas ($\sim$10 mm^2). The lithographer also has the advantage of allowing the generation of irregular, user-designed patterns that could not be easily programed with the software of the FV1000 confocal microscope.

Figure 1 shows a schematic view of the lithographer setup. In brief, the lithographer setup consists of an inverted microscope with a continuous wave diode laser (405 nm) in combination with an objective 40× NA 0.4.

The power of the laser is adjusted by inserting neutral density filters in the beam path. A shutter is interposed in the beam way to control the pattern design. The sample holder is motorized in the three axes by means of xyz stages, and the smallest step accessible for the motors is 50 nm. The control of the focus is made by means of a charge-coupled device (CCD) camera imaging the back reflection of the laser beam, and the sample focus is adjusted to minimize the image size in the CCD camera.

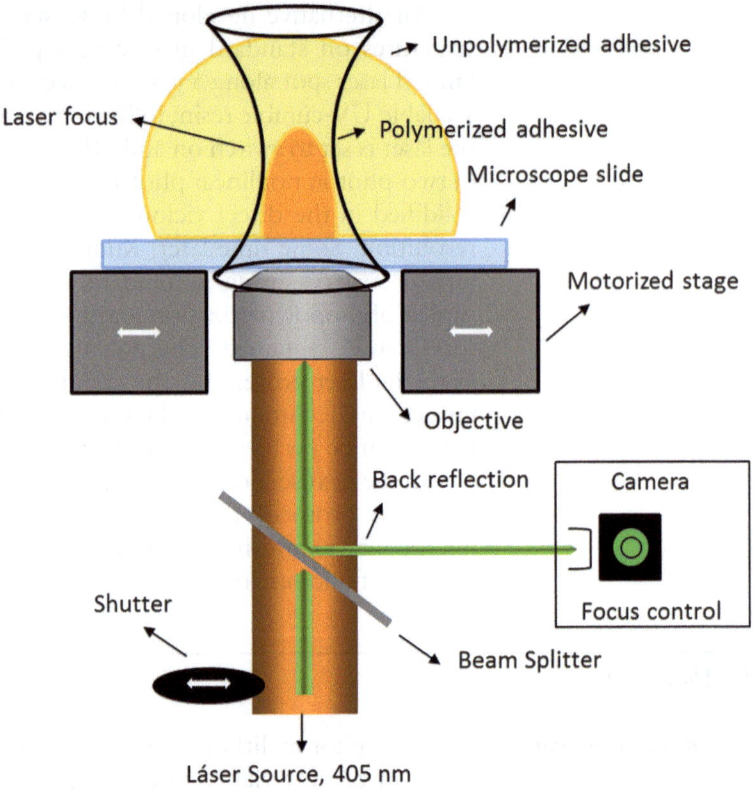

Fig. 1 Schematic diagram of the lithographer setup

2.2 Fabrication of the Master

1. Standard cover glass
2. UV-curable adhesive NOA86 (UV-curing adhesives, Norland Products, Cranbury, NJ)
3. Acetone
4. Isopropyl alcohol
5. A piece of clean transparent plastic (polypropylene)
6. Roller

2.3 Fabrication of the Stamp

1. Sylgard 184 prepolymer (Dow Corning, Midland, MI)
2. Vacuum pump
3. Oven (25–100 ± 1°C)

2.4 Printing of Biomaterial and Supported Lipid Bilayers Generation

1. Passivating protein: bovine serum albumin (BSA) 1 mg/mL or fibronectin 1 mg/mL
2. Nitrogen flux
3. Lipid mixtures: stock solutions prepared by dissolving lyophilized lipids form a reliable source (Avanti Polar Lipids, Alabaster, AL, USA) into chloroform or a 2:1 (v/v) mixture of

chloroform and methanol (*see* **Notes 1–3**). Include about 0.1% of a lipophilic fluorescent dye, for example, one of the families of long-chain dialkylcarbocyanines like DiD, DiI, or DiO (Invitrogen).

4. Phosphate-buffered saline (1× PBS Buffer): 2.7 mM KCl, 1.5 mM KH_2PO_4, 8 mM Na_2HPO_4, and 137 mM NaCl, pH 7.4. Weigh 0.2237 g KCl, 0.2041 g KH_2PO_4, 1.1357 g Na_2HPO_4, and 8.0065 g NaCl in a beaker. Fill to about 800 mL with water. Adjust the pH by adding NaOH. Add enough water to make 1 L solution. Store at room temperature.
5. 1 M calcium chloride solution. Weigh 5.5492 g $CaCl_2$ and dissolve in enough water to make 50 mL of solution. Store at room temperature.
6. Bath sonicator
7. UV-curable adhesive NOA86 or NOA88 (UV-curing adhesives, Norland Products, Cranbury, NJ).
8. Round glass coverslips of 25 mm diameter, rectangular of 22 × 40 mm or squared of 25 × 25 mm.
9. UV lamp, wavelength 250–400 nm.

3 Methods

3.1 Fabrication of the Master

1. Wash a standard cover glass with isopropyl alcohol.
2. Place a drop of the UV-curable adhesive NOA86 in the middle of the cover glass (Fig. 2a, b).
3. Cut a piece of clean transparent plastic, the same size of the cover glass, and place it over the UV-curable adhesive drop (Fig. 2c).
4. Apply pressure with a small roller to distribute the UV-curable adhesive drop homogeneously over the glass (Fig. 2d).
5. Position the cover glass in the confocal microscope or lithographer (Fig. 3) and start the polymerization process.

   For masters prepared in the confocal FV1000, arrays of dots are obtained by illuminating each dot during 1 s with a 405 nm diode laser at a laser power of 200 μW. The shutter is controlled using the microscope software.

   Arrays of lines are obtained by scanning the sample with a speed of 1,600 μm/s in the lithographer. The same laser is used at a power of 200 μW. The motors and the shutter are controlled using a PC, with custom-written programs and drivers provided by TOLKET SA.
6. Remove the piece of plastic.

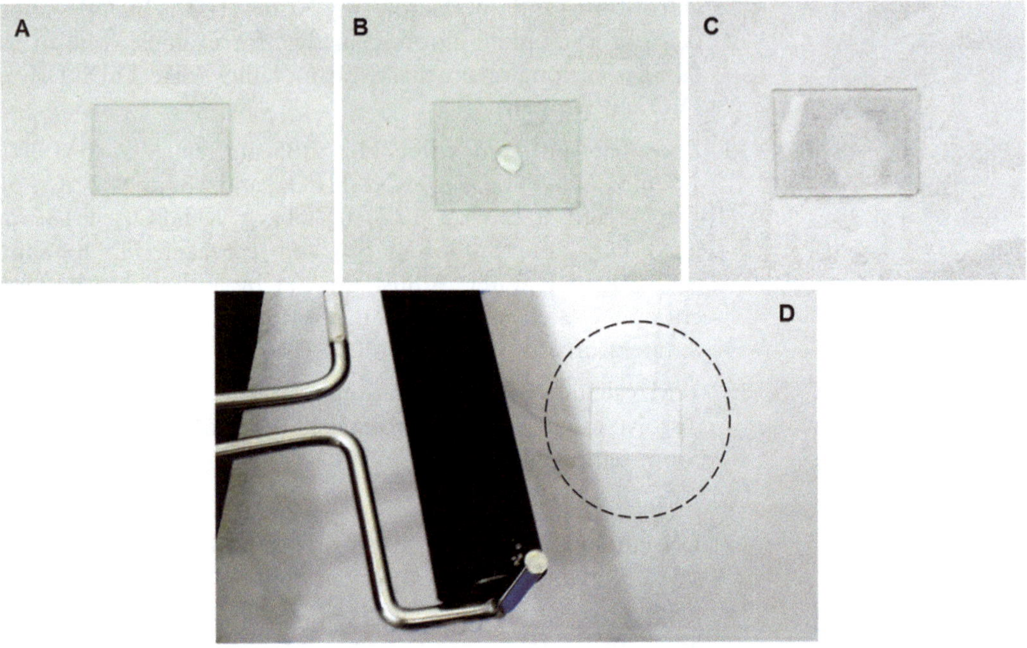

Fig. 2 (**a**) A piece of cover glass is used to fabricate the master. (**b**) A drop of the UV-curable adhesive NOA86 is placed in the middle of the cover glass. (**c**) A piece of clean transparent plastic is placed over the UV-curable adhesive drop. (**d**) The drop is distributed homogeneously over the glass by applying pressure with a small roller

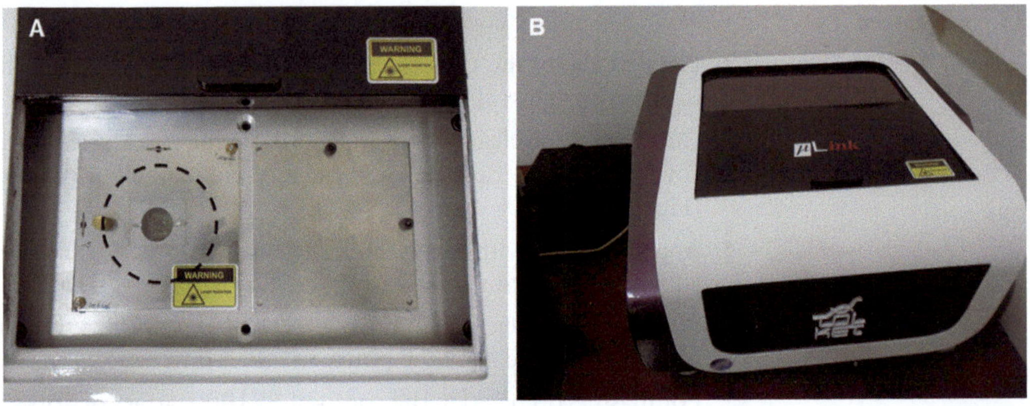

Fig. 3 (**a**) The cover glass with the UV-curable adhesive is placed in the center of the microscope or the lithographer. (**b**) After closing the cover of the lithographer, the design is set in the software, and the polymerization process begins

7. Rinse the sample first with acetone and then with isopropyl alcohol to remove residual adhesive.
8. Check the master with transmitted light in an optical microscope to check for the correct formation of the pattern.

3.2 Fabrication of the Stamp

1. To prepare PDMS stamps mix Sylgard 184 prepolymer with crosslinker in a 10:1 v/v ratio. You will notice that bubbles appear in the mixture.
2. Degas in vacuum for 30 min to remove the bubbles.
3. Place a ring of any metal on top of the glass so that the design in the master is placed inside the metal ring. This acts as a chamber to contain the Sylgard prepolymer mixture.
4. Add the mixture and cure for 4–6 h at 60°C in an oven (*see* **Note 4**).
5. Peel the elastomer off the master cover glass (*see* **Note 5**).

3.3 Preparation of Small Unilamellar Vesicles (SUVs)

1. Evaporate the solvent of the lipid mixture of interest under nitrogen flux so as to form a thin film of lipid and then subject to vacuum for 3 h to allow complete evaporation of the solvent.
2. Rehydrate to a final concentration of 1 mM in PBS. This produces a suspension of multilamellar vesicles. These can be separated into aliquots and stored at −20°C for up to 6 months.
3. Vortex the suspension to mix evenly.
4. Sonicate the suspension in aliquots of 150 μL inside a transparent bottom glass flask until the solution becomes clear. At this point SUVs are obtained (*see* **Note 6**).

3.4 Preparation of Supported Lipid Bilayer Patterns

1. Put a drop of passivating protein solution big enough to cover the entire stamp and leave at room temperature for 15 min.
2. Dry the stamp with a nitrogen flux.
3. Press the stamp against a glass coverslip for 10 min to transfer the protein onto the glass (*see* **Note 7**). This process is called passivation process (Fig. 4a, b).
4. Wash the coverslip with 1 mL of distilled water and dry with nitrogen flux (*see* **Note 8**).
5. Prepare a chamber by gluing the top of an eppendorf tube onto the coverslip with the UV-curable adhesive NOA88/NOA86 (Fig. 4c, d). Place this chamber under UV light for 15 min to allow the polymerization of the adhesive.
6. Pipette 100 μL of the clear SUV solution into the chamber and add 3 μL of $CaCl_2$ 1 M.
7. Incubate the sample at room temperature for 30 s to 1 min to allow for vesicle fusion (*see* **Note 9**).
8. Rinse several times with PBS Buffer to remove the nonfused vesicles.
9. The stamp can be reused. Sonicate the stamp in a beaker with distilled water for 5 min and then let dry (*see* **Note 10**).

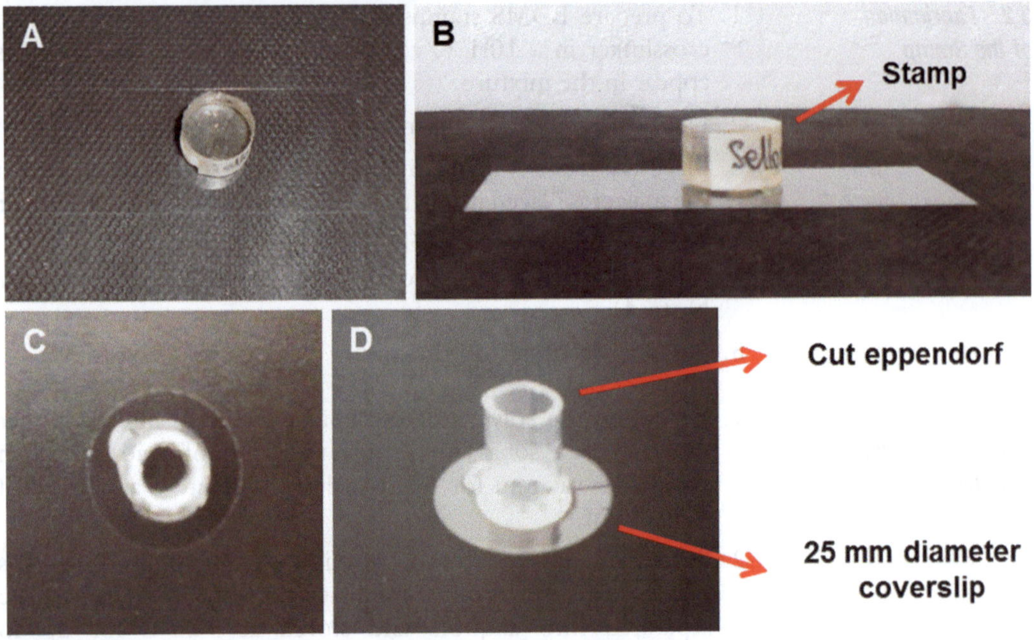

Fig. 4 (**a**, **b**) Images of the passivation process in which the protein is transferred to the coverslip. Top (*left*) and lateral view (*right*). (**c**, **d**) Images showing a homemade chamber designed for observing lipid bilayer patterns. The chamber was prepared by adhering an eppendorf tube cut to the coverslip passivated with BSA. Top (*left*) and lateral view (*right*) of the chamber. As an example we show different kind of coverslip sizes

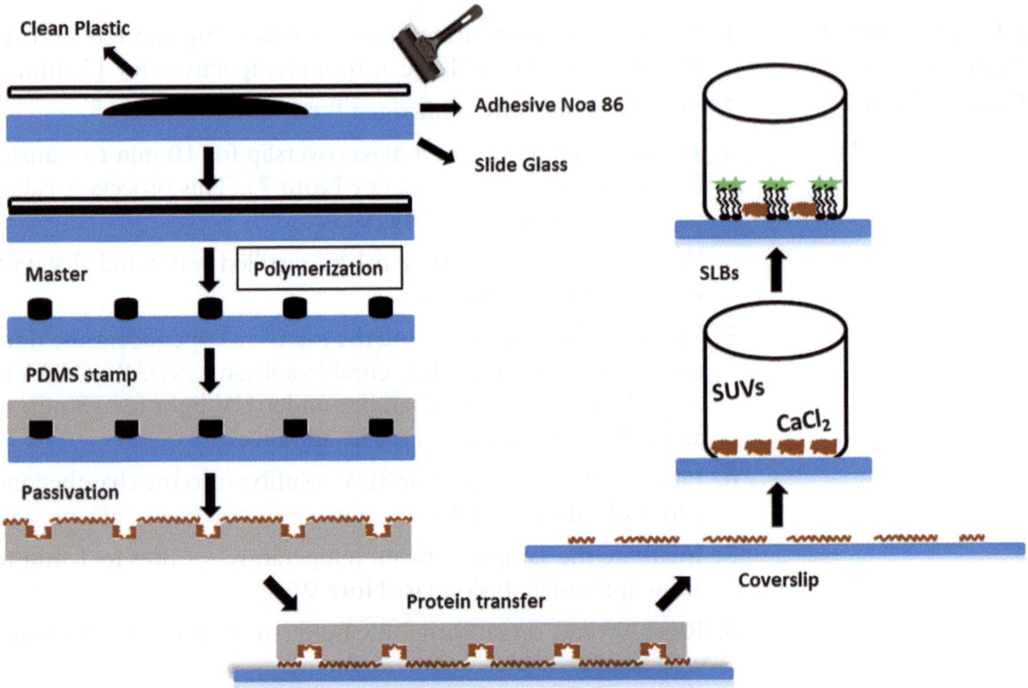

Fig. 5 Schematic diagram of the microcontact printing procedure used to generate SLB patterns

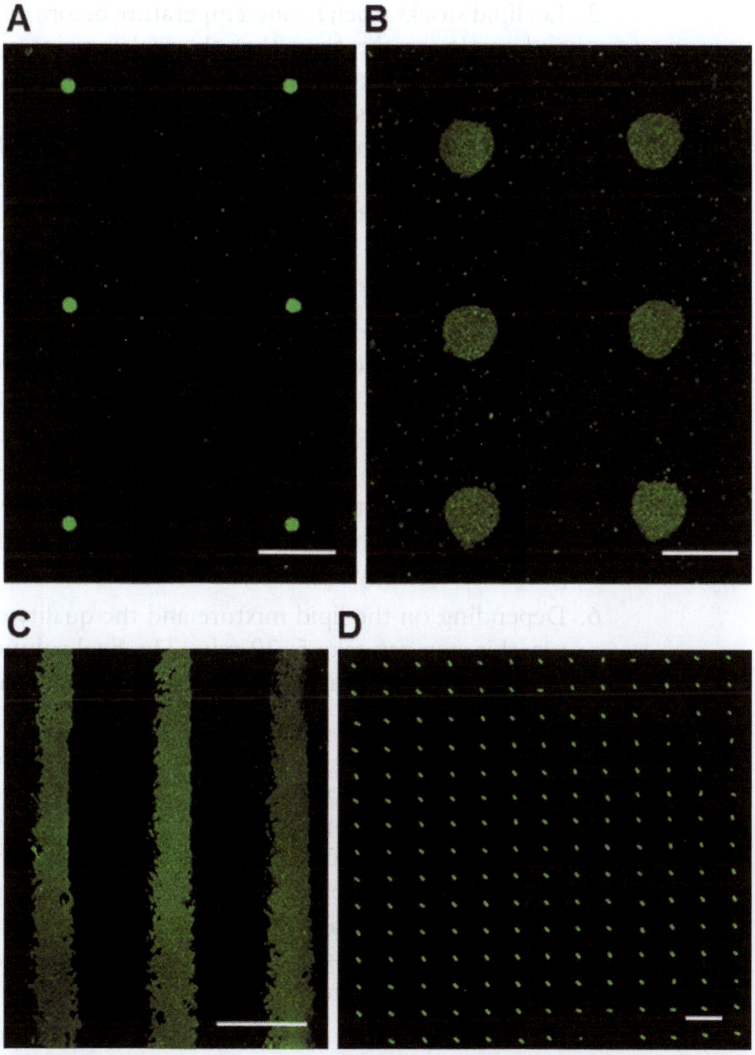

Fig. 6 Confocal images of patterned SLB of DOPC/Chol (75:25 mol/mol), 0.1 mol% DiO-C18. The size and shape of the design were controlled by the lithographer or the confocal microscope software. Scale bar, 20 μm

10. Figure 5 shows a schematic diagram of the whole process.
11. In Fig. 6, we show some examples of the patterns that can be obtained with this technique.

4 Notes

1. Avoid contact of solutions in organic solvents with plastics. Use glassware and organic solvent-resistant material for their handling.

2. Let lipid stocks reach room temperature before opening the vial and minimize the time that the vial remains open to avoid changes in lipid concentration due to solvent evaporation or pipetting mistakes due to the pipette not being calibrated to work at temperatures below room temperature. Before closing the vial again, replace the air in it by an inert gas like N_2 or Ar to avoid oxidation during storage.
3. Wear appropriate gloves to protect you from organic solvents and work under a hood.
4. Be sure that the mixture is well cured. You can touch with a scalpel the top side of the stamp, and it should have the consistency of a rubber. If you notice that it is still too sticky, you can cure for 1–2 more hours.
5. This step must be done very carefully to avoid breaking down the cover glass or the stamp. Usually it is convenient to cut well with a cutter all the periphery of the stamp and then it should peel off very easily.
6. Depending on the lipid mixture and the quality of the sonicator, this step can take 5–20 min. The final solution should be clear and transparent in comparison with the initial solution which is very milky.
7. This is a critical step in the process, and we have corroborated that the efficiency of the passivation is increased if a 100 g weight is placed on top of the stamp in contact with the coverslip for 10 min. This weight can be a small piece of metal, and as in our case the piece was much bigger than the stamp, we put old stamps under the metal to maintain a balance and achieve a homogeneous pressure over the stamp which is transferring the protein.
8. It is very useful in this step to make a rounded or squared mark with a marker on the side of the coverslip that has not been printed, on the back side. If it is done on the same side as the printing, sometimes rests of the marker can be transferred to the design (even with waterproof markers) that can contaminate the "printed area." Besides, in this way it is easy to know where is the area in which you have to prepare the chamber.
9. One of the main problems of this technique is that, sometimes after all the process, a homogeneous bilayer instead of bilayer patches is obtained. This can be due to two main reasons, first, the passivation process is failing (*see* **Note** 7) and second, the time you incubate the SUVs is longer than needed. This depends on the size of the patterns. Decreasing the incubation time of SUVs to 20–30 s can be used to improve the result. Also, a dilution of the SUVs solution can be used, for example,

diluting 30 μL of SUVs to a final volume of 100 μL with PBS and use that dilution to prepare the supported lipid bilayer.

10. To ensure that thc stamp is clean of rests of protein or dust, put it into contact with a piece of adhesive tape and make pressure over the stamp. Remove the tape gently, and it should be ready to reuse.

References

1. Groves JT, Dustin ML (2003) Supported planar bilayers in studies on immune cell adhesion and communication. J Immunol Methods 278:19–32
2. Mouritsen OG (2011) Model answers to lipid membrane questions. Cold Spring Harb Perspect Biol 3:a004622
3. Sackmann E, Tanaka M (2000) Supported membranes on soft polymer cushions: fabrication, characterization and applications. Trends Biotechnol 18:58–64
4. Tamm LK, McConnell HM (1985) Supported phospholipid bilayers. Biophys J 47:105–113. doi:10.1016/S0006-3495(85)83882-0
5. Boxer SG (2000) Molecular transport and organization in supported lipid membranes. Curr Opin Chem Biol 4:704–709
6. Nicolson GL (2014) The fluid-mosaic model of membrane structure: still relevant to understanding the structure, function and dynamics of biological membranes after more than 40 years. Biochim Biophys Acta 1838:1451–1466
7. Manz BN, Groves JT (2010) Spatial organization and signal transduction at intercellular junctions. Nat Rev Mol Cell Biol 11:342–352
8. Brian AA, McConnell HM (1984) Allogeneic stimulation of cytotoxic T cells by supported planar membranes. Proc Natl Acad Sci 81:6159–6163
9. Chan P-Y et al (1991) Influence of receptor lateral mobility on adhesion strengthening between membranes containing LFA-3 and CD2. J Cell Biol 115:245–255
10. Grakoui A et al (1999) The immunological synapse: a molecular machine controlling T cell activation. Science 285:221–227
11. Sánchez MF, Levi V, Weidemann T, Carrer DC (2015) Agonist mobility on supported lipid bilayers affects Fas mediated death response. FEBS Lett 589:3527–3533
12. Richter RP, Bérat R, Brisson AR (2006) Formation of solid-supported lipid bilayers: an integrated view. Langmuir 22:3497–3505
13. Tepass U, Tanentzapf G, Ward R, Fehon R (2001) Epithelial cell polarity and cell junctions in *Drosophila*. Annu Rev Genet 35:747–784
14. Brown D, London E (1998) Functions of lipid rafts in biological membranes. Annu Rev Cell Dev Biol 14:111–136
15. Dykstra M, Cherukuri A, Sohn HW, Tzeng S-J, Pierce SK (2003) Location is everything: lipid rafts and immune cell signaling. Annu Rev Immunol 21:457–481. doi:10.1146/annurev.immunol.21.120601.141021
16. Zhao X-M, Xia Y, Whitesides GM (1997) Soft lithographic methods for nano-fabrication. J Mater Chem 7:1069–1074
17. Rai-Choudhury P (1997) Handbook of microlithography, micromachining, and microfabrication: microlithography, vol 1. Iet
18. Dontha N, Nowall WB, Kuhr WG (1997) Generation of biotin/avidin/enzyme nanostructures with maskless photolithography. Anal Chem 69:2619–2625
19. Revzin A et al (2001) Fabrication of poly (ethylene glycol) hydrogel microstructures using photolithography. Langmuir 17:5440–5447
20. Guo LJ (2007) Nanoimprint lithography: methods and material requirements. Adv Mater 19:495
21. Torres AJ, Wu M, Holowka D, Baird B (2008) Nanobiotechnology and cell biology: micro- and nanofabricated surfaces to investigate receptor-mediated signaling. Annu Rev Biophys 37:265–288. doi:10.1146/annurev.biophys.36.040306.132651
22. Groves JT, Ulman N, Boxer SG (1997) Micropatterning fluid lipid bilayers on solid supports. Science 275:651–653. doi:10.1126/science.275.5300.651
23. Xia Y, Whitesides GM (1998) Soft lithography. Annu Rev Mater Sci 28:153–184
24. Whitesides GM, Ostuni E, Takayama S, Jiang X, Ingber DE (2001) Soft lithography in biology and biochemistry. Annu Rev Biomed Eng 3:335–373
25. Orth RN et al (2003) Creating biological membranes on the micron scale: forming

patterned lipid bilayers using a polymer lift-off technique. Biophys J 85:3066–3073
26. Kung LA, Kam L, Hovis JS, Boxer SG (2000) Patterning hybrid surfaces of proteins and supported lipid bilayers. Langmuir 16:6773–6776. doi:10.1021/la000653t
27. Hovis JS, Boxer SG (2001) Patterning and composition arrays of supported lipid bilayers by microcontact printing. Langmuir 17:3400–3405. doi:10.1021/la0017577
28. Kam L, Boxer SG (2001) Cell adhesion to protein-micropatterned-supported lipid bilayer membranes. J Biomed Mater Res 55:487–495
29. Groves JT, Boxer SG (2002) Micropattern formation in supported lipid membranes. Acc Chem Res 35:149–157. doi:10.1021/ar950039m
30. Weghuber J et al (2010) Temporal resolution of protein–protein interactions in the live-cell plasma membrane. Anal Bioanal Chem 397:3339–3347. doi:10.1007/s00216-010-3854-x
31. Bally M et al (2010) Liposome and lipid bilayer arrays towards biosensing applications. Small 6:2481–2497. doi:10.1002/smll.201000644
32. Dutta D, Kam LC (2013) Micropatterned, multicomponent supported lipid bilayers for cellular systems. Methods Cell Biol 120:53–67
33. Costantino S, Heinze KG, Martínez OE, De Koninck P, Wiseman PW (2005) Two-photon fluorescent microlithography for live-cell imaging. Microsc Res Tech 68:272–276
34. Kunik D, Aramendia PF, Martínez OE (2009) Single photon fluorescent microlithography for live-cell imaging. Microsc Res Tech. doi:10.1002/jemt.20748
35. Sánchez MF, Dodes Traian MM, Levi V, Carrer DC (2015) One-photon lithography for high-quality lipid bilayer micropatterns. Langmuir 31:11943–11950

Approaches for Preparation and Biophysical Characterization of Transmembrane β-Barrels

Bharat Ramasubramanian Iyer*, Ankit Gupta*, and Radhakrishnan Mahalakshmi

Abstract

Membrane proteins are influential members of the cellular machinery, as they are involved in a range of critical functions from immune receptors and ion transporters to enzymes and signaling molecules. These facets render membrane proteins as ideal drug targets. Despite their significance in biological systems, our knowledge of the structural and functional behavior of membrane proteins remains considerably limited, when compared with their soluble counterparts. This limitation continues to persist, owing largely to the difficulties in membrane protein purification and amenable methods for their characterization. In this review, we outline the methods for the extraction and purification of transmembrane β-barrel proteins, which will serve as a guide for generating membrane protein preparations in high amounts and of sufficient purity. Next, we provide a detailed description of the currently available protocols for efficient refolding of β-barrel membrane proteins in both lipidic and detergent systems. These protocols are designed to enable the protein to attain its native structural and functional characteristics in any chosen membrane mimetic environment. Further, we detail the current methodologies being employed for the biophysical and structural characterization of transmembrane β-barrels, using illustrative examples drawn from recent studies using various β-barrel membrane proteins. This consolidated summary of protocols and strategies will prove highly useful to membrane biologists and can be applied to the study of any uncharacterized β-barrel membrane protein.

Keywords β-Barrel folding, Biophysical characterization, Functional study, Inclusion body preparation, Membrane protein folding, Molecular spectroscopy, Protein expression and purification

Abbreviations

β-OG	*n*-Octyl-β-D-glucoside
τ_c	Rotational correlation time
BLM	Bilayer lipid membrane
CD	Circular dichroism
D7PC	1,2-Diheptanoyl-*sn*-glycero-3-phosphocholine
DDM	*n*-Dodecyl β-D-maltoside
DHPC	1,2-Dihexanoyl-*sn*-glycero-3-phosphocholine
DLPC	1,2-Dilauroyl-*sn*-glycero-3-phosphocholine
DMPC	1,2-Dimyristoyl-*sn*-glycero-3-phosphocholine

Arun K. Shukla (ed.), *Chemical and Synthetic Approaches in Membrane Biology*, Springer Protocols Handbooks, (2017) 49–116, DOI 10.1007/8623_2016_4, © Springer Science+Business Media LLC 2016, Published online: 28 September 2016

DPC	*n*-Dodecylphosphocholine
DPR	Detergent-to-protein ratio
FRET	Förster resonance energy transfer
HSQC	Heteronuclear single quantum coherence
IB	Inclusion bodies
IEX	Ion-exchange chromatography
k_q	Bimolecular quenching constant
K_{SV}	Stern-Volmer constant
LDAO	Lauryldimethylamine oxide
LPR	Lipid-protein ratio
MP	Membrane protein
MS	Mass spectrometry
NMR	Nuclear magnetic resonance
$<t>$	Average fluorescence lifetime

1 Introduction

Proteins represent a class of macromolecules made up of long polypeptide chains consisting of a specific combination of 20 different amino acids. Proteins maintain cellular homeostasis by providing structural integrity and performing a variety of functions in the complex cellular environment [1–8]. Structural proteins help build the framework on which functional proteins, including enzymes and other signaling molecules, carry out their tasks. On the basis of structural and functional requirements, proteins adopt several conformations defined by a variety of weak or strong interactions such as hydrogen bonds, ionic bonds, disulfide bonds, hydrophobic interactions, and van der Waals' interactions [9–15]. These interactions result in the formation of different secondary structures and promoting specific folds of the protein, which forms the characteristic of that particular macromolecule. The primary structure is defined by the ordered array of specific amino acid units, linked by peptide bonds. The secondary structure is a relatively more complex conformation resulting from non-covalent interactions and predominantly stabilized by hydrogen bonding. There are two major classes of secondary structures, namely, α-helix and β-sheets. By combining these structures in a myriad of ways, proteins attain their tertiary and quaternary structure [4, 9, 11, 16–18].

Proteins reside in different, but specific, cellular environments, on the basis of their functional requirement, and can be separated into two broad classes, namely, soluble proteins (including cytoplasmic, periplasmic, and secretory proteins) and membrane proteins (accounting for integral membrane proteins as well as membrane-anchored proteins). Soluble proteins typically reside in the aqueous phase – such as the cytoplasm of cells – where their interacting partners are other molecules in the intracellular milieu

[19–24]. On the other hand, membrane proteins (MPs) are surrounded by an entirely different microenvironment (the hydrophobic surroundings of the lipid membrane) and function as receptors, enzymes, and channels. Membrane proteins maintain a delicate balance between the intra-/extracellular environment of the cell. They represent ~30% of the total proteome and are involved in a variety of functions ranging from transport of ions and metabolites across the membrane to facilitating cell-cell communication. Due to their abundance and functional importance in biological systems, they are considered as an excellent target for rationalized drug design [25–32].

It is well known that MPs are involved in numerous biological processes. Sometimes, instantaneous interruption in metabolic processes affecting MPs results in the occurrence of several debilitating disease conditions. Membrane proteins have been implicated in various diseases such as deregulation of apoptosis, neurodegenerative diseases caused by protein misfolding, and aggregation such as Alzheimer's, Parkinson's, and Huntington's disease [16, 33–37]. Membrane proteins are involved in pathogenesis caused by bacterial infection (e.g., plague, diarrhea, cholera), as they are mediated by receptors or anchored proteins. Such diseased conditions can arise due to several reasons, such as mutation in polypeptide sequence, protein overexpression, underexpression of chaperones, protein modification by exogenous agents, elevated temperature, oxidative stress, and chemical or physical damages. Stress conditions may result in a change in the total architecture of the protein, or it may affect the folding or trafficking, rather than specifically affecting protein function [16]. Thus, a detailed understanding of physicochemical properties of MPs is critical to deduce their role in disease conditions.

Membrane proteins show contrasting structural characteristics compared to soluble proteins. While soluble proteins have a buried hydrophobic core, membrane proteins have exposed hydrophobic amino acids, which allows them to be accommodated in the lipid environment [38–40]. Further, soluble proteins adopt a myriad of structures, achieved by permutations of the two secondary structure elements described earlier. However, the restrained lipid milieu promotes two structural scaffolds for MPs. Hence, membrane proteins can be classified into two groups on the basis of their secondary structure: α-helices and β-barrels [41]. Helical MPs are found in the plasma membrane and inner membranes of mitochondria, chloroplast, and bacteria. Outer membranes of bacteria, mitochondria, and chloroplast have thinner membranes and usually possess β-barrel MPs. The MP structure and topology strongly depend upon the polypeptide sequence and proper insertion of the polypeptide chain into membrane bilayers [42–45]. Another crucial feature for the folding and insertion of membrane proteins is the hydrophobicity of the polypeptide sequence. Researchers have carried out a great deal of analysis on the hydropathy of individual amino acids, which in turn determines the propensity of the polypeptide chain to insert into the membrane [25, 46–56]. These studies also help us predict the secondary structural framework of the protein based on its primary sequence. Transmembrane α-helical bundles consist of single-pass or multi-pass transmembrane helices, primarily stabilized by hydrogen bonding [9, 11, 25, 46, 47]. Single-pass helices mainly function as anchoring elements to the extra-membrane functional domains. Various short-range and long-range inter-helix interactions allow the formation of complex helix bundle architecture in multi-pass helices. These proteins function as ion transporters, anchors, receptors, and enzymes and are involved in a variety of cellular processes [57–68].

β-Barrel proteins, found in the outer membranes of bacteria, mitochondria, and chloroplasts, execute a range of functions, such as transport of ions and metabolites, or they perform the role of adhesins, invasins, protein secretion pore, and protease pore-forming toxins [69–80]. Structurally, they are quite distinct from α-helical membrane proteins, since β-barrels consist of a framework made up of about 8–24 strands, connected by antiparallel hydrogen-bonding network across the strands (Fig. 1). The barrel core (interior) usually projects polar or hydrophilic residues, but can also remain hydrophobic, based on the function. The barrel exterior faces the lipid and is usually made of hydrophobic aliphatic amino acids. In past several years, several transmembrane α-helical and β-barrel proteins have been characterized. To date, the structures of 2,420 helical proteins and 338 β-barrels have been deduced using crystallography, NMR, and EM-based methods [81–84]. Of the total 2,758 structures that are currently available, only 609 structures are unique [85]. Our current understanding of β-barrel proteins is limited. Several biophysical and structural methods have been developed to further our understanding of these proteins [71, 76, 86, 87]. In this chapter, we describe biophysical approaches to characterize β-barrel proteins and how they have been applied to various bacterial and eukaryotic transmembrane barrels.

1.1 Why Are Membrane Proteins Less Characterized?

The involvement of membrane proteins in several biological processes makes it imperative to characterize them in detail; yet, studies on membrane proteins have lagged behind significantly. To understand the role of any biological molecule, deriving its structural and biophysical characteristics and gaining information regarding its interactome become important in the fields of applied science and clinical biology. Upon examination of protein structure databases and available literature, one can easily infer that our understanding of membrane proteins is limited as compared to soluble proteins. The reason behind this is the fact that membrane proteins have partially hydrophobic surfaces, resulting in problems associated with crystallization and other means of structure determination

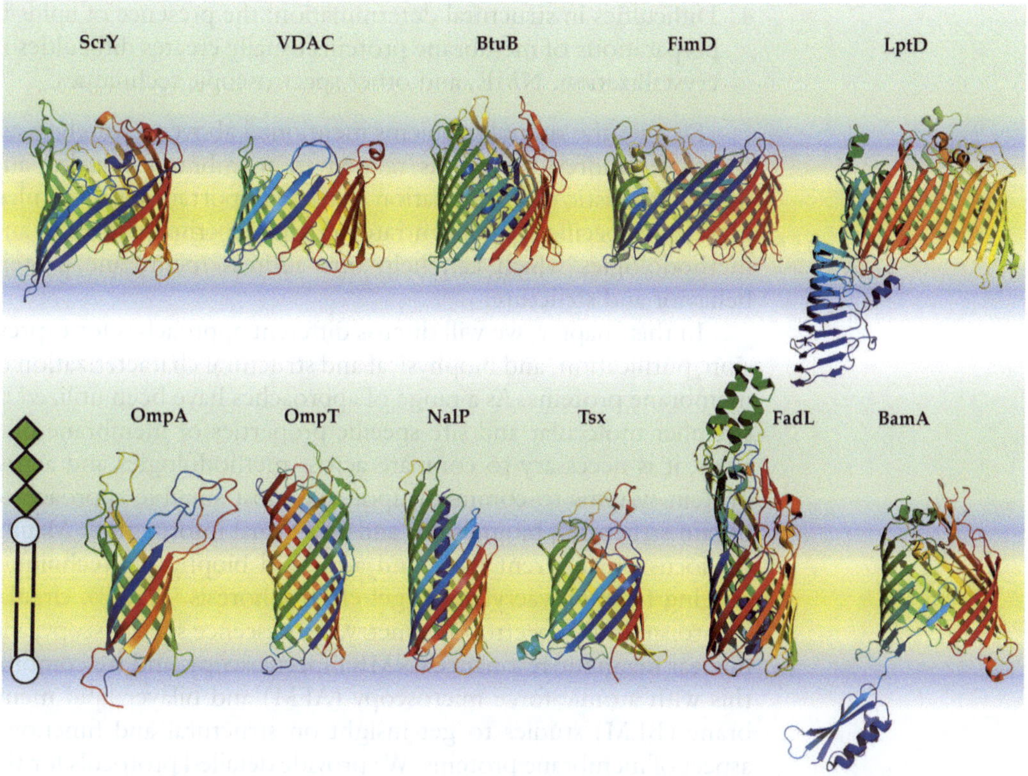

Fig. 1 Cartoon representation of transmembrane β-barrel proteins. Ribbon diagram of transmembrane (TM) β-barrels from Gram-negative bacterial (ScrY, BtuB, FimD, and LptD (*top row*), OmpA, OmpT, NalP, Tsx, FadL, and BamA (*bottom row*)) and human (VDAC, *top row*) origin. Hydrophobic region of the membrane and hydrophilic regions are demarcated by *yellow* and *blue color*, respectively. TM strands occupy the hydrophobic lipidic region, whereas loops are presented to the polar environment. TM strands are arranged in antiparallel fashion, with the exception of VDAC, where the 18th and 19th strands show parallel arrangement (notice how the *blue* and *red arrowheads* for the first and last strands point in the same direction). Despite their structural similarity, the different β-barrels perform a range of functions, such as proteases, porins, or receptors, and are involved in small ion transport, nucleoside specific diffusion, long-chain fatty acid transport, or the formation of barrel assembly machinery complex. Reproduced with permission from Kleinschmidt [76]

[81, 82]. Apart from the problems directly associated with their own chemical nature, there are several experimental hurdles in handling membrane proteins, some of which are outlined below:

1. Low expression levels: biophysical characterization needs a large amount of protein.
2. Lipid-protein interaction: membrane proteins require lipid environment to be folded into their native form.
3. Difficulties in solubilization: due to the high proportion of hydrophobic amino acid residues, membrane proteins have a high propensity to aggregate.

4. Difficulties in structural determination: the presence of lipids in preparations of membrane proteins usually creates difficulties in crystallization, NMR, and other spectroscopic techniques.

Despite the several problems mentioned above, researchers are trying to address questions related to membrane proteins and obtain atomistic characterization of these important biomolecules. Here, we specifically concentrate on the experimental tools and methodologies which can help one address membrane protein behavior and structure.

In this chapter, we will discuss different approaches for expression, purification, and biophysical and structural characterization of membrane proteins. As a range of approaches have been utilized to decipher molecular and site-specific properties of membrane proteins, it is necessary to compare across methodologies and across protein systems to comment upon the most suitable approach to obtain structural, biophysical, and functional information. Mainly, we focus on different basic and advanced biophysical techniques ranging from polyacrylamide gel electrophoresis (PAGE), circular dichroism (CD) spectropolarimetry, fluorescence spectroscopy, and nuclear magnetic resonance (NMR) spectroscopy, and we combine this with atomic force microscopy (AFM) and bilayer lipid membrane (BLM) studies to get insight on structural and functional aspect of membrane proteins. We provide detailed protocols for the different methodologies, along with the benefits and limitations of each method, which will help the reader opt for the best strategy for molecular dissection of membrane proteins.

2 Production of Membrane Proteins for Characterization

2.1 Native Purification Versus Heterologous Expression

A number of approaches have been employed for protein production by means of both homologous and heterologous expression systems (Fig. 2). Homologous expression systems designate the expression of the desired protein in the same biological host. Heterologous expression system refers to directed overexpression of the protein in cells different from the host organism, for example, human or yeast protein expressed in *E. coli* [88–92]. Protein expression can be achieved by the combination of various steps [93–95]; some of these are as follows:

1. Identification and amplification of the desired gene (corresponding to particular protein) from host genome.
2. Cloning of obtained gene product into suitable plasmids; lambda phage; cosmids; phagemids; artificial chromosomes from bacteria, yeast, or human origin (BAC, YAC, and HAC3, respectively); and transformation in any suitable prokaryotic or eukaryotic (*E. coli*, yeast, etc.) expression system.

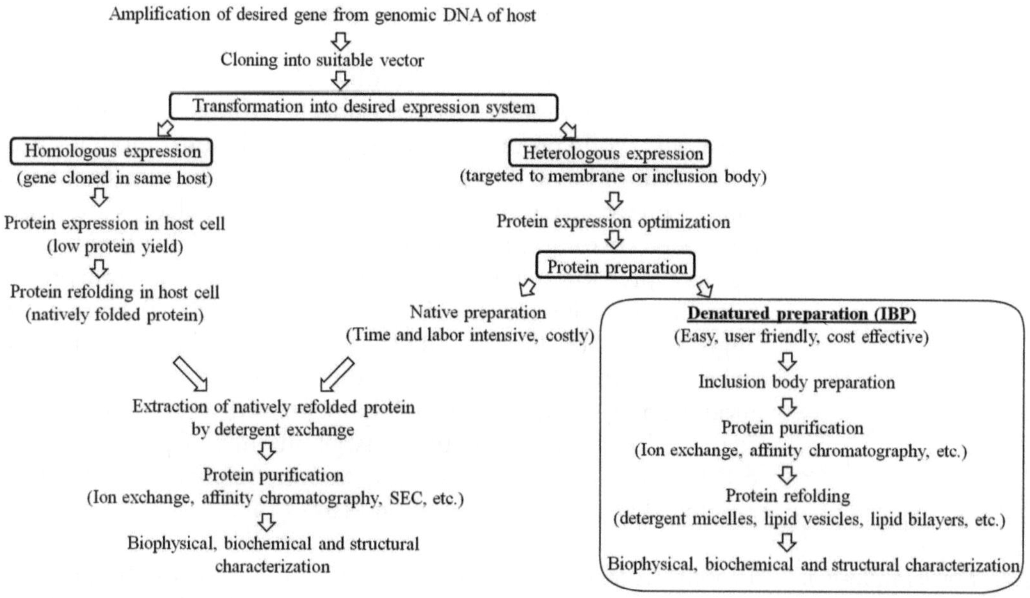

Fig. 2 Summary of steps involved in membrane protein preparation. Homologous (*left panel*) and heterologous (*right panel*) protein expression systems comprise of different procedures for protein purification, resulting in natively and in vitro refolded protein for further studies. Protein expressed using homologous and heterologous expression systems can both be purified by native preparation, mediated by detergent solubilization. The commonly used denatured preparation (*boxed, right panel*) is discussed in detail in this review. Here, protein is expressed as aggregated inclusion bodies (IBs) and refolded in vitro into detergent micelles or lipid vesicles

3. Identification of suitable condition for overexpression of the protein, followed by protein isolation.

Each step described above has to be chosen carefully, as both prokaryotic and eukaryotic systems have different advantages and disadvantages. Several strategies have been employed to expedite cloning of heterologous genes in different expression systems by simplification of the steps involved. Identification and amplification of desired genes have been simplified from genomic DNA by whole-genome amplification (WGA) principles, based on molecular biology tools such as multiple displacement amplification (MDA) and polymerase chain reaction (PCR) [96–98]. Furthermore, cloning strategies have been modified to such an extent that cloning time has been significantly reduced [99, 100]. The use of different cells or organisms to improve protein expression methodologies has been investigated to a great extent for a large variety of proteins [89, 101–105].

Both approaches (homologous and heterologous) have benefits and drawbacks and can be used on the basis of experimental requirement. Homologous protein expression systems are preferred for functional studies, due to the advantage that they provide protein folded in native conformation. The major limitation associated with this approach is that proteins are obtained in very small quantities (0.1–5 mg/l culture) [106]. On the other hand, heterologous systems (proteins expressed in bacterial cells such as *E. coli*) do not provide natively folded protein but can offer a large amount of protein. The disadvantage of using heterologous system is that most eukaryotic proteins undergo posttranslational modifications, which cannot be obtained when prokaryotic expression systems are being utilized. To overcome this barrier, several eukaryotic cell lines have been customized for protein production. While using heterologous expression systems, membrane proteins are usually expressed as inclusion bodies (IBs), and the protein obtained will represent an aggregated state (Fig. 2). Heterologous expression is usually preferred for proteins that do not require extensive posttranslational modifications to attain the folded state and for biophysical and structural studies that require a large amount of protein. Several cell-free and in vitro protein expression systems have also been employed for protein production [107–114].

2.2 Steps Involved in Obtaining Protein for Biophysical and Biochemical Characterization

Protein overexpression as inclusion bodies remains the preferred method for generating sufficient quantities (10–100 mg/l culture) of purified membrane proteins [115]. We outline the steps involved in this process, in this section:

1. *Cloning*: Amplification of the gene of interest and cloning into a suitable vector is the first critical step. The protein-coding sequence is amplified from the host genomic DNA using primers complementary to gene sequence. Occasionally, we find that codon-optimized synthetic genes provide better expression levels, particularly when bacterial systems are used to overexpress human membrane proteins. The signal sequence is not included when expression as inclusion bodies is desired. PCR-amplified genes are cloned downstream to the promoter region of commercially available vectors such as pET, pBlueScript, pUC, pBR322, and pGEX [115–119]. An affinity tag such as glutathione S-transferase (GST) and hexa-His is usually included to facilitate purification at a later stage.

2. *Expression screening*: In the last decades, several modified *E. coli* BL21(DE3) expression strains such as C41 and C43 (and their pLys versions), Rosetta, and NovaBlue are available for ready production of membrane proteins. We find that in cases wherein BL21 (DE3) cells do not provide desired protein levels, C41 cells are most conducive for rapid protein production. However, we have observed that inclusion bodies prepared using C41 cells require a more rigorous processing than the conventional DE3 cells. A usual step with transformation-positive cells is to screen for the most suitable expression condition [growth temperature, induction time and duration, concentration of inducer (isopropyl β-D-1-thiogalactopyranoside (IPTG), etc.)] for high-level protein production [101, 115, 120, 121]. As pET vectors have higher translation

efficiency, the protein produced will accumulate in the cells in the form of inclusion bodies. Most proteins can be produced in sufficient quantities after a 3-h induction when induced at a culture OD_{600} of ~0.8 [121]. However, in our laboratory, we find that a 7-h induction with IPTG is more suited for VDAC expression [122, 123].

3. Inclusion body preparation: The aim of inclusion body preparation (IBP) is to extract aggregated protein (in the form of inclusion bodies) from *E. coli* by cell lysis, with minimal contamination from other cellular proteins and lipids [124]. This process is easy to carry out, as inclusion bodies are considerably pure. Several methods are available for inclusion body preparation, with many of them being protein specific and possess many advantages and disadvantages [125–133]. In our laboratory, we have optimized the IBP such that it can be used for a wide variety of membrane proteins and can be cost effective, less laborious, and completed in the fewest possible steps [133]. We outline the procedure in this section (Fig. 3).

The IBP is divided into the following segments, and each step plays an important role in determining the final purity of the preparation:

1. Cell lysis
2. Removal of cell debris
3. Elimination of protein impurities
4. Removal of lipidic impurities
5. IB protein recovery and estimation of purity

The complete protocol is described in the form of a flowchart comprising of detailed information related to each of these steps [133]. The description also includes specific information pertaining to incubation times as well as centrifugation conditions and, therefore, provides a complete guideline for obtaining sufficiently purified IBs. It is important that each of these steps be carried out carefully, since minute errors in any of the steps can lead to a considerable reduction in final protein purity. We have successfully tested this method on several transmembrane β-barrels from both bacteria and humans, with reproducible results [133]. This protocol can be utilized for characterizing novel proteins, without the need for modifications.

4. Protein purification and storage: Purified protein for biophysical, biochemical, structural, and functional studies can be directly obtained from native purifications (wherein homologous expression is used) or in the unfolded/ aggregated state from IBPs [126].

Membrane proteins are classified on the basis of their interaction with the lipid bilayer into peripheral and integral membrane proteins. Generally, MPs are associated with the lipid membrane through covalent, non-covalent, or hydrophobic interactions. Low

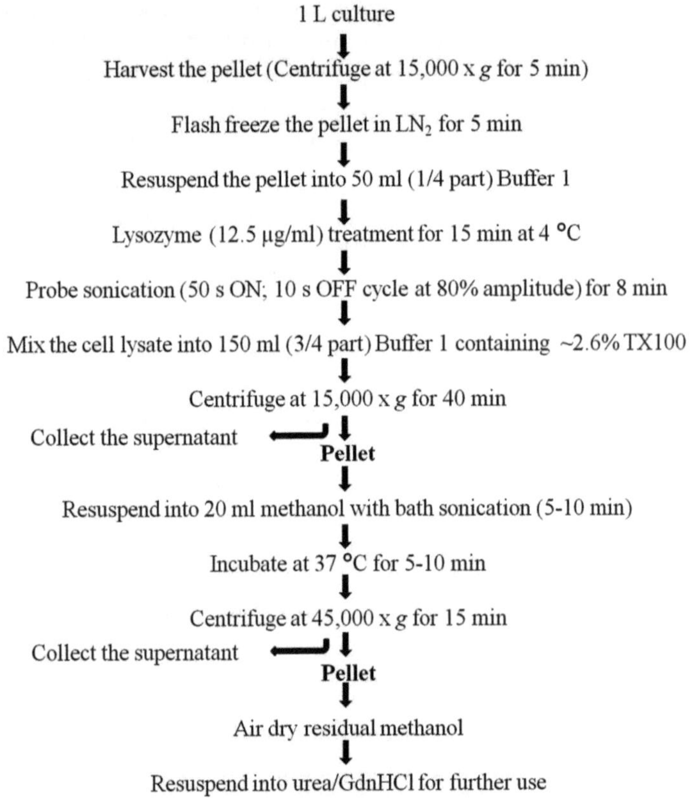

Fig. 3 Protocol for modified rapid inclusion body preparation. Steps involved in inclusion body preparation (IBP) of a 1 L bacterial culture are shown. The steps mentioned above need to be followed strictly for efficient removal of protein and lipid impurities, as minor changes in the protocol can affect the final purity of the IB pellet. The protocol has been tested for a large variety of transmembrane β-barrel proteins and can be used for higher culture volumes without significantly increasing the processing time. Mentioned centrifugation speeds, wash volumes, detergent, and DTT concentrations are optimal to obtain clean inclusion bodies, but can be adjusted carefully when used for extraction of novel proteins. All the supernatant (washes) and final IB pellet can be checked on SDS-PAGE, to ensure minimal protein loss and confirm purity of the preparation. The extracted IB pellet can be purified using denaturant purification (after solubilizing in urea or GdnHCl), and the purified protein can be used for biophysical and functional studies. Redrawn from [133] with permission from the Royal Society of Chemistry

expression of membrane proteins is the primary challenge in the field of membrane protein purification. The extraction of native protein from an organism requires disruption of phospholipid bilayer and solubilization into detergent micelles [134–136]. Membrane proteins tend to become highly unstable and prone to aggregation, when they are outside their native environment. Therefore, it becomes important to solubilize them using appropriate detergent conditions, which helps them retain their structural integrity and biological function. These properties make it difficult to generate homogenous preparations of membrane protein in sufficiently high amount and purity [91]. Thus, purification of MPs without disruption of native structure is a challenge [102, 120].

We discuss native protein purification using the human VDAC protein as the example and compare the results with VDAC generated from inclusion bodies. The voltage-dependent anion channel, or VDAC, is present in the outer mitochondrial membrane in three different isoforms, VDAC-1, VDAC-2, and VDAC-3 [137–143]. The purification of human VDAC has been attempted from the mitochondrial outer membrane (Fig. 4). The purification process comprises of different steps, starting with the isolation of mitochondria, followed by solubilization in mild detergents and chromatographic separation. First, the rat liver tissue is homogenized in cold isotonic sucrose solution. The cell lysate is then subjected to differential centrifugation to remove cell debris and isolate mitochondria [144]. The mitochondrial membrane is then solubilized into nonionic detergents (triton X-100 or Genapol X-100), followed by multiple rounds of ion-exchange chromatography (Fig. 4). HTP columns are particularly useful to purify VDAC, as it binds the refolded protein based on the ionic strength of the buffer used [145, 146].

Native purification of human VDAC using detergent exchange has been employed for functional assays as well as structure determination. Other ways of purifying VDAC using affinity chromatography have also been utilized [147]. However, in the past decade, it has been accepted that protein prepared in vitro also possesses similar properties as the natively purified protein [148, 149]. Furthermore, in vitro preparations are less laborious, save time, and provide a higher protein yield required for a large number of structural and functional studies [133, 149]. For expressing the full-length VDAC in *E. coli* as inclusion bodies, we find that BL21 (DE3) cells provide us with protein preparations that contain less contaminating lipid impurities than C41 cells [122]. Inclusion bodies enriched with VDAC can be purified using ion-exchange chromatography and affinity columns. Human VDAC-2 isoform, for example, can be purified in a single step of anion-exchange chromatography using the chaotropic agent urea as the denaturant and an NaCl gradient for protein elution. On the other hand, human VDAC-3 requires two purification steps – anion-exchange followed by Ni-NTA affinity chromatography to obtain protein preparations that are >95% pure [150].

VDAC purifications are carried out under highly alkaline buffer conditions (20 mM Tris-HCl pH 9.5) as the theoretical pI of VDACs lies between ~7.5 and 8.5. Concomitantly, anion-exchange matrices are most suitable for these buffers. Although VDACs

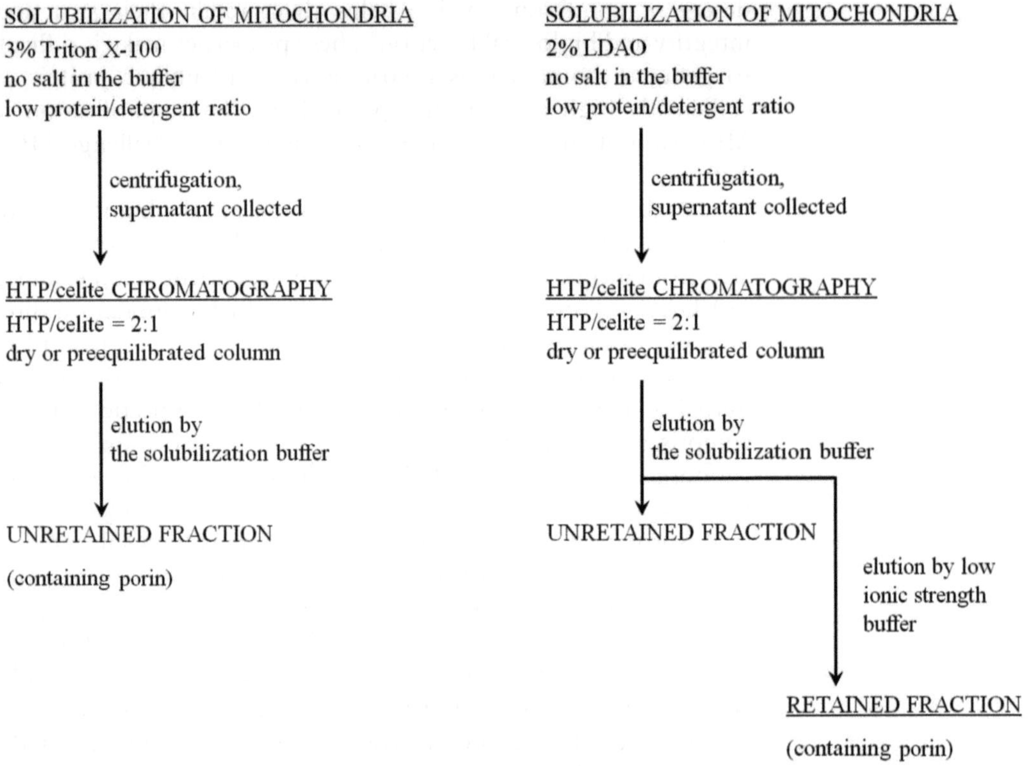

Fig. 4 Flow chart describing native purification of porin from mitochondria. Schematic representation of VDAC purification using detergent solubilization in triton X-100 (*left panel*) and LDAO (*right panel*). The mitochondrial membrane is directly solubilized in detergent micelles and subjected to multiple steps of ion-exchange purification, using hydroxyapatite (HTP) and celite columns. Porin purification by using triton X-100 (*left panel*) is fast and is a simple purification procedure and allows the preparation of high amounts of pure porin in a single chromatographic step. On the other hand, when LDAO-mediated purification is attempted, porin shows high-binding affinity to the chromatographic material. The bound porin can then be specifically eluted by the addition of a buffer of moderate ionic strength. The binding of porin to the hydroxyapatite/celite column in the presence of LDAO enables on-column detergent exchange, which facilitates porin crystallization. Redrawn with permission from Palmieri and De Pinto [146]

contain cysteine residues (human VDAC-1 has two, VDAC-2 has nine, and VDAC-3 has six cysteines), we find that the use of reducing agents for purification is not required [122].

Similar to hVDAC-2, the bacterial outer membrane β-barrels PagP (from *E. coli*, *Salmonella typhimurium*), OmpX (from *E. coli*), and Ail (from *Yersinia pestis*) can be purified from inclusion bodies using ion-exchange chromatography [121, 151–153]. The theoretical pIs of PagP, OmpX, and Ail are ~5.5, 5.1, and 7.8, respectively. On the basis of their different pIs, these proteins are purified using buffers of pH 8.5 (for PagP and OmpX) and 5.0 (for Ail). These proteins are eluted in their pure forms using an NaCl gradient, applied onto anion-exchange (for PagP and OmpX) or cation-exchange (for Ail) matrices. A representative IEX purification

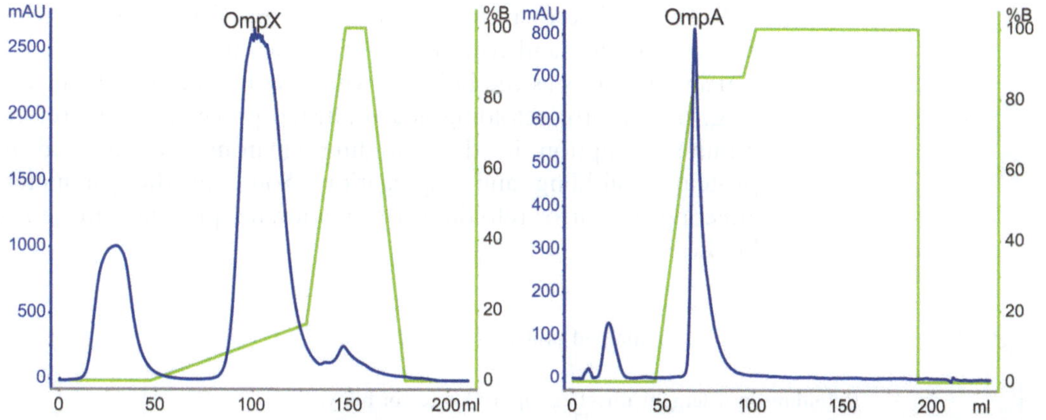

Fig. 5 Purification profiles of denatured barrels using ion-exchange chromatography. Denaturant-based purification of OmpX (*left*) and OmpA (*right*) in buffer containing 8.0 M urea using anion-exchange column. The *green line* represents the salt (NaCl) gradient used for purification, and the *blue line* indicates the absorbance recorded at 280 nm. Note how the gradients are different for each protein. The elution corresponding to the protein of interest is indicated. Reproduced with permission from Maurya, Chaturvedi, and Mahalakshmi [121]

profile is provided in Fig. 5. Fractions containing pure protein is usually dialyzed against water to remove salts and urea and lyophilized to obtain a white protein powder, for long-term storage. We find that IEX offers advantages over affinity chromatography (such as His-tag-based purifications), since it can effectively remove residual lipids that are strongly associated with the protein. Affinity columns can be effective to remove protein impurities (as we see for VDAC-3 preparations), but they are ineffective in providing lipid-free preparations [150].

Both native and denatured purification procedures have their advantages and disadvantages, and the choice of purification method strictly depends on the study being carried out [147]. For heterologous protein expression systems, wherein the protein is expressed as inclusion bodies, denaturant-based purification is preferred, since inclusion bodies (aggregated form of protein) cannot be dissolved directly in any solvent without the addition of chaotropic agents [126]. However, this method is limited to the generation of those proteins which can successfully be refolded to the near-native state. We now look into known procedures for membrane protein refolding.

5. *In vitro refolding*: In vivo, transmembrane β-barrel proteins require the assistance of several translocons, holdases, and chaperones (SecYEG, Skp, DegP, SurA, and BAM) for translocation and insertion in a functional conformation in the outer membrane [55, 154–158]. On the other hand, in vitro refolding is largely unassisted. It is achieved by the rapid dilution of unfolded protein prepared in a denaturant, into lipid mimetic systems that can support a functional and stable fold of the protein [40, 126,

159–165]. These include detergent micelles, lipid vesicles, bilayers, bicelles and nanodiscs. Here, the chaotropic denaturant usually functions as the holdase, while the folding is spontaneous (Fig. 6). In vitro refolding is a multistep process, and therefore, minute disruption in the refolding environment can lead to protein misfolding and aggregation. Some of the parameters affecting in vitro refolding of membrane proteins are listed below:

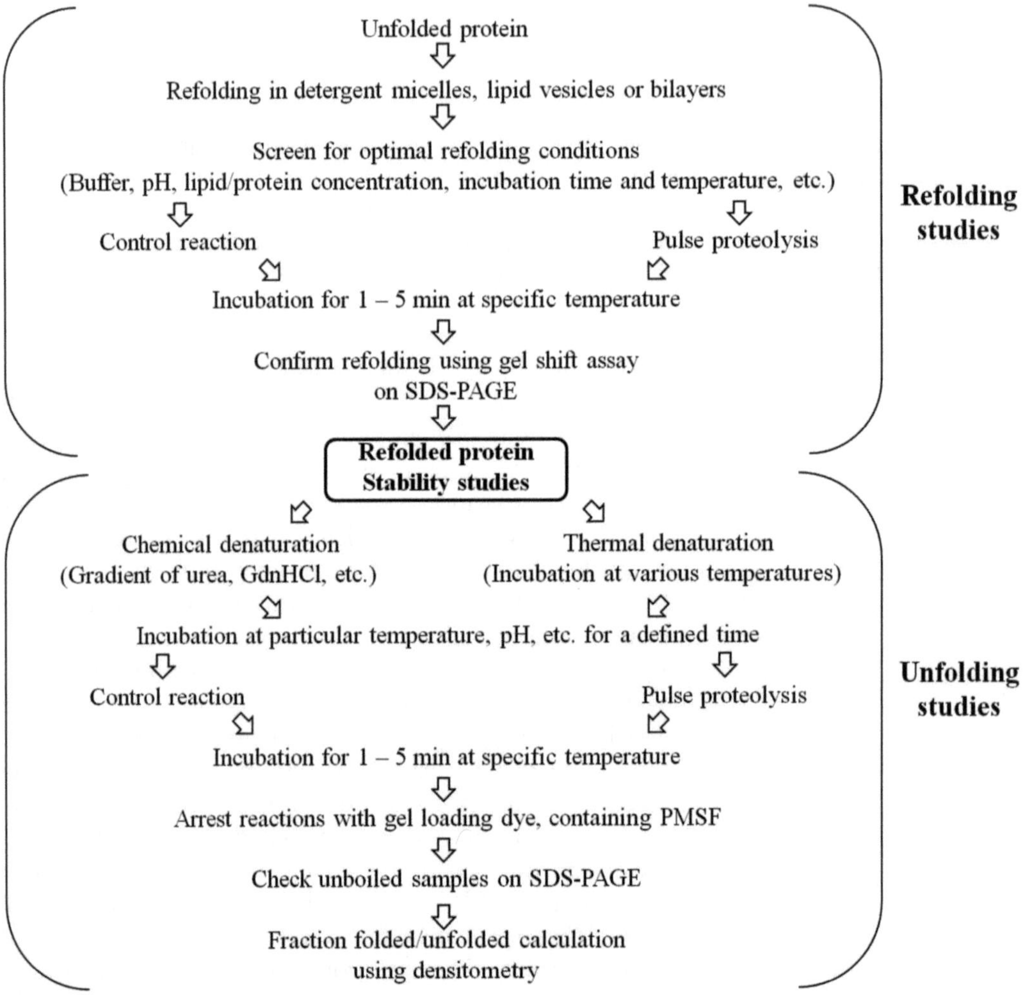

Fig. 6 Folding-unfolding studies of transmembrane barrels using SDS-PAGE. Refolding screens of membrane proteins under different conditions (pH, temperature, detergent or lipid concentrations, etc.) can be carried out using SDS-PAGE gel mobility shift assay (*upper panel*). The unfolded protein is diluted into detergent micelles or lipid vesicles to attain the folded conformation. The refolded protein can then be subjected to protease digestion to confirm the refolded state of the barrel. Once the protein is folded, the stability of folded barrel in the presence of chemical or thermal denaturant can be measured, followed by a brief exposure to a protease (pulse proteolysis). The susceptibility of the unfolded protein to protease digestion can be checked on SDS-PAGE gels and is then used to differentiate the folded and unfolded states of the protein by densitometry analysis

1. Choice of denaturant: Chemical (urea, guanidine hydrochloride) or detergent (tetraethylene glycol monooctyl ether (C8E4), 3-(*N,N*-Dimethylmyristylammonio)propanesulfonate (SB3-14))
2. Detergent properties: Critical micelle concentration, aggregation number, size and shape of micelles, and headgroup characteristics
3. Lipid characteristics: Chain length, chain saturation, size of vesicles, doping with saturated and unsaturated lipids or different headgroup or cholesterol, etc.
4. Packing defects in the lipid vesicle or membrane
5. Molecular composition of folding microenvironment: Buffer, pH, salt concentration, and oxidoreductive environment
6. Physical parameters: Temperature, physical vortexing or mixing, dilution rate, etc.
7. Protein and lipid concentration or lipid-protein ratio (LPR)
8. Dilution procedure (stepwise or rapid) of unfolded protein into lipid milieu
9. Presence of cofactor, substrate, or other small molecules

These parameters are crucial for proper refolding of the membrane protein. It is also necessary to choose conditions that preserve the folded state of the refolded protein during storage. Refolding conditions in artificial systems have been screened and successfully optimized for a variety of transmembrane β-barrels, such as hVDAC-2, OmpX, Ail, PagP, OmpA, OmpLA, OmpW, FhuA, etc. [69, 121, 122, 152, 153, 166–174]. We will discuss the refolding methods employed for β-barrel membrane proteins in detergent micelles and lipid vesicles, using hVDAC2, PagP, OmpX, and Ail as examples.

hVDAC-2 (human VDAC isoform 2) is a 19-stranded β-barrel, abundant in cysteine residues. Hence, refolding of hVDAC-2 is carried out in a reducing environment to prevent the formation of nonspecific disulfides, which can lead to protein oligomerization and aggregation [122, 138, 175, 176]. Hence, we achieve refolding of hVDAC-2 by first dissolving the denatured protein in 6.0 M guanidine hydrochloride (GdnHCl) containing a reducing agent such as 10 mM DTT (dithiothreitol). Complete reduction of all nonspecific disulfides can also be achieved by a brief exposure of the protein to incubation at marginally elevated temperatures. This step also facilitates complete solubilization of any dissolved aggregates. Refolding of hVDAC-2 is achieved when the GdnHCl-denatured reduced protein is rapidly diluted tenfold in a concentrated solution of detergent micelles (such as LDAO (lauryldimethylamine oxide) or DDM (n-dodecyl β-D-maltoside)) prepared in 50 mM phosphate buffer, pH 7.2, containing 100 mM NaCl, at a temperature of 4 °C. Low temperatures prevent the formation of unfolded hVDAC-2 aggregates. The rapid dilution step allows the formation of protein-micelle complexes. Further incubation with gentle mixing for 4–5 h at 4 °C promotes complete refolding of the protein. This method yields protein preparations that are >95% refolded [122, 177]. Further purification is not required, which is an added advantage.

In vitro folding of bacterial outer membrane β-barrels is easier to accomplish than their eukaryotic counterparts. For example, PagP, OmpX, and Ail are three eight-stranded transmembrane β-barrel proteins of bacterial origin, residing in the outer membranes of *S. typhimurium*, *E. coli*, and *Y. pestis*, respectively [121, 151–153]. PagP is nearly ubiquitous in all Gram-negative bacteria, while OmpX and Ail are believed to be homologous proteins. We achieve refolding of PagP and OmpX by carrying out a tenfold dilution of the protein denatured in 8.0 M urea, into the refolding reaction containing detergent micelles, prepared in 20 mM Tris-HCl pH 9.5 at 4 °C. Immediately after the rapid dilution, we administer a "heat shock" to the refolding mixture by transferring it from 4 °C to 70 °C. This heat shock is administered for 3 min, following which OmpX is incubated at 4 °C for 10 min, whereas the incubation time for PagP is increased to ~12 h. Through these simple steps, we obtain complete refolding of the barrel [151, 153]. We have shown that alteration in the lipid dynamics, induced by a change in temperature, can help in β-barrel membrane protein folding. The mechanism accelerating the folding is lipid independent and can be used for screening of protein folding conditions and generation of stable and functional barrels.

A similar heat shock method can effectively refold OmpX and OmpA (also from *E. coli*) in phosphocholine vesicles (e.g., DLPC and DMPC SUVs) [121]. However, the refolding procedure for PagP into lipid vesicles cannot be achieved by the procedure described above, and minor modifications are employed to achieve complete folding of the barrel. PagP refolding is achieved by a 100-fold rapid dilution of denatured PagP (0.4 mM) prepared in 8.0 M GdnHCl, into a final concentration of 5 mM lipid vesicles in 50 mM phosphate buffer pH 8.0 containing 7 M urea, at 25 °C. Then, we incubate the refolding reaction further at 25 °C for ~12 h [153, 161].

Despite possessing a similar barrel structure, the refolding procedure for Ail is achieved without the need for heat shock or incubation. Rapid tenfold dilution of denatured Ail prepared in 8.0 M GdnHCl, into detergent micelles (LDAO) prepared in 20 mM Tris-HCl pH 8.5 at 25 °C, is sufficient to achieve Ail refolding [152]. Once complete refolding is achieved, it is important to remove all the partially refolded, misfolded, and aggregated fractions. For this, aggregated protein is removed by high-speed centrifugation at 15,500×*g* for 1 h at the respective temperatures (4 °C for hVDAC-2, PagP, and OmpX; 10 °C for Ail). The final refolded protein prepared in the desired DPR (detergent-to-protein ratio) or LPR (lipid-protein ratio) is checked for refolding (as described in next sections) and can be used for biophysical, biochemical, and functional characterization [121–123, 151–153].

It is evident from this section that different transmembrane β-barrel proteins require specific refolding environments and protocols to achieve proper refolding of the barrel. As we can see, the refolding procedure for PagP differs when we move from detergent micelles to lipid vesicles, suggesting that refolding microenvironment strongly influences protein folding [153]. Similarly, when we compare refolding of two structurally homologous barrels OmpX and Ail, we find that the protein sequence drastically affects the folding of the barrel [151, 152]. We have also noticed that each protein requires different DPRs or LPRs and refolding microenvironment, further highlighting the importance of the folding milieu. For example, in LDAO micelles, optimal refolding efficiency is obtained for hVDAC-2 only in DPRs of 2,600:1–13,000:1 [122]. Beyond this range, we find that the stability of the folded protein is considerably affected. Hence, refolding conditions require careful handling and a great extent of optimization for biophysical, biochemical, and functional characterization. Owing to the elaborate screening process involved in identifying conditions that support a near-native structure for membrane proteins, it has been advocated that native preparations be used to study these important biomolecules [138, 175, 176].

3 Tools to Study Membrane Proteins

3.1 Polyacrylamide Gel Electrophoresis

Polyacrylamide gel electrophoresis (PAGE) is the most common and inexpensive technique used to characterize transmembrane β-barrel proteins on the basis of gel mobility shifts. Generally, PAGE is performed in native (e.g., native PAGE, blue native PAGE (BN-PAGE)) or denaturing conditions (e.g., SDS-PAGE, 2D-gel electrophoresis) [178, 179]. Native PAGE is the method of choice for characterizing protein-protein interactions, calculation of the native mass of the protein of interest, and determination of oligomeric states [178, 179]. However, the limitation with native PAGE is the tedious process of optimizing the buffer conditions and maintaining ideal environments to retain the native protein state. On the contrary, denatured gel electrophoresis is run in relatively harsh conditions where SDS and β-mercaptoethanol or DTT are used along with boiling at 95–100 °C for 3 min, as denaturing agents.

A revolutionary finding in the field of membrane protein folding was the discovery of anomalous migration exhibited by unboiled samples of refolded β-barrel proteins on SDS-PAGE gels, in comparison with their unfolded counterparts. This is generally referred as the gel shift assay for protein folding [180–182], where a difference in band migration is seen after protein folding in lipidic environments (micelles, vesicles, bilayers, nanodiscs, etc.). Here, the insertion of the β-barrel protein into the lipid milieu leads to a change in protein compaction, thus affecting the gel migration [69, 153, 181, 183–188]. In unboiled samples, it is believed that the lipid or detergent environment is replaced by SDS molecules, which preserves the folded state of the barrel and gives rise to the difference in mobility. It has been observed that for proteins with molecular weight less than ~18–20 kDa, the folded species shows retarded mobility on Laemmli gels, compared to the unfolded protein. This phenomenon is reversed in tricine gels [189] and when the molecular weight exceeds 20 kDa. In the case of helical membrane proteins, SDS binding and variation in acrylamide concentration result in migration change [180, 181]. The gel shift assay has been extensively used to monitor the folding of various transmembrane β-barrels such as OmpX, Ail, PagP, OmpLA, OmpA, Opa, PagL, FadL, OmpT, OmpW, etc., in the presence of different lipids and detergents [61, 73, 78, 152, 153, 161, 173, 183, 187, 190–195].

Another interesting observation in the case of membrane proteins is their ability to resist proteolysis when folded. Several studies with refolded transmembrane β-barrels report the persistence of the refolded protein on SDS-PAGE gels, after brief exposure to several proteases including subtilisin [196, 197], proteinase K [121], trypsin [198, 199], and thermolysin [200–202]. Unfolded and aggregated counterparts, on the other hand, can be readily proteolyzed. It is believed that the folded barrel assembly is preserved by the surrounding lipid or detergent milieu, as the protease cannot access the transmembrane region. However, the loop regions are accessible to the protease, as seen for Ail, refolded in LDAO micelles [152]. Proteolysis of the loop regions does not affect the integrity of the transmembrane region, as they are stabilized by interstrand hydrogen bonds, strong protein-lipid, and intra-protein hydrophobic interactions. Therefore, we observe a gel mobility shift for the proteolyzed folded protein, similar to its undigested counterpart, despite the digestion of loop segments by the protease. Gel mobility shift assays are usually combined with protease digestion experiments to establish the folded state of the β-barrel protein [121, 152, 153].

Protein stability to chemical (urea, guanidine hydrochloride, etc.) and thermal (heat) denaturation has also been investigated with the combination of gel shift assay and pulse proteolysis (Fig. 6). Here, the refolded protein is first subjected to increasing gradient of denaturant to promote protein unfolding, following which the reaction is exposed to proteases for a short pulse of time [121, 152, 202–204]. Incubation with proteases, such as proteinase K, subtilisin, and trypsin, leads to digestion of unfolded and partially folded protein species, whereas the refolded protein remains inaccessible to proteases, as the transmembrane region remains occluded by the surrounding lipid or detergent. We use this selective digestion of unfolded and partially folded protein as a screen for transmembrane β-barrel refolding. The folded fractions (% folded species) can be easily calculated for protein folding and unfolding, or stability studies, using the following formula [121]:

$$\text{Fraction folded} = \text{Folded band intensity}/(\text{Folded band intensity} + \text{Unfolded band intensity})$$

We have also used this approach to calculate the percentage oligomer population of Ail and its tryptophan mutants refolded in different concentrations (DPRs) of LDAO micelles. Here, we can differentiate between the oligomerization propensities of mutant Ail proteins. We find that the W42 → F mutant of Ail has a lesser oligomerization propensity than Ail WT in all the DPRs monitored. Similarly, we are able to establish using gel mobility shifts coupled with densitometry studies that the oligomerization propensity proportionally decreases with increase in DPR, for all the mutants.

Another protein prone to oligomerization is the human mitochondrial β-barrel VDAC. We have investigated the role of cysteines on the oligomerization of the hVDAC-2 isoform, for native (hVDAC-2 WT) and Cys-less (hVDAC-2 C0) mutant proteins, using SDS-PAGE. We observe that both WT and C0 constructs form ~40% oligomeric species, and ~60% remain monomeric, suggesting that oligomer formation is independent of cysteines in hVDAC-2. Furthermore, we observed that oligomerization tendency is independent of the DPR. Formaldehyde cross-linking studies examined using SDS-PAGE reveal that hVDAC-2 also has the tendency to form multimeric protein species or higher-ordered aggregates. Furthermore, refolded hVDAC-2 also shows a unique digestion pattern when treated with trypsin [122], providing insight on protein behavior in various DPRs of detergent. Hence, information on protein folding and oligomerization can be readily obtained using SDS-PAGE methods, when coupled with protease digestion experiments, cross-linking using chemical agents such as formaldehyde or glutaraldehyde and densitometry studies.

In our laboratory, we have employed the widely used technique of gel mobility shift to perform rapid refolding screens of different transmembrane β-barrels [121, 133, 152, 153]. A similar approach has previously used by the Fleming group [183]. As discussed in the previous section, using the β-barrel protein OmpX as the model system, we have identified temperature-induced refolding of transmembrane β-barrels (Fig. 7). The efficacy of "heat shock" refolding can be readily screened in different detergent micelles (LDAO, lauryldimethylamine oxide; β-OG, *n*-octyl-β-D-glucoside; DPC, *n*-dodecylphosphocholine), lipid vesicles (DHPC,

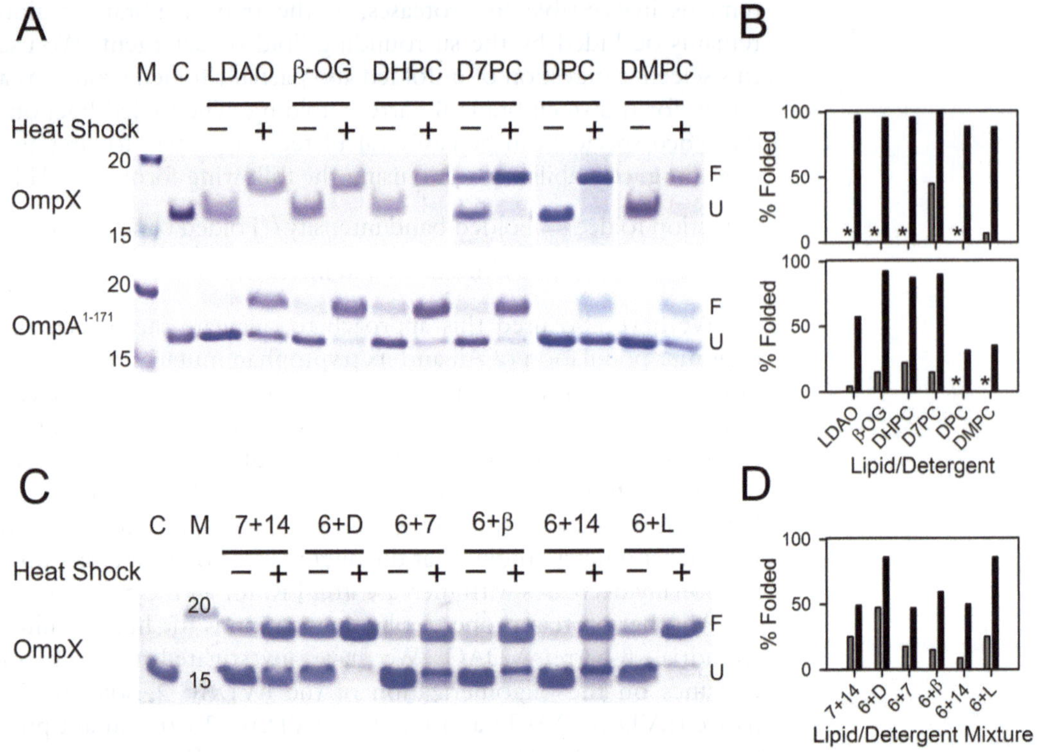

Fig. 7 Use of heat shock for refolding membrane proteins. (**a**) Refolding efficiency of OmpX and OmpA (residues 1–171, corresponding to the transmembrane region) on SDS-PAGE. (**b**) Histograms representing the folded fractions (% folded) of OmpX and OmpA in the presence of various lipids and detergents. (**c**) Refolding screen of OmpX in various detergent-micelle mixtures, on SDS-PAGE. (**d**) Histogram comparing the percent folded OmpX in mixtures of detergent and lipids. Here, both OmpX and OmpA fold efficiently when a heat shock (*plus*; indicated above the gel) for 3 min is given at 70 °C. The anomalous gel mobility shift (shift of ~3.0 kDa) upon refolding of the OmpX and OmpA barrels is evident in all the gels. The ~20 kDa band corresponds to the refolded protein, whereas the lower, ~15 kDa band corresponds to the unfolded protein. Reproduced with permission from Maurya, Chaturvedi, and Mahalakshmi [121]

1,2-dihexanoyl-*sn*-glycero-3-phosphocholine; D7PC,7, 1,2-diheptanoyl-*sn*-glycero-3-phosphocholine; DMPC, 1,2-dimyristoyl-*sn*-glycero-3-phosphocholine), and mixtures of detergent and lipids, using gel shift assay and densitometry (Fig. 7). Hence, gel mobility shifts serve as an inexpensive and rapid method to perform refolding screens in a multitude of conditions as well as identify the refolding efficiency in various detergents and lipids [121].

SDS-PAGE finds its use in a variety of applications for β-barrel membrane proteins, including the determination of protein purity and integrity, confirmation of β-barrel folding and stability, estimation of oligomeric species, monomer to oligomer formation or oligomer to monomer dissociation, and estimation of lipid adsorbed (partially folded) species. The major advantage of time, minimal sample requirement, and simplicity of the technique has rendered this method as one of the most popular approaches to address more sophisticated questions. The most recent applications include the determination of factors that modulate β-barrel folding [155, 187], protein folding, and unfolding studies [61, 205].

3.2 Circular Dichroism

Circular dichroism (CD) spectropolarimetry is a widely used technique in the field of structural biology and biophysics. The chirality of non-Gly amino acids in proteins renders CD as a very effective tool to monitor changes in the structural and behavioral properties of proteins [206]. In the field of membrane protein biology, this technique has been employed in a variety of model systems to analyze protein structure, characterize protein folding events, monitor protein-protein interactions, and unravel several other molecular details [207, 208]. The basic principle of CD deals with the differential absorption of left- and right-handed circularly polarized light, mediated by chiral molecules such as proteins. It enables us to decipher valuable information pertaining to the secondary structure (from far-UV CD) and tertiary organization (from near-UV CD) of these macromolecules. Despite lagging significantly behind their soluble counterparts, membrane proteins have increasingly been characterized using CD [161, 193, 209]. In this section, we will focus on the recent developments in studies of β-barrel membrane proteins using CD spectropolarimetry.

Transmembrane β-barrel proteins show a characteristic negative ellipticity centered at ~215 nm, similar to soluble β-rich structures. The ellipticity at 215 nm can, therefore, be used to monitor the rate of protein folding and unfolding, as well as the secondary structure content. For example, in the case of human VDAC-2 (hVDAC-2), we have monitored thermal denaturation of the protein refolded in LDAO micelles in low and high DPR conditions using far-UV CD at 215 nm (Fig. 8) [122]. Similarly, to decipher the effect of a chemical denaturant, we have monitored the thermal denaturation of the wild-type protein (WT) and its cysteine-less mutant (C0) in the presence of increasing amounts of urea and obtained some very interesting results. In low DPR, both proteins remain largely unaffected by an increase in the denaturant concentration. But, in high DPR conditions, C0 shows a drastic reduction in both secondary structure content and T_m values, whereas no appreciable change in T_m values was observed for the WT protein. Using CD, we were able to demonstrate that the effect of denaturant on protein stability is possibly a resultant of the differences in protein-lipid interactions (Fig. 8).

CD can also be used for bicelles and lipid vesicles in low concentrations. For example, to obtain a clear understanding of protein-lipid interactions in human VDAC-2, under the mixed micelle conditions, we have titrated varying amounts of the vesicle-forming lipid DMPC into the protein-micelle complex (Fig. 8). We observe a high sensitivity on the part of both proteins (WT and C0 constructs) to the curvature stress induced by DMPC

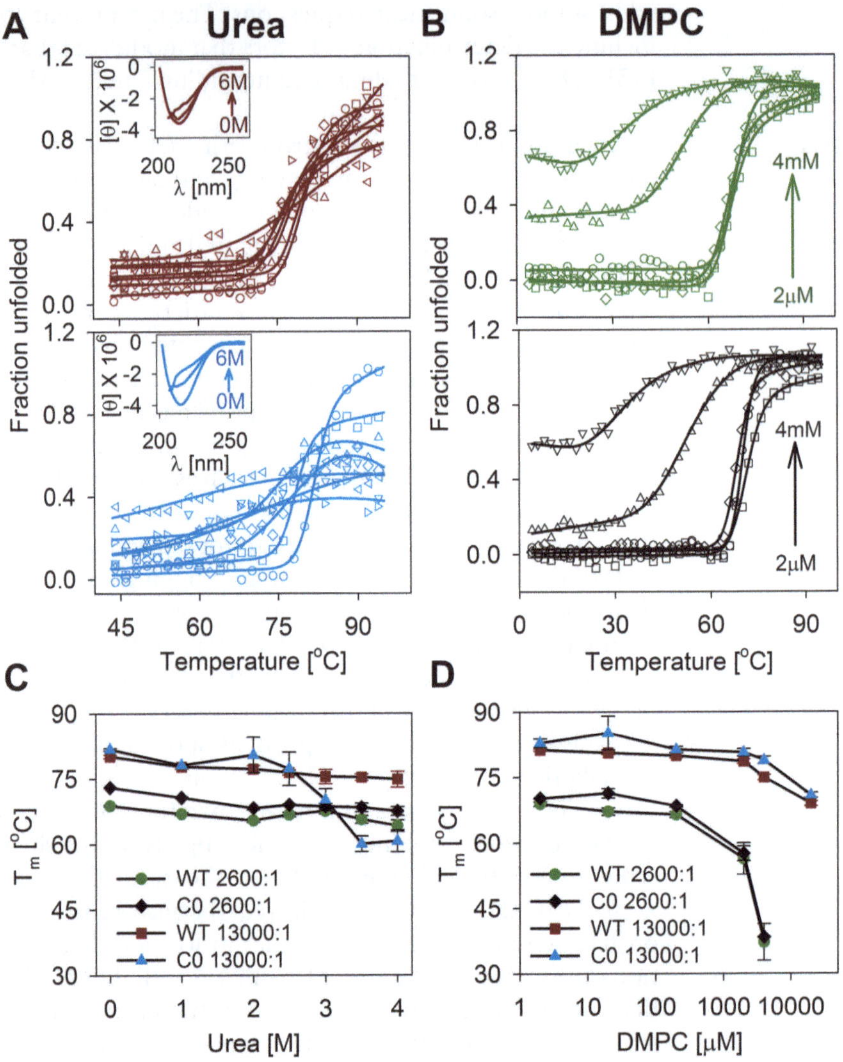

Fig. 8 Thermal denaturation profiles of refolded hVDAC-2 monitored using CD. Shown here is an example of how CD can be used to monitor the effect of urea and DMPC on the stability of WT (native) and C0 (Cys-less) mutants of hVDAC-2, refolded in LDAO micelles. Refolded hVDAC2 (WT and C0), in the presence of increasing concentrations of (**a**) urea and (**b**) DMPC, is subjected to thermal denaturation, and the folded fractions are calculated at each temperature. The midpoint of thermal denaturation (T_m), in the presence of (**c**) urea and (**d**) DMPC, is plotted for hVDAC-2 refolded in LDAO micelles. (**c**) Using CD, we find that in high DPR conditions, C0 is destabilized to a greater extent when the urea concentration is increased. (**d**) Increasing DMPC affects hVDAC-2 stability by inducing a curvature stress on the protein, as deduced by the lowering of T_m. This research was originally published in *The Journal of Biological Chemistry*. Maurya and Mahalakshmi. Modulation of Human Mitochondrial Voltage-dependent Anion Channel 2 (hVDAC-2) Structural Stability by Cysteine-assisted Barrel-lipid Interactions. The Journal of Biological Chemistry. 2013; 288, 35, 25584–25592. © The American Society for Biochemistry and Molecular Biology

molecules. Specifically, in low DPR conditions, both WT and C0 display higher destabilization upon increase in DMPC amounts, as compared to high DPR. This destabilization is evident from the rapid decrease in T_m values derived from ME_{215} data. The T_m values remain stable up to 100 μM DMPC after which, sudden destabilization occurs for both proteins. This effect may be due to a marginal hydrophobic mismatch between the barrel structure and the bilayer thickness presented by the vesicles formed by DMPC [122].

Several parameters that can be derived from thermal denaturation measurements using CD include the T_m, $T_{m\text{-start}}$, $T_{m\text{-end}}$, ΔT_m, and ΔH_{app}. Here, T_m, midpoint of thermal denaturation, is derived from fitting the unfolding profile to a two-state equation. Additionally, $T_{m\text{-start}}$ and $T_{m\text{-end}}$ denote starting and end points of unfolding transition, respectively. Furthermore, ΔT_m is calculated from the difference between $T_{m\text{-start}}$ and $T_{m\text{-end}}$. Finally, the slope of the unfolding transition curve is represented as ΔH_{app}. The T_m and $T_{m\text{-start}}$ values indicate the stability of the folded protein state, while ΔT_m and ΔH_{app} provide important information on the cooperative nature of the aggregation process and enthalpic changes associated with the unfolding of solvent-exposed segments of the β-barrel [177].

Proteins such as VDAC-2 undergo complete unfolding and aggregation, when subjected to thermal denaturation. Hence, CD can be used to derive the kinetic stability of membrane proteins such as VDAC-2, which undergo irreversible unfolding. The activation energy derived from such experiments gives information about energy barrier separating the folded and unfolded (aggregated) state of the protein. By monitoring the rate of unfolding as a function of temperature and DPR, we find that in the case of VDAC-2, the rate of unfolding proportionally increases with a rise in temperature and decreases as we increase the detergent concentration (and DPR) (Fig. 9). Activation energy values obtained from Arrhenius plots can also be used to demarcate the contribution of a mutant to the kinetic barrier. For example, we obtain a ~5 kcal/mol difference in the activation energy for aggregation between hVDAC-2 WT and C0 constructs [123]. Similarly, we have investigated the role of cysteine residues on the kinetic stability of hVDAC-3 by measuring the rate of protein unfolding and aggregation, as a function of temperature. We observe a slow unfolding followed by aggregation for hVDAC-3 WT and all the cysteine mutants. Cysteine removal drastically affects the activation energy values, and hVDAC-3 C0 mutant shows an increase of ~5–7 kcal/mol when compared to WT protein (unpublished data). These, and other examples [210–212], demonstrate the application of CD to quantify kinetic barriers separating the refolded protein from the aggregated state.

All membrane proteins do not exhibit unfolding upon thermal denaturation. For example, the *E. coli* proteins OmpX and PagP are considerably resistant to denaturation by heat [121, 151, 153, 161]. Hence, in our studies, we monitor the variations in the near-UV or far-UV CD profiles in response to changes in

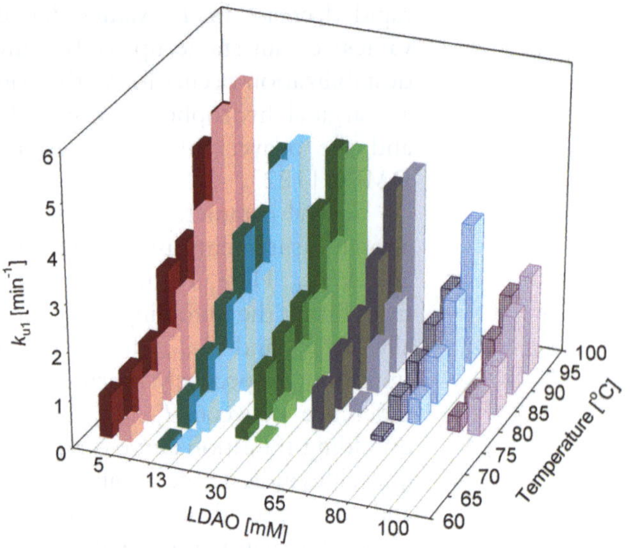

Fig. 9 Comparison of unfolding and aggregation rates for hVDAC-2 using CD. hVDAC-2 WT (*dark bars*) and C0 (*light-colored bars*) refolded in different concentrations of LDAO are subjected to thermal denaturation, and the process can be monitored using CD. The fast rate of unfolding (k_{u1}) is plotted against different concentrations of LDAO, at various temperatures. The data provides information on how the rates of unfolding are strongly influenced by LDAO concentration and cysteine residues. We infer from the CD experiments that the removal of cysteine (C0) stabilizes the barrel in lower LDAO concentrations, whereas at higher LDAO concentration, C0 exhibits faster unfolding. Reproduced from Maurya and Mahalakshmi [123], under the Creative Commons Attribution (CC BY) license

temperature, to understand the tertiary interactions in the protein, and derive information on the effect of mutations (Fig. 10) [153]. OmpX neither exhibits complete protein unfolding nor does it possess a characteristic tertiary CD. Hence, normalized unfolding profiles monitored at 215 nm can be qualitatively compared across mutants, to obtain insights on the folding and unfolding pathways [121, 151].

By monitoring changes in the tertiary CD at 231 nm, we can deduce the effect of point mutations in the outer membrane enzyme PagP from *S. typhimurium* and *E. coli* [153]. The molecular mechanism behind this tertiary interaction (Fig. 11) has been previously characterized in detail [193]. This protein exhibits a sigmoidal unfolding profile, allowing us to readily deduce the T_m, $T_{m\text{-start}}$, and $T_{m\text{-end}}$. PagP is one of the few outer membrane proteins which also functions as an enzyme. It transfers a palmitoyl chain from outer membrane phospholipids to the lipid A component of lipopolysaccharides (LPS) [213]. This protein displays an unusual structure for an integral membrane protein, with a highly dynamic barrel domain tilted with respect to the membrane

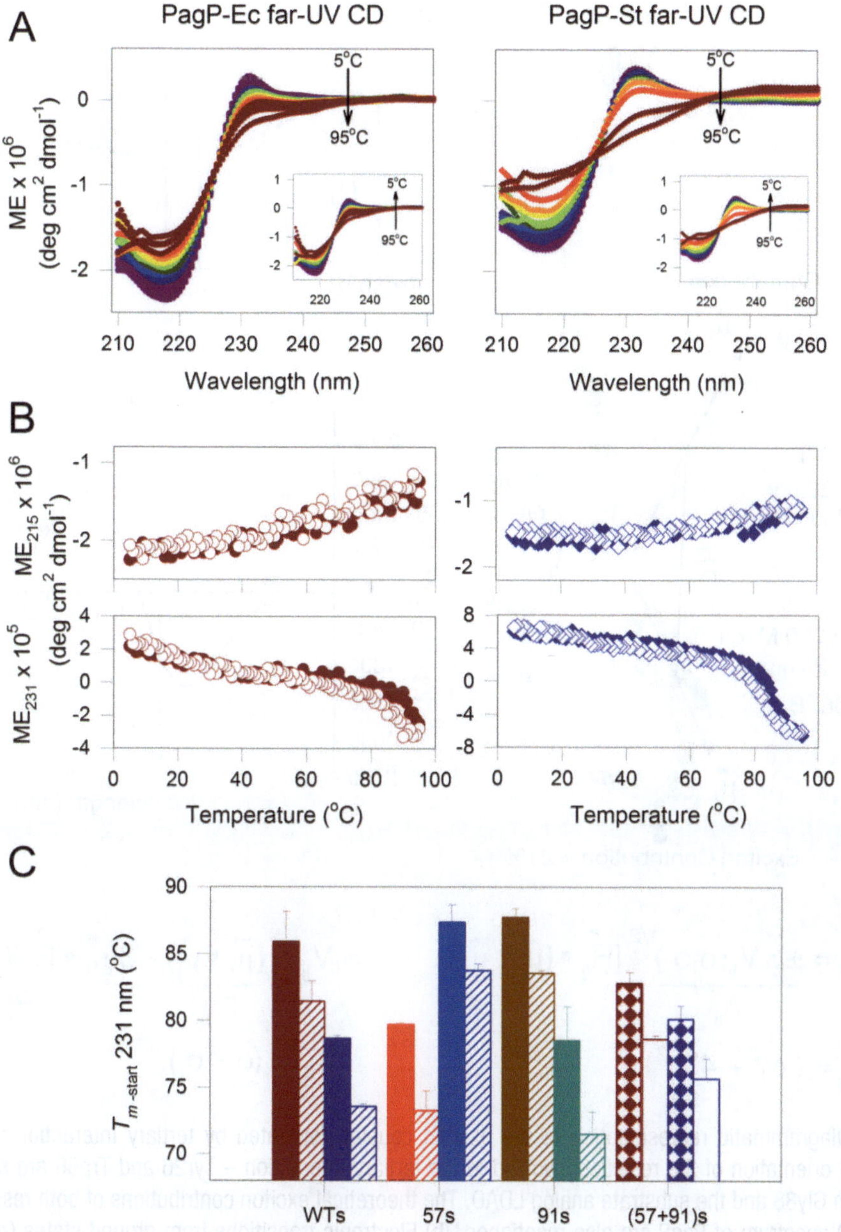

Fig. 10 Thermal denaturation of PagP and its mutants monitored by CD. Here, the stability characteristics of two PagP variants with near-identical primary sequence are studied using CD. (**a**) Far-UV CD scans of PagP derived from *E. coli* (PagP-Ec, *left*) and *S. typhimurium* (PagP-St, *right*) as a function of increasing temperature show changes in molar ellipticity values at both 215 nm (ME_{215}) as well as 231 nm (ME_{231}). (**b**) The reduction in secondary structure (ME_{215}, *upper panel*) and the tertiary interaction (ME_{231}, *lower panel*) is displayed for the melting (*filled symbols*) and recovery (*empty symbols*) profiles. (**c**) Histograms displaying the point at which the tertiary interaction begins to drop (denoted as $T_{m\text{-start}}$) show that PagP-Ec is more stable than PagP-St with a difference of about ~8°C. Reprinted with permission from Iyer and Mahalakshmi, Biochemistry, 2015, 54, 5712–5722. Copyright (2015) American Chemical Society

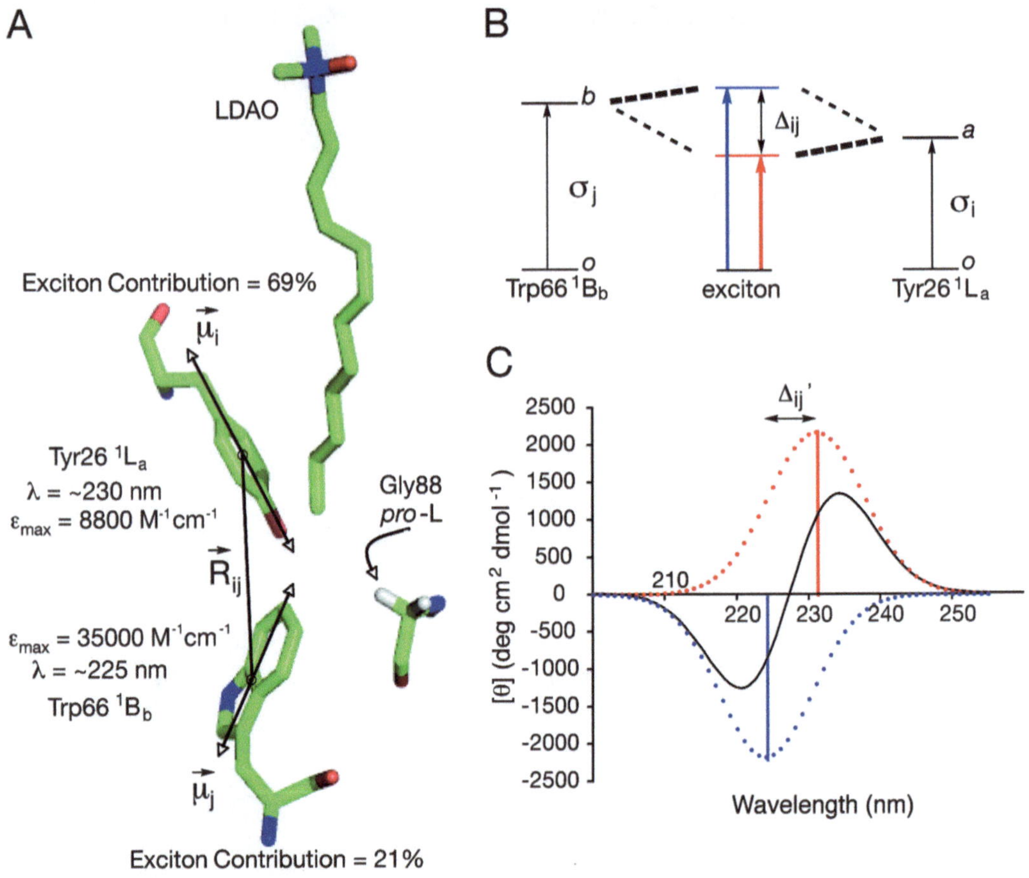

D

$$R_o = \pm \frac{\pi V_{ij}(\sigma_i \sigma_j)^{1/2}}{\Delta_{ij}} [\vec{R}_{ij} \bullet (\vec{\mu}_i \times \vec{\mu}_j)] \qquad V_{ij} \approx \frac{(\vec{\mu}_i \bullet \vec{\mu}_j)}{R_{ij}^3} - \frac{3(\vec{\mu}_i \bullet \vec{R}_{ij})(\vec{\mu}_j \bullet \vec{R}_{ij})}{R_{ij}^5}$$

$$\Delta_{ij} = (\delta_{ij}^2 + 4V_{ij}^2)^{1/2} \qquad \delta_{ij} = (\sigma_j - \sigma_i)$$

Fig. 11 Diagrammatic representation of the exciton couplet generated by tertiary interaction in PagP. (**a**) Structural orientation of the residues involved in the tertiary interaction – Tyr26 and Trp66 are shown here, along with Gly88 and the substrate analog LDAO. The theoretical exciton contributions of both residues to the far-UV CD spectrum of PagP are also mentioned. (**b**) Electronic transitions from ground states (*o*) to excited state, which generate the CD cotton effect, are depicted for Tyr26 (*a*) and Trp66 (*b*). (**c**) The exciton couplet produces an effect in equal magnitude but opposite directions, leading to contributions to the 215 nm and 230 nm region of far-UV CD. Reprinted with permission from Khan et al., Biochemistry, 2007, 46, 4565–4579. Copyright (2007) American Chemical Society

normal, and that it contains an N-terminal amphipathic helix. The dynamic structure helps PagP switch between two distinct states – one facilitating easier substrate access and the other favorable for the catalytic reaction [214]. PagP is a well-characterized protein, and it has also been extensively used as a model system to understand the principles that drive β-barrel membrane protein folding [171, 190, 215, 216].

The major drawback in CD experiments is observed when higher concentrations of lipids are used, wherein a considerable contribution from scattering affects the conclusions. Further, high salt concentrations interfere in CD measurements specifically for membrane proteins, due to the added contribution of lipids and detergents used in the refolding mix. Finally, denaturation carried out using GdnHCl cannot readily be followed using far-UV CD and relies heavily on the presence of a tertiary CD (near-UV CD) in the protein.

3.3 Steady-State Fluorescence

Fluorescence is a phenomenon where absorbance of light at specific wavelengths causes excitation of a fluorophore, and return to the ground state is achieved by emission of light of longer wavelength and lower energy. The high extinction coefficient and prolonged excited state lifetime of tryptophan (as against Phe and Tyr) allow us to use the indole as an intrinsic probe to obtain information about local changes in the vicinity of the fluorophore. Changes in fluorescence behavior of tryptophan have been used to understand β-barrel membrane protein folding (protein anchoring to lipids, insertion into membrane, rearrangement of the barrel), unfolding, and stability both in vivo and in vitro, for a variety of proteins such as hVDAC-2, OmpX, PagP, Ail, OmpA, OmpF, etc. [122, 152, 153, 184, 217–220]. Here, we will discuss the studies carried out on the role of tryptophan and other interfacial residues on the stability of transmembrane β-barrels, using Trp fluorescence as a probe.

A shift in the tryptophan fluorescence spectrum upon chemical or thermal de(re)naturation of the macromolecule is the preferred way to understand protein un(re)folding and stability. Here, change in fluorescence intensity is a result of a change in the local vicinity of fluorophore caused by a conformational and structural change in the protein. Along with steady-state fluorescence measurements, Stern-Volmer constant calculation from quenching experiments (K_{SV}), tryptophan anisotropy (r), average lifetime ($<t>$), bimolecular quenching constant (k_q) measurements, and rotational correlation time (τ_c) calculations further help us to derive solvent accessibility and conformational rigidity of the fluorophore [53, 122, 153, 217, 221].

We have used Trp as the fluorescence probe to investigate the contribution of cysteines to the overall stability of hVDAC-2 and to study the influence of Trp on (un)folding pathways and stabilities of Ail and OmpX. hVDAC-2, Ail, and OmpX are transmembrane β-barrels containing four, two, and two tryptophan residues, respectively. All the tryptophans are located at the interface and form a part of the aromatic girdle that anchors and stabilizes membrane proteins [185, 222]. As we mentioned earlier, membrane proteins require an optimal detergent or lipid-protein ratio for proper refolding of the barrel, a lack of which may lead to protein misfolding or aggregation. Folding kinetics measurements carried out for native hVDAC-2 (WT) and Cys-less hVDAC-2 (C0) proteins reveal that unlike other transmembrane proteins, hVDAC-2 shows rapid refolding in LDAO micelles (Fig. 12). We are able to obtain folding rates in the scale of 0.385 ± 0.037 min^{-1} for WT and 0.185 ± 0.015 min^{-1} for C0 using changes in Trp anisotropy as the means for measurement. Our measured rates from anisotropy suggest that removal of cysteines in C0 construct has increased the rate of folding compared to WT protein [122]. Data obtained from Trp quenching experiments using acrylamide as the quencher of fluorescence reveals valuable information on the role of mutations in overall protein structure and stability. Compared to hVDAC2 WT, the C0 mutant (wherein all cysteines are mutated) has a lower K_{SV}, which suggests that the Trp residues are more buried; yet the lower anisotropy values than WT in both DPR conditions indicate a less rigid environment for the indoles (Fig. 12). Hence, fluorescence measurements can provide considerable information on subtle changes in the local folded environment of the β-barrel [122].

Ail possesses two tryptophans W42 and W149 at opposite faces of the barrel. We have carried out equilibrium folding and unfolding studies for Ail in LDAO micelles using Trp fluorescence. This allows us to not only understand the role of detergent concentration in barrel folding and stability but also the importance of tryptophans at the interfacial position. As expected for membrane proteins, we observe that high detergent-to-protein ratios (DPR) stabilize the barrel. The observed increase in stability may arise due to strong protein-micelle interaction, at high concentrations of the detergent (Fig. 13). Interestingly, while refolding is DPR independent, the process shows a W42-dependent folding intermediate. Substitution of Trp 42 to Phe stabilizes the barrel but slows down the barrel folding by the formation of an on-pathway intermediate. On the other hand, substitution of Trp 149 to Phe destabilizes the barrel, as compared to native (WT) Ail protein (Fig. 13). The contribution of both indoles to the stability of the barrel can be obtained by following the refolding and unfolding processes using the fluorescence properties of the single-Trp Ail mutants [152]. Thus, we can measure a differential contribution of both the tryptophans in the folding and stability of Ail.

Similarly, the homologous transmembrane barrel OmpX also carries two tryptophans (W76 and W140) at the interfacial position of the barrel [217]. To understand the positional significance and their contribution to protein folding and stability, we carried out chemical denaturation (folding-unfolding equilibrium measurements) of OmpX and its Trp mutants in DPC micelles. We follow the contributions of the two tryptophans to OmpX stability as well as deduce information on the local environment using fluorescence

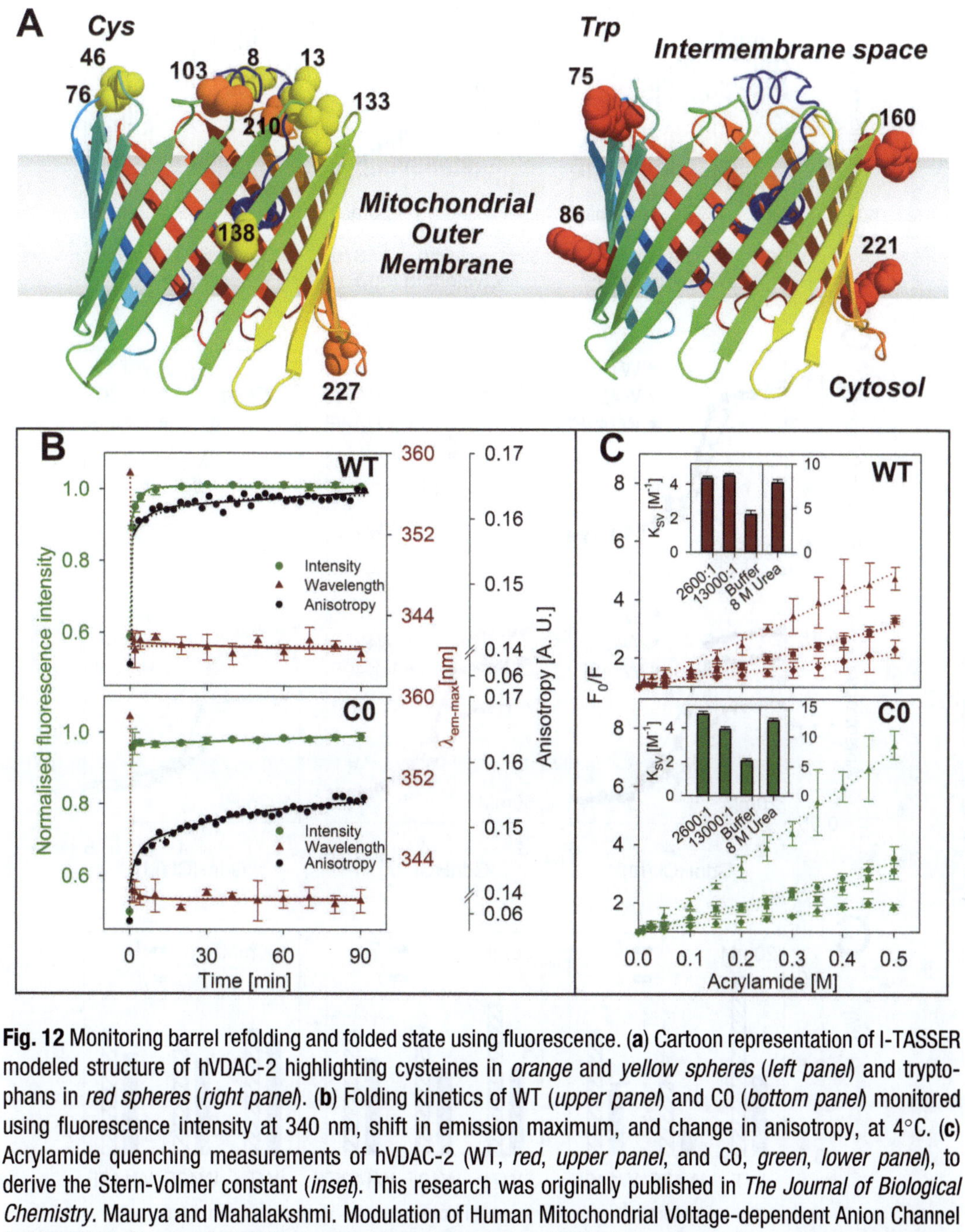

Fig. 12 Monitoring barrel refolding and folded state using fluorescence. (**a**) Cartoon representation of I-TASSER modeled structure of hVDAC-2 highlighting cysteines in *orange* and *yellow spheres* (*left panel*) and tryptophans in *red spheres* (*right panel*). (**b**) Folding kinetics of WT (*upper panel*) and C0 (*bottom panel*) monitored using fluorescence intensity at 340 nm, shift in emission maximum, and change in anisotropy, at 4°C. (**c**) Acrylamide quenching measurements of hVDAC-2 (WT, *red*, *upper panel*, and C0, *green*, *lower panel*), to derive the Stern-Volmer constant (*inset*). This research was originally published in *The Journal of Biological Chemistry*. Maurya and Mahalakshmi. Modulation of Human Mitochondrial Voltage-dependent Anion Channel 2 (hVDAC-2) Structural Stability by Cysteine-assisted Barrel-lipid Interactions. The Journal of Biological Chemistry. 2013; 288, 35, 25584–25592. © The American Society for Biochemistry and Molecular Biology

quenching and thermal and chemical denaturation measurements [217]. We observe that W76 is relatively rigid and helps in folding, whereas W140 is comparatively flexible and contributes to the overall stability of the protein (Fig. 14). The properties of these

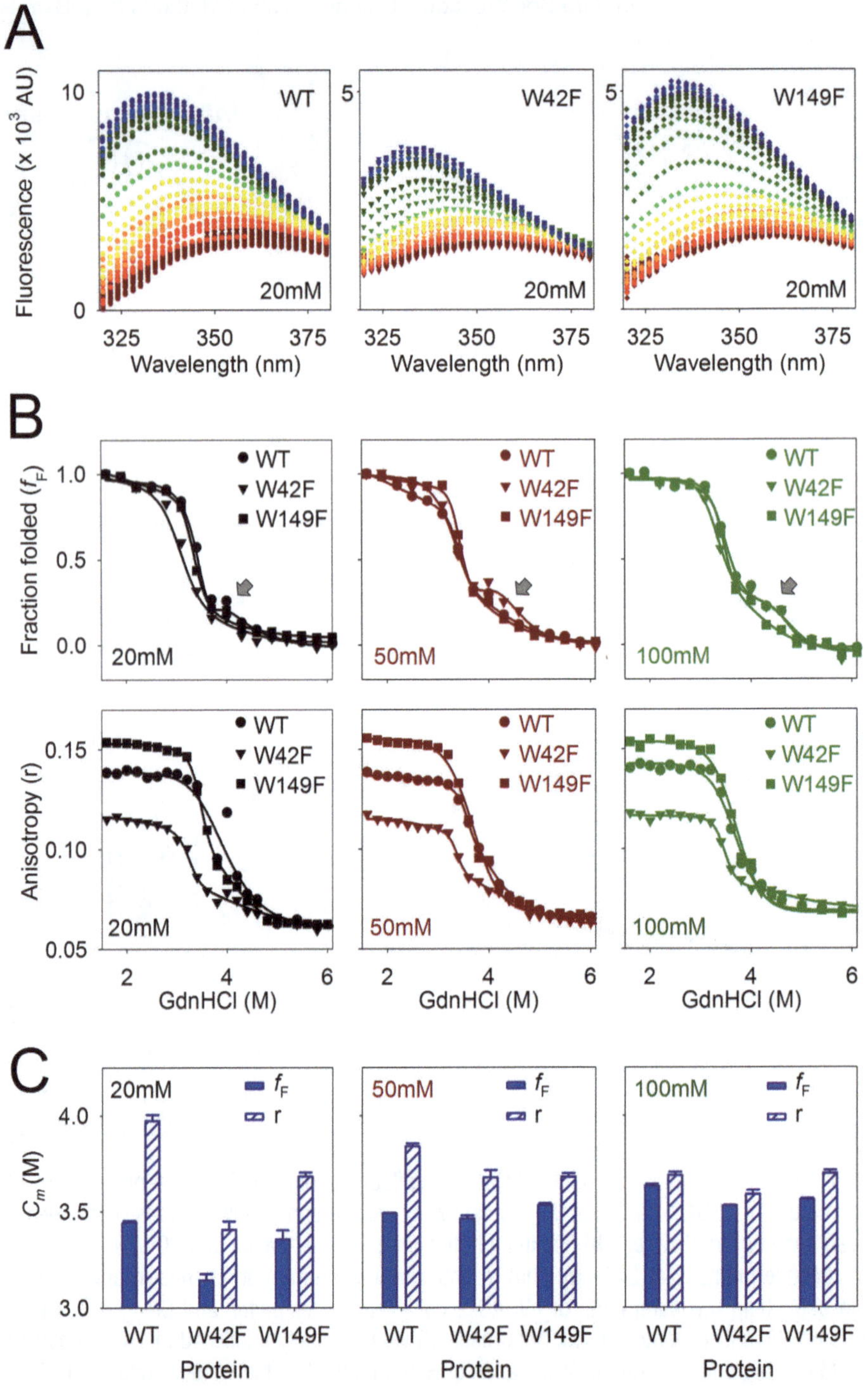

Fig. 13 Monitoring the effect of DPR on refolding of Ail and its mutants using tryptophan fluorescence. (**a**) Tryptophan fluorescence emission profiles in the presence of different concentrations of GdnHCl for Ail WT, W42F, and W149F, refolded in 20 mM LDAO. (**b**) Fraction folded (f_F) (*upper panel*) and anisotropy (*r*) values calculated from Trp emission intensity at 330 nm at increasing GdnHCl gradient for all three mutant proteins, at various LDAO concentrations (20, 50, and 100 mM). (**c**) C_m values obtained from fraction folded and anisotropy data, at various LDAO concentrations (20, 50, and 100 mM). Fluorescence data indicate that WT and W149F show two-state folding, whereas we observe a three-state folding for W42F construct, suggesting the presence of a prominent folding intermediate (indicated with an *arrow*). Reproduced with permission from Gupta, Zadafiya, and Mahalakshmi [152]

tryptophans can be investigated in detail by performing quenching experiments using different quencher molecules such as acrylamide and iodide. While acrylamide can quench buried tryptophans by associating with hydrophobic regions of the protein or micelles, iodide shows accessibility to only the solvent-exposed indoles. Hence, using data from multiple quenchers, we can obtain detailed information on the local tryptophan environment. Additionally, by measuring the average Trp lifetime as a function of quencher concentration, we can deduce the quenching mechanism. By means of this array of experiments, we were able to conclude that in OmpX,

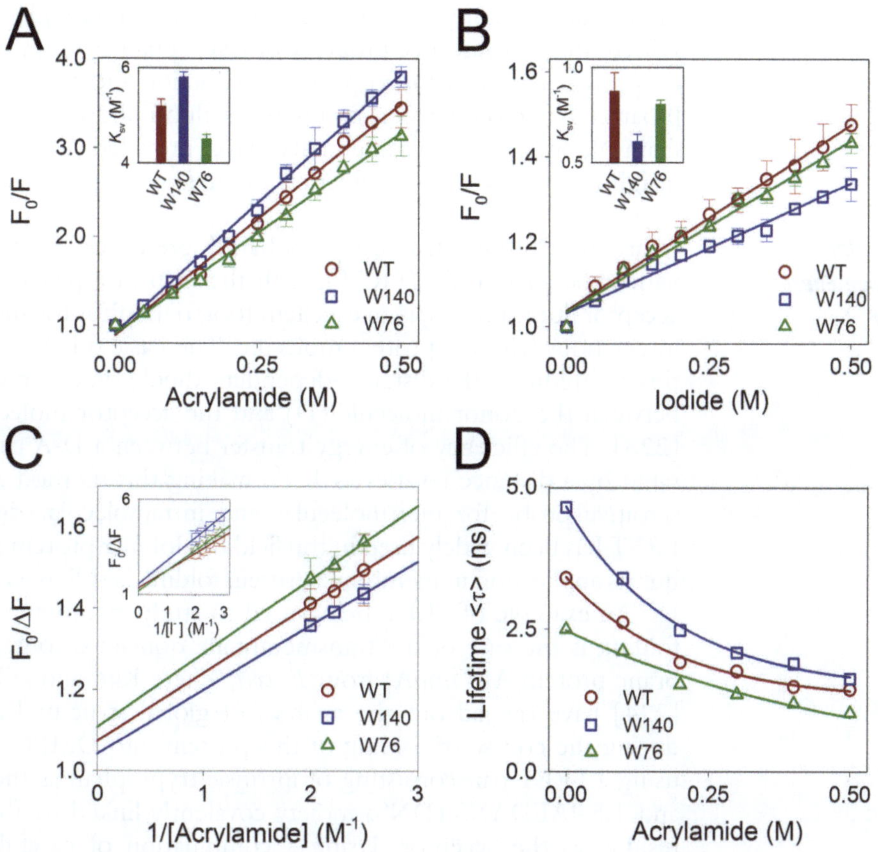

Fig. 14 Assessment of local tryptophan environment of OmpX using coupled quenching and lifetime measurements. (**a**) Acrylamide quenching measurements for WT OmpX and the W140 and W76 tryptophan mutant proteins. Stern-Volmer constant (K_{sv}) values are compared in the *inset* across the mutants. (**b**) K_{sv} values derived from iodide quenching measurements for WT and the single tryptophan mutants. Note how the K_{sv} values are different in (**a**) and (**b**), suggesting that the measured K_{sv} values depend on the accessibility of the fluorophore by the quencher. (**c**) Accessible fraction for tryptophans to the quencher acrylamide can be derived from the *y*-intercept obtained from plots of acrylamide concentration against the change in fluorescence intensity. (**d**) Comparison of average Trp lifetime ($<\tau>$) for all the mutants in the presence of various concentrations of acrylamide provides information on the local tryptophan vicinity. Reproduced with permission from Chaturvedi and Mahalakshmi [217]

W140 is embedded in a hydrophobic environment near the micelle core, whereas W76, albeit rigid, is present in a solvent-exposed and an interfacial environment [217].

Acrylamide quenching and anisotropy experiments that we carried out for Ail refolded in different detergent concentrations suggest that W42 is solvent exposed and flexible, whereas W149 is buried and rigid in nature. Similarly, we observed that W76 is relatively rigid and helps in folding, whereas W140 is buried and contributes to the overall stability of OmpX [217]. Hence, fluorescence measurements from both Ail and OmpX suggest that Trp residues differentially contribute to overall folding and stability of the protein, further maintaining optimal scaffold of the barrel with intermediate folding efficiency and moderate stability [152, 217]. Differential fluorescence property of tryptophan strongly depends on protein-lipid or protein-protein interaction and packing of the β-barrel. Based on the local nature of the fluorophore, observations from fluorescence measurements have provided us with new insights on local barrel architecture and the overall protein scaffold.

3.4 Förster Resonance Energy Transfer

Apart from using intrinsic tryptophan fluorescence, Förster resonance energy transfer (FRET), with the help of a pair of donor-acceptor fluorophores, is an excellent tool to monitor the molecular interactions within a protein molecule. The basic principle behind this technique is the distance-dependent dipole-dipole interaction between the donor molecule (D) and the acceptor molecule (A) [223]. The efficiency of energy transfer between a D-A pair separated by a distance r varies as $1/r^6$, making this method a highly sensitive probe for intermolecular and intramolecular distances. FRET has been widely used in the field of globular protein folding, but its application in membrane protein folding is still in its infancy.

An example of FRET being used to study membrane protein folding is the case of the transmembrane domain of outer membrane protein A (OmpA) from *E. coli*, where Kim and colleagues [224] have carried out the analysis of global structural changes during the course of folding of this protein into DMPC vesicles, using a FRET pair consisting of intrinsic tryptophan as the donor and 1,5-IAEDANS (DNS) moiety covalently linked to a cysteine residue as the acceptor. Using a combination of carefully constructed mutants, they carried out folding kinetics experiments to study the folding mechanism of OmpA [224]. Results indicate that the most rapid changes in FRET efficiency occur for the mutants reporting on pore formation. The changes occurring in intermediate timescales are for the mutants involved in bilayer traversal. Finally, the slowest exchanges are observed for mutants monitoring the strand extension across the bilayer. These FRET-based measurements suggest that tertiary structure formation occurs parallely during the course of folding and dictates the final strand extension and complete secondary structure formation for this protein. The qualitative information gained from this study can serve as a platform for further FRET-based analyses of membrane protein folding [224].

Recently, we have characterized the conformational switch from β-hairpin to α-helix of the first transmembrane domain (TM1) of mycobacteriophage D29 holin using FRET between the intrinsic tryptophan side chain as the donor and Alexa Fluor 350 or dansyl chloride as the acceptor (Fig. 15) [225]. The donor-acceptor distance in the α-helical conformation, observed for wild-type $TM1^{PG}$ (TM1 containing a central Pro-Gly dipeptide unit) in high-detergent conditions and the TM1 analog designed to take up helical structure – $TM1^{AA}$ (TM1 with a central dialanine unit) – is expected to be ~34 Å, due to which the FRET efficiency is low (Fig. 15). Upon structural transition to the β-hairpin form, the distance reduces drastically, which is evident from the high FRET

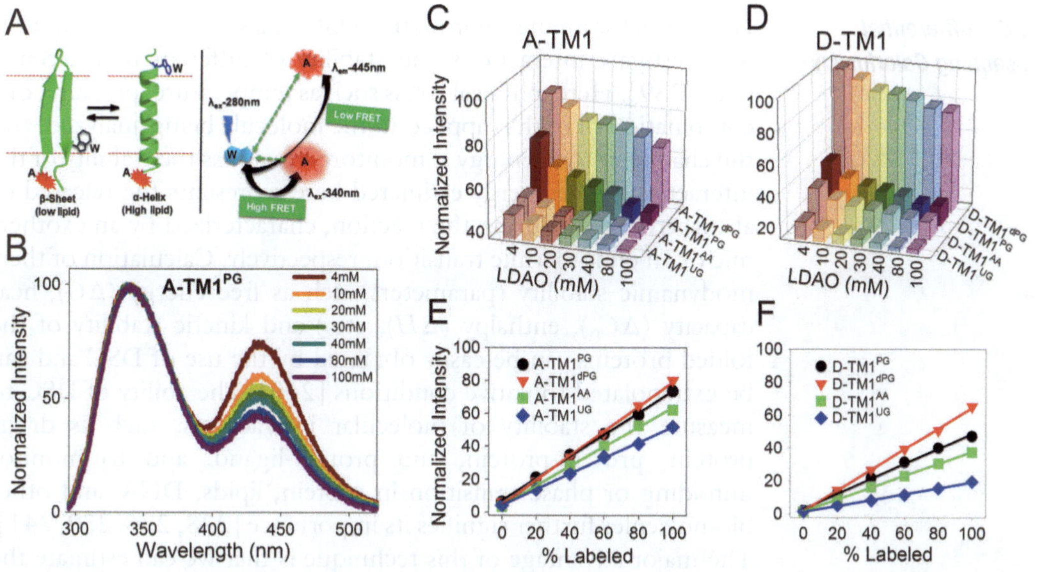

Fig. 15 The use of FRET to monitor conformation switch in a transmembrane peptide. (**a**) Schematic representation of conformational interconversion predicted to occur in the $TM1^{PG}$ peptide, also highlighting the FRET donor (Trp)-acceptor (Alexa Fluor 350; **a**) pair. Here, the presence of helical structure is anticipated to show low FRET efficiency due to the greater distance between the FRET donor and acceptor, whereas the formation of β-hairpin leads to increase in FRET and increased acceptor fluorescence. (**b**) Change in fluorescence intensity of Alexa Fluor-labeled $TM1^{PG}$ peptide (A-$TM1^{PG}$) with increasing concentration of LDAO. Increased acceptor fluorescence suggests the structural interconversion from α-helical structure to β-hairpin. (**c**) Use of FRET to study the effect of different concentrations of LDAO on the structural conversion of Alexa Fluor 350-labeled A-$TM1^{PG}$ and its peptide analogs. (**d**) Comparison of the acceptor (dansyl chloride; **d**) fluorescence across the four peptides, with increasing LDAO concentration. Dependence of FRET efficiency on the concentration of (**e**) Alexa Fluor 350-labeled and (**f**) dansyl chloride-labeled peptide in the presence of 4 mM LDAO. Linear increase in the observed FRET is suggestive of structural interconversion without the formation of peptide aggregates. Reprinted with permission from Lella et al., ACS Chemical Biology, 2016, 11 (4), 910–920. Copyright (2016) American Chemical Society

efficiency observed for $TM1^{PG}$ in lower micelle concentration, as well as for the $TM1^{dPG}$ peptide (TM1 containing a central ^{D}Pro-Gly dipeptide unit) designed to take up a β-hairpin structure [225]. We also validated that the observed change in FRET efficiency was indeed due to intramolecular energy transfer between the donor-acceptor pair (Fig. 15). By obtaining a linear dependence of FRET intensity as a function of labeled peptide concentration, we were able to ascertain that the structural transition for $TM1^{PG}$ was observed for conversion from helical to monomeric β-hairpin form (Fig. 15) [225].

FRET-based measurements hold considerable promise in the field of membrane protein conformational studies and in deducing the folding-unfolding pathways. With the availability of modified fluorophore libraries for site-specific protein tagging [226, 227], it should be now possible to extend FRET-based studies to other membrane proteins that are difficult to otherwise characterize.

3.5 Differential Scanning Calorimetry

Differential scanning calorimetry (DSC) is a widely used technique, to investigate interactions and stability of different biomolecules [228–239]. Here, physical stress such as temperature, pressure, or a combination of both is applied to the molecule being analyzed, and the change in total energy is monitored to assess the stability of the interactions. The energy estimated here represents the released or absorbed energy during the reaction, characterized by an exothermic or an endothermic transition, respectively. Calculation of thermodynamic stability (parameters such as free energy (ΔG), heat capacity (ΔC_p), enthalpy (ΔH), etc.) and kinetic stability of the folded proteins can be easily obtained by the use of DSC and can be extrapolated to native conditions [240]. The ability of DSC to measure the stability of molecular interactions, such as drug-protein, protein-protein, and protein-ligand, and to monitor unfolding or phase transition in protein, lipids, DNA, and other biomolecules further signifies its importance [228, 236–239, 241]. The major advantage of this technique is that we can estimate the thermal behavior of the proteins above the boiling point of solvents (e.g., 100°C for water), as the experiments can be run above atmospheric pressure. In this section, we will discuss the use of DSC in studying protein unfolding, stability, and aggregation.

The study of protein unfolding and lipid phase transitions is extremely challenging, as the transitions are very fast and involve minute energetic changes. Increased sensitivity of DSC enables the investigation of fast reactions, but in parallel, it requires careful sample preparation as well as data processing and interpretation [237, 238, 241]. The two major components of DSC-based study are:

A. Preparation of suitable buffer. The buffer should contain all the components identical to sample, except the protein.

B. Sample preparation. A homogenous sample with a single population of analyte in reference buffer is ideal for DSC studies. To achieve homogeneous sample preparations, high-speed centrifugation, filtration, and degassing are usually preferred, followed by estimation of sample integrity using SDS-PAGE, CD, or any other chromatographic or spectroscopic techniques.

We have used DSC for thermal denaturation studies of native OmpX and its mutant protein (H100 → N), to probe the effect of the mutation on the protein (Fig. 16). Furthermore, we have also investigated the effect of the refolding procedure on protein stability. Here, we refolded OmpX in LDAO micelles using slow refolding and heat shock refolding procedures, as described previously [121, 151]. Both refolding methods display the anticipated gel mobility shifts and protection against nonspecific proteases such as proteinase K. Further, OmpX does not undergo complete thermal denaturation at the temperatures achievable using CD [121, 151]. Hence, DSC serves as an ideal method to study the mutant proteins. After confirming the refolding, we unfolded the OmpX WT and mutant proteins using temperature as a denaturant.

The midpoint of thermal denaturation (T_m), heat capacity (ΔC_p), and enthalpy of unfolding (ΔH) can be readily calculated by fitting the unfolding profile to a two-state unfolding equation. In the case of OmpX, the observed melting temperatures were similar for all the preparations, and ΔC_p also did not show significant differences [151]. We find that the WT protein and H100N mutant shows similar enthalpy of unfolding (276 ± 2.49 and 272 ± 1.97 kcal/mol, respectively) when refolded using the slow refolding procedure. But, when both the proteins are refolded using heat shock method, we observe a drastic increase in the enthalpy of unfolding. We observe that heat shock refolding significantly stabilizes the OmpX barrel, and the degree of stabilization follows the order of H100N > WT mutant protein, wherein the enthalpy for H100N increases by ~200 kcal/mol, whereas it increases ~50 kcal/mol for WT protein [151]. Such differences in protein stability, which appear to have a high impact on the overall protein behavior, are difficult to measure using other biophysical methods.

Other than thermodynamic studies, DSC is also the preferred method to estimate kinetic stability and aggregation behavior of the proteins and other biomolecules [228–231, 233–235, 237–239, 241]. For estimation of kinetic stability, proteins in varying concentrations and at different ramp rates (scan rates) are titrated against temperature, and the change in T_m is compared across the conditions. A proportional increase in the T_m with an increase in scan rate is suggestive of kinetic stabilization of the protein. As we mentioned earlier, DSC can measure energy changes of both endothermic and exothermic reactions; the extent of exothermic transitions and their appearance in the thermal profiles can be compared to understand temperature-dependent aggregation of the proteins.

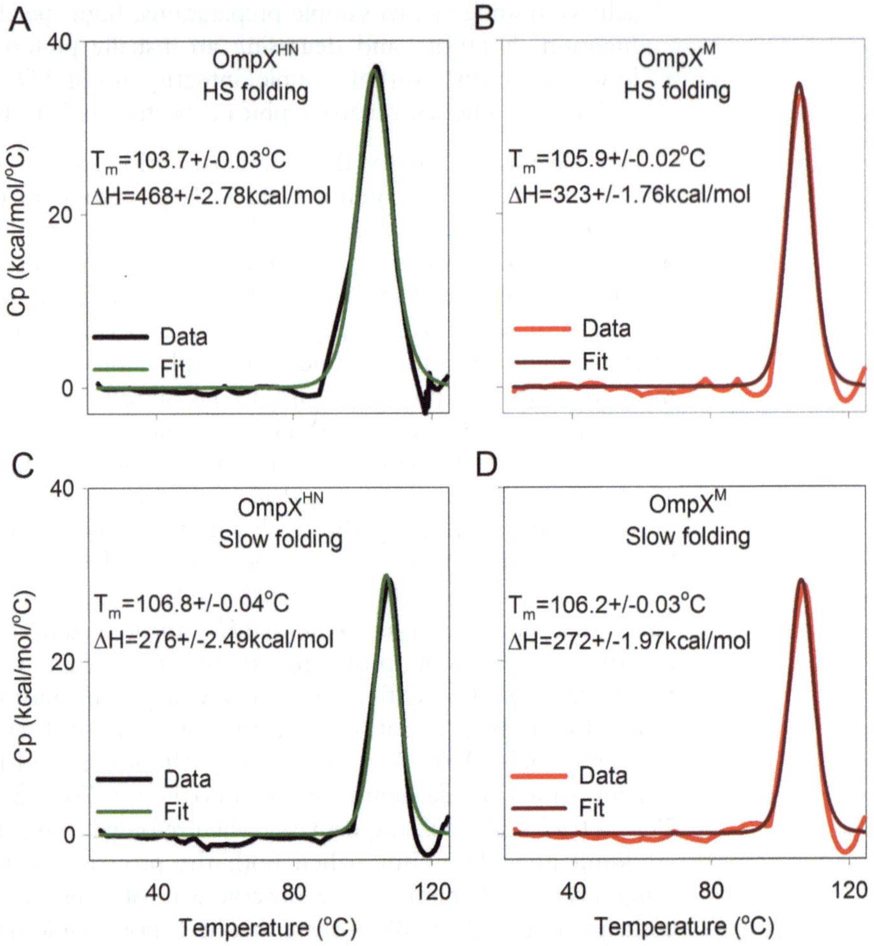

Fig. 16 Thermal denaturation of OmpX monitored using differential scanning calorimetry. Thermal denaturation DSC curves of wild-type OmpX ($OmpX^{HN}$; **a** and **c**; *black* and *green*) and the Met-less mutant ($OmpX^{M}$; **b** and **d**; *red* and *brown*), refolded in 50 mM LDAO generated using heat shock folding (**a** and **b**) or slow refolding (**c** and **d**) methods, respectively. Different parameters are derived from data by fitting to a two-state function to obtain T_m and ΔH, as mentioned in each panel. DSC is a highly sensitive method that can accurately detect variations in the enthalpy of the system, particularly when such differences are not evident from other biophysical methods. Reproduced from Chaturvedi and Mahalakshmi [151], under the Creative Commons Attribution (CC BY) license

This has been successfully carried out for soluble proteins [240, 242–244], and we have been able to extend this to transmembrane β-barrel proteins as well (Fig. 17).

As lipids serve as host for membrane proteins, the estimation of phase transitions, thermal stability, and understanding lipid dynamics and protein-lipid interactions is equally important. Since lipids display very sharp phase transitions, and it is very difficult to capture these transitions by means of spectroscopic methods, DSC serves as an excellent tool to study these reactions [241, 242, 245, 246]. In a

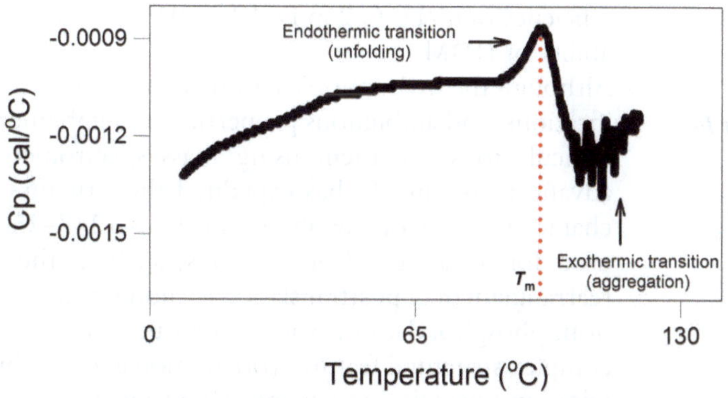

Fig. 17 Thermal denaturation and aggregation of Ail monitored by DSC. Representative raw DSC thermogram of Ail refolded in LDAO micelles. Ail shows both endothermic and exothermic transitions, corresponding to unfolding and aggregation, respectively. The temperature at the midpoint of unfolding (T_m) is marked with a *dotted red line*

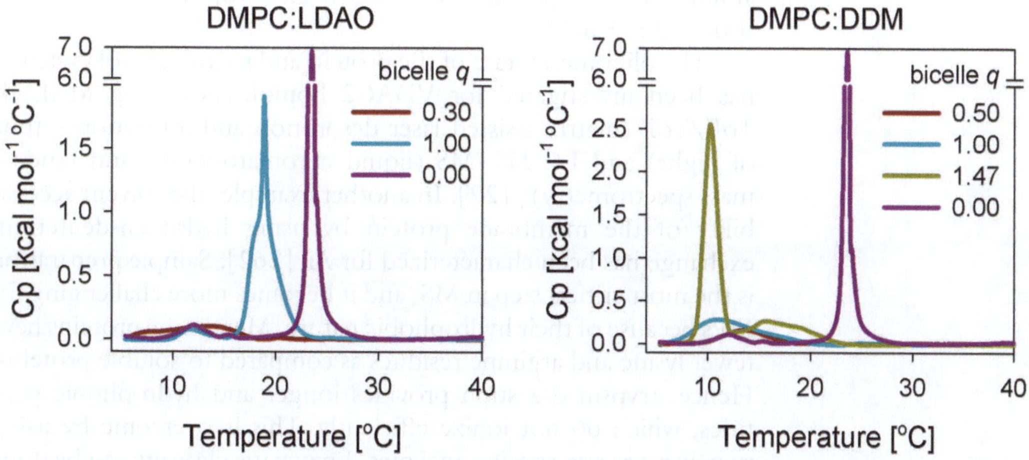

Fig. 18 Lipid transitions monitored using DSC. Comparison of DSC profiles of pure DMPC vesicles with empty DMPC:LDAO (*left panel*) and DMPC:DDM bicelles (*right panel*), respectively. The nature of the doped detergent drastically affects the phase transition of the lipid and can be monitored using DSC. This research was originally published in *The Journal of Biological Chemistry*. Maurya and Mahalakshmi. N-helix and Cysteines Inter-regulate Human Mitochondrial VDAC-2 Function and Biochemistry. The Journal of Biological Chemistry. 2015; 290, 51, 30240–30252. © The American Society for Biochemistry and Molecular Biology

recent study, we investigated the phase transition temperatures of DMPC vesicles and DMPC:LDAO and DMPC:DDM bicelles using DSC [177]. We observed higher temperature of phase transition for empty DMPC vesicles than bicelles, as anticipated (Fig. 18). Surprisingly, we also found that the nature of the doped detergent also drastically affects the phase transition of lipid. We observed that DDM forms better bicelles when mixed with DMPC, as compared to LDAO [177]. With the aid of this information, we were able to attribute the better refolding efficiency of hVDAC-2 in DMPC:DDM due to the bicelle-forming ability of DDM.

3.6 Mass Spectrometry for In Vitro Disulfide Mapping

Although the hydrophobic nature, complex posttranslational modifications, and ambiguous properties of membrane proteins make it difficult to study them using mass spectrometry (MS), recent advancements in MS has expedited the structural and proteomic characterization of membrane proteins [247–250]. MS is being used for a variety of applications, such as the study of minor rearrangements, posttranslational modifications (e.g., glycosylation, phosphorylation, oxidation reduction, etc.), investigation of complex structural features (orientation and stoichiometry) of proteins, and protein sequencing. Due to increased separation, sensitivity, and accuracy of MS, it becomes a powerful tool for mapping subtle structural changes or modification in membrane proteins [251–256]. The most significant application of mass spectrometry in recent years is the identification of non-annular lipids associated with helical transmembrane proteins and the competitive binding affinities in the gas phase [257–260]. This work has been recently reviewed in [261].

The oligomeric state of the protein and redox state of cysteines has been investigated for VDAC2 homologues using MALDI-ToF/ToF (matrix-assisted laser desorption and ionization – time of flight) and LC-MS/MS (liquid chromatography and tandem mass spectrometry) [122]. In another example, the solvent accessibility of the membrane protein by using hydrogen-deuterium exchange has been characterized for Ail [262]. Sample preparation is the most critical step in MS, and it becomes more challenging for MPs because of their hydrophobic nature. Membrane proteins have fewer lysine and arginine residues as compared to soluble proteins. Hence, trypsin digestion provides longer and hydrophobic peptides, which do not ionize efficiently. This is overcome by using two or more site-specific enzymes. Enzymatic cleavage can be done in two ways – in solution digestion and in-gel digestion – depending on whether the folded state of the membrane protein is being mapped or the characterization is at the level of the sequence. In the case of membrane proteins, the folding milieu contains a mixture of lipid, buffer, and denaturants, and so, removal of such components is necessary in order to obtain a good signal-to-noise ratio in the mass spectrum. For this reason, in-gel digestion serves as the best option. Shotgun methods have been developed for soluble proteins as a powerful alternative to gels [250, 251, 263]. Examples wherein this is extended to membrane proteins are, however, currently limited.

Recently, we have investigated the occurrence of disulfide patterns in hVDAC2 using MALDI-ToF/ToF. hVDAC-2 has nine cysteine residues that are distributed primarily across the loop regions connecting the transmembrane strands and are amenable to disulfide bond formation upon protein folding [122]. Here, we separated hVDAC-2 refolded in LDAO micelles on a 12% SDS-PAGE gel. The band corresponding to hVDAC2 was excised, and we prepared iodoacetamide-treated samples using reported protocols [264]. After digesting the protein sample with trypsin at 37°C, the extracted fragments were subjected to peptide mass fingerprinting by MALDI-ToF/ToF. Figure 19 represents the mass spectrometric profiles of in-gel trypsin-digested refolded hVDAC-2 with and without treatment with iodoacetamide. We observed that most of the cysteines of hVDAC-2 remain in the reduced state after the protein is folded. We obtained a minor population of Cys76-mediated disulfide bond formation with Cys43 and Cys103 (shown by three low-intensity peaks) in untreated samples [122]. The cysteines involved in disulfide formation, identified by MS analysis, are shown in the hVDAC-2 sequence and structure in Fig. 19. Our result indicates that disulfides are inherently not present in the refolded hVDAC2 and are transient entities. Further, we have discussed in the previous section that hVDAC-2 shows the existence of higher-order oligomers. The data from MS studies, when considered along with the results from hVDAC-2 C0 construct, confirm that the cysteines in this protein are present in reduced state (MS results) and reveal the fact that oligomerization of hVDAC-2 is not driven by disulfide formation [122].

Similar to hVDAC-2, the redox state of the six cysteines of recombinant as well as native hVDAC-3 (isoform 3 of human VDAC) proteins has been investigated using LC-MS/MS. Interestingly, in hVDAC-3, we could observe different oxidation states, such as sulfonic, sulfinic, and disulfide bond and SH reduced form for C2, C8, C36, C65, and C229 [150]. In particular, C36 was observed in all possible oxidative states (reduced, sulfonic, and sulfinic). hVDAC-3 isolated from rat liver mitochondria indicated that nearly all cysteines were in the reduced state; C65 and C165 could be irreversibly oxidized in a fraction of the population, and only the sulfonic acid form of C229 could be detected [150]. Unlike hVDAC-2, cysteines of the isoform 3 can exhibit multiple oxidation states, allowing us to surmise that VDAC-3 may specifically be involved in sensing cellular redox levels. Our findings validate the work of Dodo, Sodeoka and coworkers, who investigated the role of cysteine-mediated disulfide bond formation on channel activity of hVDAC-3 [265], using planar bilayer measurements (described in the next section). They were able to show using a carefully constructed cysteine mutant library and by the addition of reducing agents such as DTT and S-nitrosoglutathione that Cys122 of hVDAC-3 participates in disulfide bond formation with either Cys2 or Cys8 [265]. These results reaffirm our observations from MS studies of hVDAC-3 that the cysteines of this VDAC isoform can exist in multiple oxidation states and regulate voltage-dependent gating of the channel. Hence, MS can serve as an excellent method to map modifications in membrane proteins.

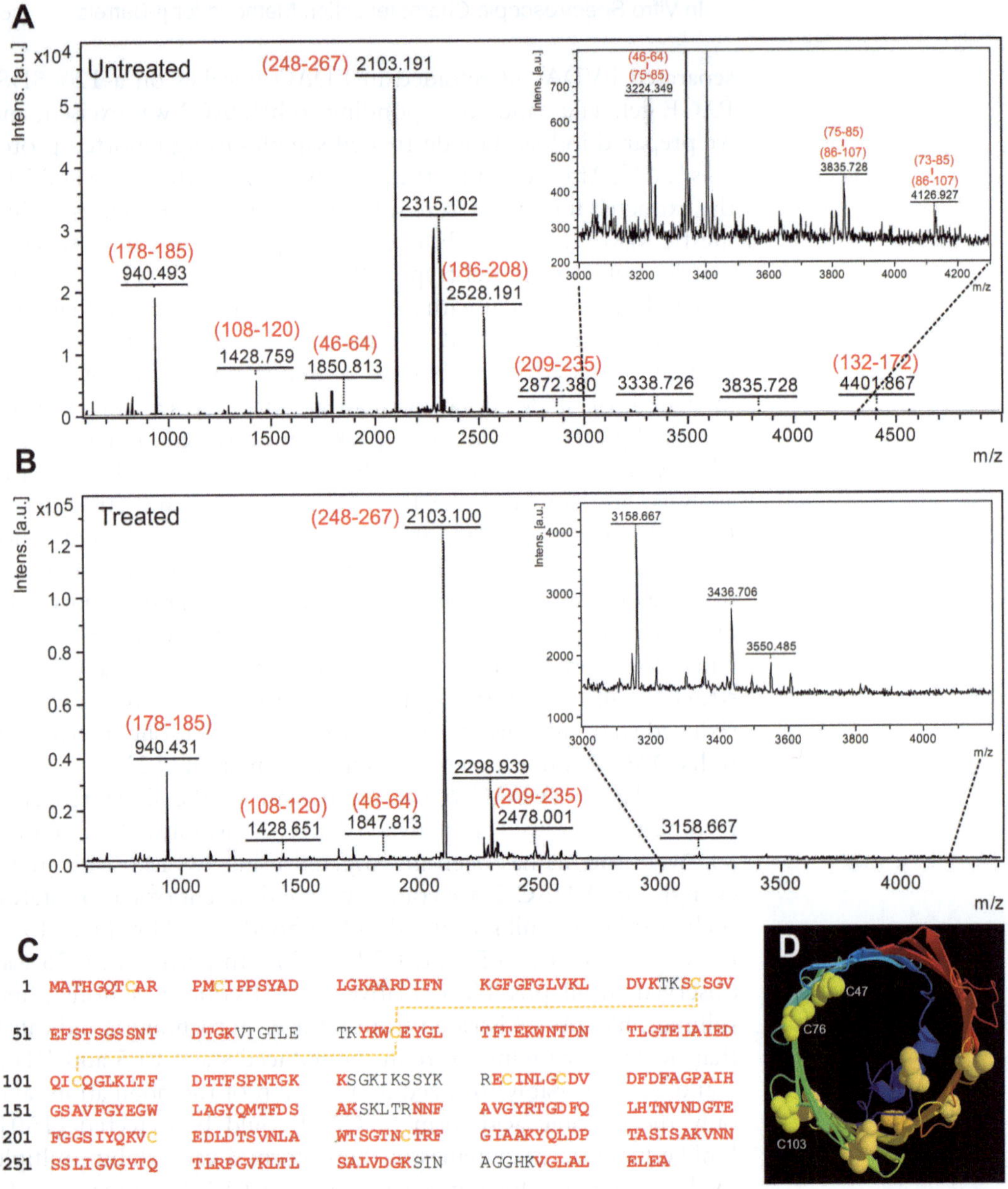

Fig. 19 Determination of oxidation state of cysteines using mass spectrometry. Representative MS profiles of in-gel trypsin-digested samples of WT hVDAC-2 refolded in LDAO, without (**a**) and with (**b**) iodoacetamide treatment. (**a**) In the untreated sample, we observe a minor disulfide-bonded hVDAC-2 population, from the presence of three low-intensity peaks. These peaks correspond to two disulfide bonds involving residue Cys76 with Cys47 and Cys103 (*inset*), which are not present in the iodoacetamide-treated samples ((**b**), *inset*). (**c**) Sequence of hVDAC-2 WT indicating the disulfide bond connectivity detected in MS (*dotted yellow lines*). Segments of the protein identified in the mass spectrometric profile have been marked in *red*. (**d**) Ribbon diagram of hVDAC-2 WT barrel highlighting three cysteine residues (*yellow spheres*) forming disulfide bonds (denoted as C47, C76, and C103). This research was originally published in *The Journal of Biological Chemistry*. Maurya and Mahalakshmi. Modulation of Human Mitochondrial Voltage-dependent Anion Channel 2 (hVDAC-2) Structural Stability by Cysteine-assisted Barrel-lipid Interactions. The Journal of Biological Chemistry. 2013; 288, 35, 25584–25592. © The American Society for Biochemistry and Molecular Biology

3.7 Planar Lipid Bilayer Measurements of Channel-Forming Barrels

Recent development in single-molecule studies made it possible to study folding-unfolding, function, or other biophysical characteristics of a single-protein molecule without interference from complex microenvironment. The revolution in the field of single-channel functional studies was the discovery of patch clamp technique, where channel behavior of a single-protein molecule can be monitored [266–268]. Outer membrane proteins reside in a lipid bilayer environment and play a central role in the transport of different ions, metabolites, energy currencies (ATP, NADPH, etc.), and other macromolecules. Black lipid membrane (BLM) system, which presents a planar lipid bilayer, works as a host matrix for membrane proteins [269, 270]. In BLM studies, a single-protein molecule is inserted in the membrane, which separates two buffer-filled chambers. The transport of various ions and molecules across the membrane from one chamber to the other is monitored at defined voltages by the change in current. Channel activity of a wide variety of proteins, such as VDACs and its homologues, OmpF, OmpA, OmpX etc., has been investigated using BLM in a controlled system [148, 271–275]. Here, we will discuss channel behavior of VDACs and other OMPs using BLM.

VDACs are abundantly present in the outer mitochondrial membrane and transports ions and metabolites across the mitochondrial membrane. Recently, we investigated the role of cysteine residues and the unique N-terminal extension sequence for hVDAC-2, by generating the Δ^{1-11} and Δ^{1-32} deletion mutants of the hVDAC-2 WT proteins (Fig. 20) [177]. We successfully refolded all the constructs in LDAO micelles containing 0.1% cholesterol and 1% triton X-100. In BLM measurements, the refolded protein is added to the *cis* side of the planar lipid bilayer composed of DiPhPC (1,2-diphytanoyl 3-phosphatidylcholine). Once the refolded protein is added, the channel insertion events are recorded at a holding voltage of 10 mV [177]. The BLM system can be used to calculate the conductance (G/G_{max}), open probability (P_{open}) plots, steepness of the voltage dependence (n), and the voltage at which half the channels are open (V_0).

Using the BLM setup and multichannel voltage ramp measurements, wherein the voltage is ramped in multiple cycles from –60 mV to +60 mV at defined ramp rates, we were able to establish differences between the WT and deletion constructs of hVDAC-2. hVDAC-2 WT shows asymmetric voltage gating, whereas we observe a symmetric voltage gating for hVDAC-2 Δ^{1-11} construct (Fig. 21). Cysteine deletion and deletion of the N-terminal extension affect the voltage sensing of hVDAC-2, wherein the deletion of N-terminal extension changes the voltage dependence for WT and C0 mutants in the positive and negative arm, respectively [177]. On the other hand, complete removal of the N-terminal extension and N-terminal helix causes a complete loss in the voltage dependence of the barrel. This loss in voltage gating cannot be

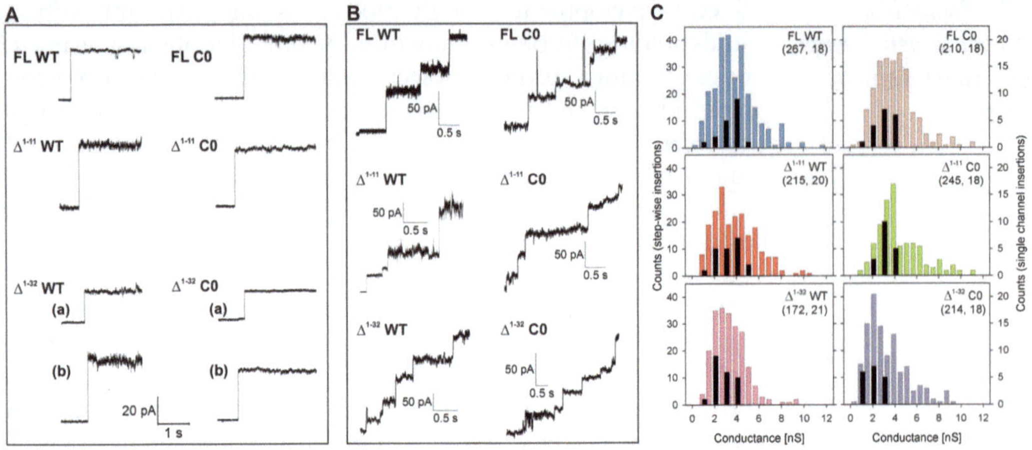

Fig. 20 Monitoring single-channel activity in planar lipid bilayers. Representative (**a**) single-channel and (**b**) stepwise insertion events of hVDAC-2 (WT and C0) and its truncations. The recordings are carried out in a DiPhPC membrane containing 0.1% cholesterol, at a holding voltage of 10 mV. The hVDAC-2 Δ^{1-32} WT mutant lacking the N-terminal helix shows insertion at both the subconductance (~2 nS) and open (~4 nS) states. (**c**) Histograms derived from stepwise insertion events (*colored bars*) and single-channel insertion events (*black bars*) for hVDAC-2 WT and its mutants. The number of insertion events, considered for deriving the histograms, is mentioned in the brackets for stepwise and single-channel insertion events, respectively. We observe from planar bilayer measurements that the removal of the N-terminal helix promotes the channel insertion in the subconductance state. This research was originally published in *The Journal of Biological Chemistry*. Maurya and Mahalakshmi. N-helix and Cysteines Inter-regulate Human Mitochondrial VDAC-2 Function and Biochemistry. The Journal of Biological Chemistry. 2015; 290, 51, 30240–30252. © The American Society for Biochemistry and Molecular Biology

recovered by supplementing a synthetic N-terminal helix peptide in the *cis* chamber. In line with these observations, we also observe two different conductance states of 2 and 4 nS for WT and C0 constructs during single-channel monitoring measurements, which remains similar for hVDAC-2 full length and the Δ^{1-11} constructs. Interestingly, we observe two conductance states for hVDAC2 Δ^{1-32} WT construct, but the removal of cysteines hampers the open state of the barrel for C0 mutants lacking residues 1–32. These results suggest that open state of hVDAC-2 is cysteine dependent and can be observed when residues from the N-terminal segment are removed [177]. Overall, BLM measurements allow us to establish that the 11 residue extension in the N-terminal helix influences the voltage sensing of hVDAC-2 (Fig. 21).

Channel conductance measurements using natively refolded (prepared by detergent exchange as described in the previous section) and in vitro refolded (in C8E4 (tetraethylene glycol monooctyl ether) micelles) OmpA, reconstituted into DiPhPC bilayers, reveal the presence of two types of channels [148]. After insertion into lipid bilayer, native OmpA shows three different conductance states: (1) closed channel, (2) small channel with the conductance of 50–80 pS, and (3) large channels with the conductance of

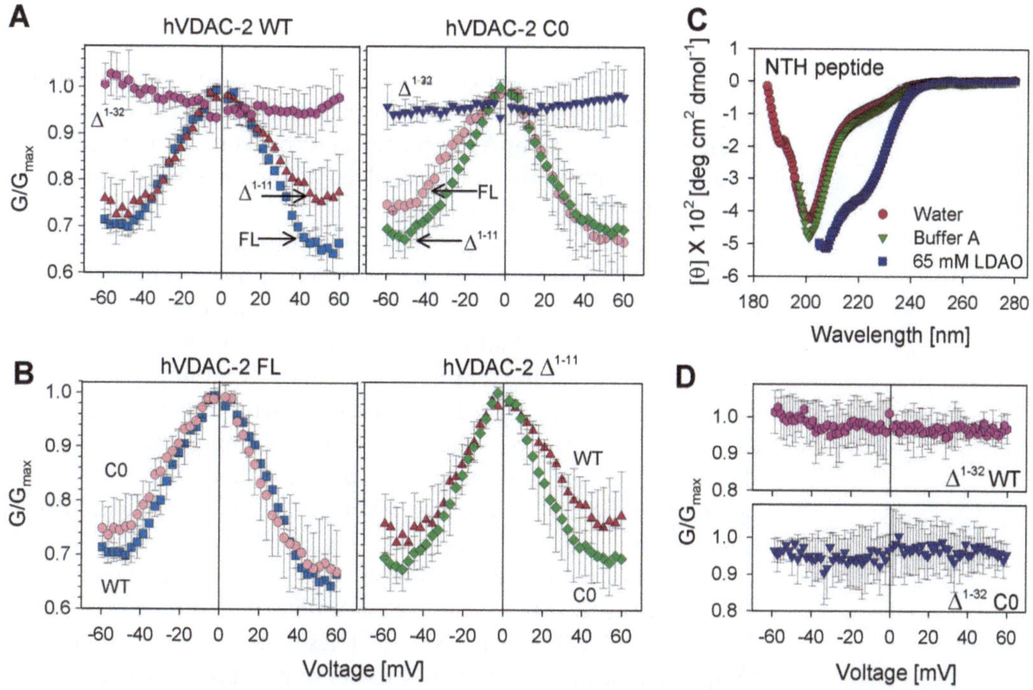

Fig. 21 Multichannel activity measurements in planar bilayers using BLM. (**a**) Normalized conductance (G/G_{max}) plots at a voltage gradient ranging from +60 mV to −60 mV in DiPhPC + 0.1% cholesterol membrane, for hVDAC-2 WT proteins (*left panels*) and C0 constructs (*right panels*), and their N-helix truncation mutants. (**b**) Comparison of the G/G_{max} plots of hVDAC-2 full length (FL; *left panel*) and hVDAC-2 Δ^{1-11} (*right panel*) of WT and C0 mutants. (**c**) Secondary structure content of synthetic N-helix peptide in various conditions, determined using far-UV CD. (**d**) Voltage-dependent G/G_{max} recording for hVDAC-2 WT Δ^{1-32} and hVDAC-2 C0 Δ^{1-32} in the presence of exogenously supplemented N-helix peptide at the protein-peptide ratio of 1:1,000. BLM measurements demonstrate that the N-terminal helix in hVDAC-2 regulates channel conductance by sensing the applied voltage, only when present in *cis* (**b**) and not when supplemented exogenously (**d**). This research was originally published in *The Journal of Biological Chemistry*. Maurya and Mahalakshmi. N-helix and Cysteines Inter-regulate Human Mitochondrial VDAC-2 Function and Biochemistry. The Journal of Biological Chemistry. 2015; 290, 51, 30240–30252. © The American Society for Biochemistry and Molecular Biology

260–320 pS. Small channels were observed with the highest frequency, whereas a low frequency of larger channels was observed, in 1 M KCl solution at a +100 mV membrane potential. Channel behavior of in vitro refolded protein, reconstituted in bilayers, also shows similar conductance pattern (close, small, and large channels) as observed for natively folded protein [148]. This indicates that the artificial refolding environment presented in BLM experiments also provides a similar level of the structural and functional scaffold to OmpA. Similarly, the importance of tryptophan residues on channel activity was investigated. It was observed that the channel behavior was largely unaffected by single tryptophan substitution for OmpA [148]. However, similar measurements on single-Trp mutants of hVDAC-2 from our laboratory indicate that the indoles do bear an influence on barrel opening and closing events (unpublished results).

Another group has investigated the temperature dependence of single-channel conductance and open probability for outer membrane protein A (OmpA) in planar lipid bilayers, as increased temperature is known to elevate molecular dynamicity of the barrels [274]. Here, OmpA refolded in C8E4 was reconstituted in DiPhPC bilayer, and single-channel conductance was recorded at +100 mV. Channel activity was recorded at five temperature levels from 15 °C to 37 °C, for 20 min at each temperature. The mean conductance of small and large channels at different temperatures varies with increasing temperature. At 15 °C, OmpA shows only very short openings to small channels with mean conductance amplitude of 30 pS, whereas at higher temperatures, OmpA shows both short and large openings [274]. OmpA shows small openings of 58 pS, 76 pS, and 105 pS at 22 °C, 25 °C, and 30 °C, respectively, and do not show any small opening at 37 °C. On the other hand, large channels with mean conductance amplitude of 130 pS, 180 pS, 250 pS, and 350 pS could be observed at 22 °C, 25 °C, 30 °C, and 37 °C. These observations suggest that OmpA is a flexible protein and exists in two temperature-sensitive conformations [274]. OmpA favors a smaller-channel conformation at lower temperatures, and larger channels become evident when the temperature is increased.

The studies explained here represent the local behavior of a single-protein molecule, independent of molecular overcrowding and environmental factors. Thus, they can be marked as protein-specific responses due to minor variation in the experimental procedure or conditions. Such microscopic differences in channel behavior monitored via BLM studies have significant application in correlating structure to function of the transmembrane protein [148, 274, 276, 277].

3.8 Nuclear Magnetic Resonance for Complete Structure and Screening

Structure determination using NMR spectroscopy is another preferred methodology when it comes to obtaining structural information for macromolecules [278–282]. A major obstacle in the field of integral membrane proteins is the lack of a universal protocol – involving buffer conditions, lipids or detergents, refolding temperature and pH, etc. – that can be used for a range of membrane proteins for characterization using NMR. Each protein requires its own process of optimization of conditions making it amenable to NMR characterization. Although complete structure determination has not been possible for a great number of β-barrel membrane proteins using NMR (the few successful examples include PagP, OmpX, OmpA, human VDAC-1, BamA [75, 283–286]), this technique has been used to gain insightful information regarding the molecular dynamics of membrane proteins. In this section, we will discuss the recent NMR studies carried out on β-barrel membrane proteins that have been able to yield valuable information on their molecular properties.

In the case of the outer membrane enzyme PagP, Kay and coworkers [214] have identified a structural interconversion between two distinct states, termed as relaxed (R) and tense (T) states, using ^{1}H-^{15}N NMR spectroscopy. With the help of a combination of TROSY-based experiments and another set of CPMG (Carr-Purcell-Meiboom-Gill) pulse train relaxation dispersion experiments, they were able to characterize the thermodynamics of the transition, yielding a ΔH and ΔS of −10.7 cal/mol and −37.5 cal/mol/K, respectively [214]. The structural changes were also annotated to the specific region of the protein implicated in the catalytic site formation. Specifically, the L1 loop residues and the nearby barrel region, containing the active site His, undergo major changes in their chemical shifts during the transition from one conformer to the other [214]. These changes are expected to be brought about by the need of this particular enzyme to switch between a "substrate binding" relaxed state and a catalytically active tense state.

The effect of amino acid substitution on the folding and structure of a transmembrane β-barrel protein OmpX was investigated using NMR spectroscopy [151]. HSQC-TROSY spectrum of methionine-less mutants ($OmpX^{M}$) and substitution of histidine to asparagine at position 100 ($OmpX^{HN}$) are shown in Fig. 22. HSQC-TROSY experiments have helped us in deriving information regarding the folded state and structural variations at the level of hydrogen bond pattern and backbone conformation [151]. Here, $OmpX^{M}$ refolded in 150 mM 6:0 PC (6:0 diether PC; 1,2-dihexanoyl-*sn*-glycero-3-phosphocholine) in 20 mM Tris-HCl at pH 9.5 showing well-dispersed resonances indicates that Omp^{XM} adopts a well-folded conformation when folded in 6:0 PC micelles [151]. Similarly, well-dispersed and superimposable resonances obtained for $OmpX^{HN}$ and $OmpX^{M}$ in LDAO also indicate that methionine substitution does not alter the barrel scaffold.

In line with the mutational studies described earlier, temperature-induced changes in folded state and structure of OmpX were also investigated by monitoring temperature-dependent chemical shifts in NMR spectrum [217]. We observe similar dispersion in NMR spectrum for WT, W76 (W140 → F), and W140 (W76 → F) mutants, suggesting a similar level of folding for all Trp mutants. Upon temperature scanning, we do not observe any change in indole resonance corresponding to Trp140, but the resonance corresponding to Trp76 disappears at the temperatures above 65 °C [217]. Such behavior suggests that the local environment of W140 is unaffected by temperature, whereas W76 is conformationally flexible. Interestingly, the marginal downfield shift of Ala10 and Gln11 was observed in the OmpX W140 mutant construct. These amino acids reside in the vicinity of W140, and the

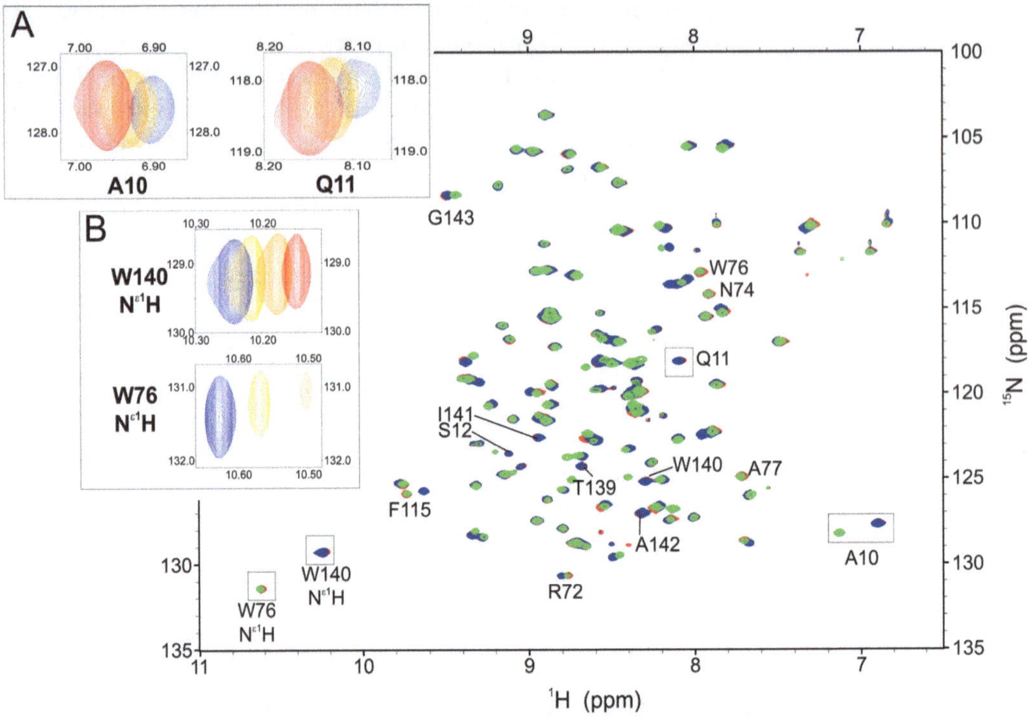

Fig. 22 Temperature-induced conformational changes on folded OmpX monitored using NMR. Overlay of the ^{1}H–^{15}N HSQC spectra of OmpX WT (*red*), and the single-Trp mutants W76 (*green*) and W140 (*blue*) refolded in 100 mM DPC. Resonances showing temperature-induced chemical shift are shown in *boxes* and further expanded in the *inset* display. *Inset*: temperature-dependent chemical shift variation of (**a**) Ala10 and Gln11, and (**b**) the indole amides, for OmpX WT, at different temperatures [40°C (*blue*), 50°C (*yellow*), 60°C (*orange*), and 70°C (*red*)]. Reproduced with permission from Chaturvedi and Mahalakshmi [217]

downfield shift proportionally increases with increases in the temperature [217]. Our observed results indicate that Trp140 maintains strong local contacts with the protein backbone that is preserved even at high temperatures, which destabilizes upon a Trp → Phe substitution.

Similar to these studies, exposed and buried amides present in refolded Ail were investigated using the hydrogen-deuterium NMR exchange experiment [262]. Incubation of refolded Ail (in DHPC micelles prepared in water) in D_2O results in an exchange of few amide protons, but some of the amides resist the exchange, even for several days. The amides participating in proton exchange can be solvent exposed, whereas the amides resisting the proton exchange can be explained as buried amides, representing the buried amino acids in vesicles or transmembrane segment of the barrel [262]. Recently, Muller, Hiller, and coworkers investigated the formation of interstrand hydrogen bonds during the folding process of OmpX in detergent micelles using hydrogen-deuterium (H/D) exchange in combination with NMR spectroscopy [287]. Refolding of OmpX follows a two-state mechanism, where the unfolded polypeptide chain undergoes a hydrophobic collapse and associates with detergent molecules, followed by permanent structural rearrangement giving rise to formation of a folded barrel. Such detailed information of hydrogen bond formation during folding of a β barrel membrane protein helps us understand membrane protein folding-unfolding pathways [287]. Hence, in addition to complete structure determination, NMR has been used to characterize local and global dynamics of membrane proteins.

3.9 Microscopy and Single-Molecule Unfolding Measurements

Recent development in single-molecule studies has made it possible to study membrane protein unfolding measurements using atomic force microscopy (AFM) [288–293]. Here, a single molecule is mounted on a tip or bead, and in the presence of mechanical forces, changes in protein architecture is monitored with respect to the initial conformation. Further development in the single-molecule study was single-molecule force spectroscopy (SMFS), where folding-unfolding and stability studies of the protein molecule can be performed in lipid bilayers under near-native conditions [291, 293]. On the other hand, microscopic techniques, such as SEM (scanning electron microscopy), TEM (transmission electron microscopy), and confocal microscopy, help in understanding topology and localization of the proteins. In the cellular microenvironment, it is difficult to investigate molecular interactions between macromolecules. Thus, single-molecule studies provide a platform to investigate effects of molecular chaperones on the folding trajectories of soluble proteins. Here we discuss AFM and SMFS, used in studying protein folding-unfolding and stability.

To understand the folding of FhuA, Hiller, Muller, and coworkers reconstituted the full-length protein in *E. coli* lipid membranes. According to AFM topographs, densely packed FhuA assemblies in proteoliposomes were used for mechanical unfolding of FhuA using SMFS, where reconstituted FhuA was adsorbed on freshly cleaved mica. Once the protein is attached to the mica surface, the unbounded fraction was removed by repeated washing with buffer solution. The calibrated AFM tip is pushed to the membrane and a force of ~600 pN for 500 ms is applied. The force-distance curve is then recorded by pulling back at 1 μm s^{-1} for 180 nm to partially unfold FhuA. To relax the unfolded polypeptide chain, the cantilever is again pushed toward the membrane and kept at a distance of 10 ± 2 nm above the membrane surface. The pulling and pushing back of AFM cantilever at different forces is repeatedly performed to record unfolding and refolding experiments, and force-distance curves are recorded.

Data corrected for deflection sensitivity and force offset for both unfolding and refolding experiments can now be analyzed to understand FhuA folding-unfolding. Fitting of force peak pattern with extensions (fully unfolded and stretched FhuA polypeptide) leads to the determination of lengths and positions of the peptide stretches that unfolded in each step. Force peaks observed for FhuA suggested the presence of well-defined unfolding intermediates in transmembrane β-barrel unfolding. The force-distance curve recorded by SMFS corresponded to the unfolding of individual β-hairpins, where unfolding of the plug domain is followed by stepwise unfolding of each of the 11 single β-hairpins H1–H11. In line with unfolding experiments, the stability of the partially unfolded β-barrel in the membrane was confirmed by first partially unfolding the FhuA, followed by checking the stability of the remaining β-hairpins in the membrane.

After studying the unfolding of FhuA, for studying the refolding of FhuA, only the N-terminal of the plug domain is unfolded, and the AFM tip is brought back in proximity to the membrane (~10 nm) to allow the polypeptide folding for 1 s. The comparison of the native fingerprint of folding, with the observed experimental fingerprint, reveals that both the force peaks do not match with each other and, therefore, must have arisen from nonnative, misfolded structures. But, when SurA (the natural holdase chaperone) was used during refolding experiments, individual force peaks at positions of the native fingerprint pattern were observed. This observation suggests that SurA has altered the folding trajectory of the FhuA polypeptide and allows the reinsertion of single β-hairpins into the lipid membrane. Unfolding experiments reveal that FhuA shows a stepwise unfolding with the presence of unfolding intermediates, which cannot be reversibly folded. SMFS-based folding-unfolding studies have, therefore, highlighted the importance of chaperones in the folding of FhuA.

Similar to FhuA, conformational flexibility of hVDAC-1 (isoform 1 of human VDAC) has also been recently investigated by Muller, Zweckstetter, and coworkers, with a combination of SMFS and NMR [294]. Several molecular interactions stabilize different functional states of single hVDAC-1 channels by forming molecular interactions and larger interaction networks, mediated by Ca^{2+} and Mg^{2+} (Fig. 23). Here, recombinantly generated hVDAC-1 was reconstituted in DMPC and nonspecifically attached to the AFM cantilever tip. Once immobilized, the mechanical force (pulling and pushing) is applied to the protein molecule, to understand hVDAC-1 unfolding [294]. Interestingly, hVDAC-1 shows stepwise unfolding with the presence of various intermediates, composed of β-hairpins. The stepwise unfolding observed for hVDAC1 represents a mechanism similar to that observed for FhuA and other transmembrane proteins. The use of single-molecule spectroscopy helps in capturing the intermediates in the unfolding process, which is not possible by other spectroscopic methods [294].

Other than protein unfolding studies, AFM has also been used for imaging of biological molecules in dry or solution state. Here, the sample is scanned with the suitable cantilever with scanning or tapping mode to get higher-resolution images. AFM enables us to image particles from micrometers to nanometer range, with a higher resolution [295–297]. Similar to all other techniques

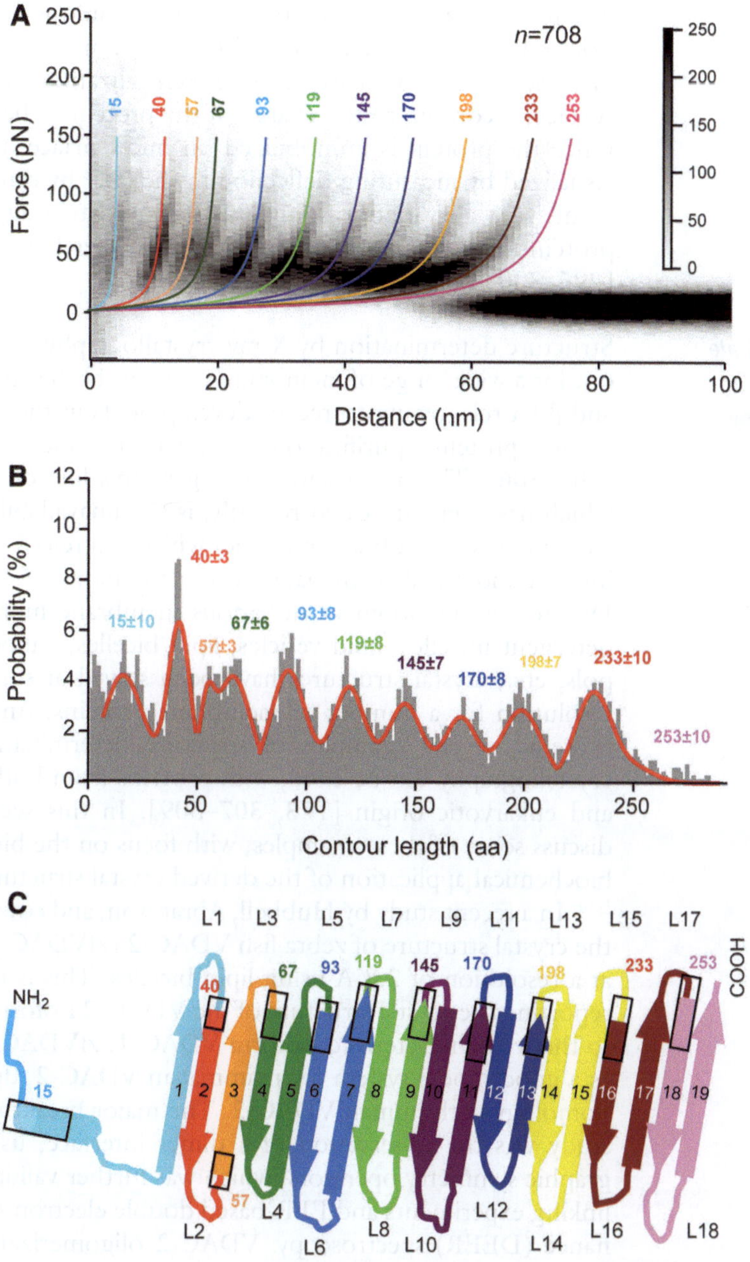

Fig. 23 Unfolding of a transmembrane protein measured using AFM. (**a**) Superimposition of 708 force-distance curves of a single hVDAC-1 molecule. *Colored lines* indicate the worm-like chain curves fitting individual force peaks. The numbers denote contour lengths (amino acids) of the unfolded polypeptide chain. (**b**) Probability of detecting the unfolding force peaks at certain contour lengths. (**c**) Schematic highlighting the structural segments of hVDAC-1 that exhibit unfolding in single steps (marked with the similar color to the force peaks). Reproduced with permission from Ge et al. [294]

described above, the sample preparation is a challenging process, because of the presence of lipid bilayers or detergent micelles and high hydrophobicity of membrane proteins. The most preferred way to visualize membrane proteins is to immobilize the preparation on a mica surface. Repulsion from mica surface due to increased hydrophobicity requires neutralization of the charged surface, according to the charge of the protein to be investigated. Once the protein is immobilized on mica surface, the protein is visualized by measuring deflection in the laser by cantilever movement. A wide variety of amyloid fibrils, protein aggregates, and protein-lipid assemblies have been visualized by using AFM [295–300].

3.10 Solid-State Structures Using Crystallography

Structure determination by X-ray crystallography has been carried out for a wide range of membrane proteins, both α-helical bundles and β-barrels, owing to recent developments in the field of membrane protein purification, extraction, and crystallization [301–306]. The major shortcoming in this line of investigation, which has been addressed recently, is the unavailability of a sufficient range of membrane mimetics which can refold the protein of interest and are also amenable to the technique of crystallization. Due to the development in various membrane mimetics such as detergent micelles, lipid vesicles, lipid bicelles, nanodiscs, amphipols, etc., crystal structures have been solved at sufficiently high resolution for a number of membrane proteins. Among β-barrel proteins, several examples of structure determination by X-ray crystallography can be cited, with proteins from both prokaryotic and eukaryotic origin [173, 307–309]. In this section, we will discuss some of these examples, with focus on the biophysical and biochemical application of the derived crystal structures.

In a recent study by Hubbell, Abramson, and coworkers [307], the crystal structure of zebra fish VDAC-2 (zfVDAC-2) was solved at a resolution of 2.8 Å using lipid bicelles. This is the first study reporting the crystal structure of the VDAC-2 isoform, as opposed to the well-characterized isoform VDAC-1. zfVDAC-2 also serves as a good model system for mammalian VDAC-2, due to its high homology with human VDAC-2. The major breakthrough in this study was the observation of a dimer interface, using crystallographic symmetry operators, which was further validated by cross-linking experiments and EPR-based double electron-electron resonance (DEER) spectroscopy. VDAC-2 oligomerization has been associated with its cellular function [307], making this observation an insightful development in our understanding of VDACs and their role in the mitochondrial outer membrane.

Prive, Bishop and coworkers derived the crystal structure of the lipid A palmitoyltransferase PagP, isolated using the detergent LDAO, at a resolution of 1.9 Å. They were able to crystallize this protein with one of the detergent molecules bound to the interior of the barrel, which paved the way for functional analysis of this protein. The LDAO molecule was inserted in the cavity, with its acyl chain interacting with hydrophobic side chains from the transmembrane region and its polar moiety projected outward. The active site residues, present in the extracellular region, are oriented in such a manner that their side chains are facing the bound detergent molecule. This led to the inference that the LDAO molecule can be considered as a substrate analog. The structure further gave rise to a detailed investigation into the mechanism of action of this lipid metabolizing enzyme, as we know from subsequent studies on the molecular properties of PagP [153, 171, 190, 216, 310, 311].

Structure determination and molecular analysis of the transmembrane pore-forming staphylococcal α-hemolysin is a prime example of the structural information gained by employing crystallographic methods (Fig. 24) [312, 313]. The crystal structure, obtained at a resolution of 1.9 Å, shows a mushroom-shaped architecture formed by a homo-oligomeric heptamer complex, with each monomer contributing two β-strands to the transmembrane pore-forming region. The transmembrane domain of each monomer, also known as the stem domain, combines to form an aqueous pore and therefore presents bilayer-spanning antiparallel β-barrels as a pore-forming motif for bacterial toxins [314, 315].

Another groundbreaking discovery has come from the structural analysis of the bacterial assembly machinery (BAM) complex

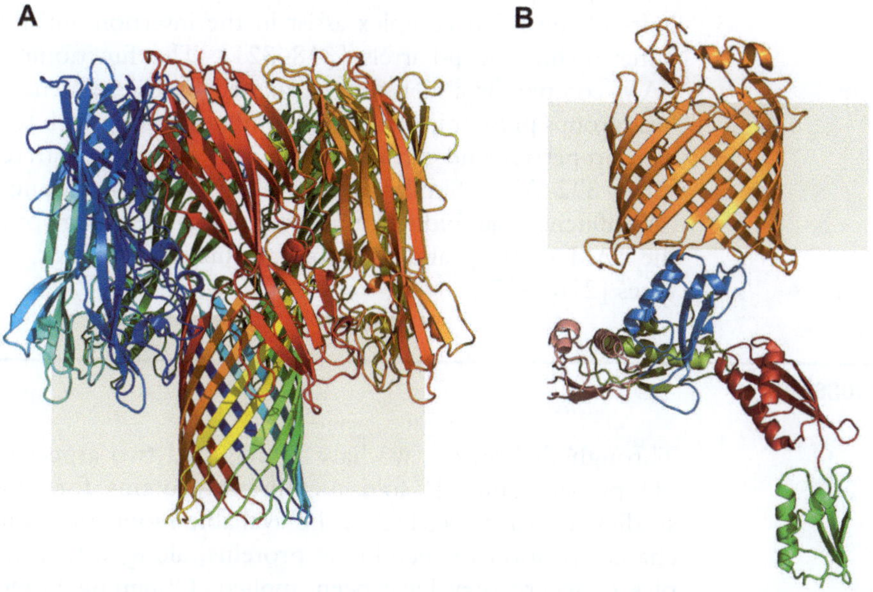

Fig. 24 Transmembrane β-barrel protein structures determined using X-ray crystallography. (**a**) Structure of α-hemolysin from *Staphylococcus aureus* (PDB ID: 7AHL, refolded in micellar environment and crystallized at a resolution of 1.8 Å. The protein takes up a homo-oligomeric heptamer, with the transmembrane region attaining the pore-forming activity. (**b**) Crystal structure of BamA from *Neisseria gonorrhoeae* (PDB ID: 4K3B) displaying the 16-stranded transmembrane β-barrel region and the N-terminal domain consisting of five identical POTRA domains. It facilitates insertion and folding of outer membrane proteins by means of the transmembrane β-barrel region, whereas the N-terminal region acts as a recognition motif, which stabilizes the complex

performed by Buchanan, Noinaj, and coworkers (Fig. 24) [75, 77, 316]. The high-resolution structures of BamA subunit, the central component of the BAM complex, from different source organisms, have revealed a 16-stranded transmembrane β-barrel domain along with an N-terminal periplasmic region comprising of five homologous polypeptide translocation-associated (POTRA) domains. Furthermore, lateral opening and exit pore formation upon the barrel structure was revealed by a combination of structural analysis and molecular dynamics simulations performed on a homology model of BamA from *E. coli*, based on the obtained crystal structures of *Haemophilus ducreyi* BamA lacking the first three POTRA domains and *Neisseria gonorrhoeae* BamA. The role of the other subunits in this multi-protein complex was brought to the fore when the structural analysis of BamB subunit revealed that it binds to the POTRA domain 3 of the BamA subunit [317]. Most recently, the crystal structure of the BAM complex from *E. coli*, solved at a resolution of 3.4 Å, has yielded invaluable information regarding the regulation of this assembly process. The binding of the BamCDE complex appears to exert an allosteric effect by modulating the conformation of the BamA subunit and effectively regulating the assembly machinery [317]. Essentially, these results, in combination with structural analyses reported by other groups, reveal the mechanism by which the ancillary subunits of the BAM complex assist in the insertion and folding of outer membrane β-barrels [318–321]. The functioning of the BAM complex is also tightly regulated by the influence of lipid headgroups present in the outer membrane, along with the molecular properties and structural integrity of the substrate OMPs [155, 322–324]. Similar factors, along with membrane fluidity, can influence the folding of mitochondrial membrane proteins by the SAM (sorting and assembly machinery) complex, in eukaryotes [276, 325].

4 Discussion

Through this review, we have highlighted two aspects: how to (1) prepare and (2) fold membrane proteins for biophysical studies. Further, we discuss the available biophysical studies for characterization of membrane proteins, along with brief examples of where they have been applied. Obtaining homogenous preparations of natively folded protein in sufficient amount and purity is the major bottleneck in these studies. It has been overcome to a great extent by recent advancement in protein expression methodologies and in vitro refolding approaches. Despite these developments, the predicted and actual numbers of solved structures, for MPs, have a glaring mismatch, as of today.

In combination with structure determination, biophysical studies complement our understanding of the overall architecture, scaffold, and functional behavior of the protein. In this review, we have discussed the various techniques employed to characterize β-barrel membrane proteins. The use of different basic biophysical techniques such as SDS-PAGE, CD, fluorescence, and DSC with combination of advanced methodologies (NMR, X-ray and AFM) provides detailed information about local and global changes in structure in the course of β-barrel membrane protein folding and unfolding. These studies can be used to understand protein (un) folding pathways, as well as to derive information on the stability of the transmembrane β-barrels. The evaluation of posttranslational modifications and functional variation due to environmental factors, variation in the protein sequence, and overall fold can also be obtained using advanced methodologies such as MS and BLM.

Availability of improved biophysical methods for characterization will allow us bridge the gap between soluble and membrane proteins. It will further our understanding of how these proteins fold, how they are stabilized, and factors that govern their function and regulation. Knowledge of membrane protein behavior will provide us with new avenues to engineer these molecules and understand disease progression and in novel therapeutic development.

Acknowledgments

B.R.I. thanks the University Grants Commission, Govt. of India, for Senior Research Fellowship. A.G. thanks IISER Bhopal for Senior Research Fellowship. R.M. is a Wellcome Trust/DBT India Alliance Intermediate Fellow. This work is supported by the Wellcome Trust/DBT India Alliance award number IA/I/14/1/501305, Department of Biotechnology, Govt. of India award numbers BT/01/IYBA/2009 and BT/HRD/35/02/2006, and the Science and Engineering Research Board of the Department of Science and Technology, Govt. of India award number SR/FT/LS-47/2010 to R.M.

References

1. Orengo CA, Todd AE, Thornton JM (1999) From protein structure to function. Curr Opin Struct Biol 9(3):374–382. doi:10.1016/S0959-440X(99)80051-7
2. Sackmann E (2015) How actin/myosin crosstalks guide the adhesion, locomotion and polarization of cells. Biochim Biophys Acta 1853(11 Pt B):3132–3142. doi:10.1016/j.bbamcr.2015.06.012
3. Kitchen P, Day RE, Salman MM, Conner MT, Bill RM, Conner AC (2015) Beyond water homeostasis: diverse functional roles of mammalian aquaporins. Biochim Biophys Acta 1850(12):2410–2421. doi:10.1016/j.bbagen.2015.08.023

4. Keskin O, Tuncbag N, Gursoy A (2016) Predicting protein-protein interactions from the molecular to the proteome level. Chem Rev. doi:10.1021/acs.chemrev.5b00683
5. Fanelli F, Felline A, Raimondi F, Seeber M (2016) Structure network analysis to gain insights into GPCR function. Biochem Soc Trans 44(2):613–618. doi:10.1042/BST20150283
6. Chap H (2016) Forty five years with membrane phospholipids, phospholipases and lipid mediators: a historical perspective. Biochimie. doi:10.1016/j.biochi.2016.04.002
7. Schwartz TU (2016) The structure inventory of the nuclear pore complex. J Mol Biol. doi:10.1016/j.jmb.2016.03.015
8. Shrestha L, Young JC (2016) Function and chemotypes of human Hsp70 chaperones. Curr Top Med Chem. doi:10.2174/1568026616666160413142028
9. Joh NH, Min A, Faham S, Whitelegge JP, Yang D, Woods VL, Bowie JU (2008) Modest stabilization by most hydrogen-bonded side-chain interactions in membrane proteins. Nature 453(7199):1266–1270. doi:10.1038/nature06977
10. Dong H, Sharma M, Zhou HX, Cross TA (2012) Glycines: role in alpha-helical membrane protein structures and a potential indicator of native conformation. Biochemistry 51(24):4779–4789. doi:10.1021/bi300090x
11. Chung HS, Piana-Agostinetti S, Shaw DE, Eaton WA (2015) Structural origin of slow diffusion in protein folding. Science 349 (6255):1504–1510. doi:10.1126/science.aab1369
12. Min D, Jefferson RE, Bowie JU, Yoon TY (2015) Mapping the energy landscape for second-stage folding of a single membrane protein. Nat Chem Biol 11(12):981–987. doi:10.1038/nchembio.1939
13. Makwana KM, Mahalakshmi, R (2015) Implications of aromatic–aromatic interactions: From protein structures to peptide models. Protein Sci 24 (12):1920-1933. doi:10.1002/pro.2814
14. Haney CM, Werner HM, McKay JJ, Horne WS (2016) Thermodynamic origin of alpha-helix stabilization by side-chain cross-links in a small protein. Org Biomol Chem. doi:10.1039/c6ob00475j
15. Smith AE, Zhou LZ, Gorensek AH, Senske M, Pielak GJ (2016) In-cell thermodynamics and a new role for protein surfaces. Proc Natl Acad Sci U S A 113(7):1725–1730. doi:10.1073/pnas.1518620113
16. Sanders CR, Myers JK (2004) Disease-related misassembly of membrane proteins. Annu Rev Biophys Biomol Struct 33:25–51. doi:10.1146/annurev.biophys.33.110502.140348
17. Gold V, Kudryashev M (2016) Recent progress in structure and dynamics of dual-membrane-spanning bacterial nanomachines. Curr Opin Struct Biol 39:1–7. doi:10.1016/j.sbi.2016.03.001
18. Touw WG, Joosten RP, Vriend G (2016) New biological insights from better structure models. J Mol Biol 428(6):1375–1393. doi:10.1016/j.jmb.2016.02.002
19. Akutsu M, Dikic I, Bremm A (2016) Ubiquitin chain diversity at a glance. J Cell Sci 129 (5):875–880. doi:10.1242/jcs.183954
20. Geeves MA (2016) The ATPase mechanism of myosin and actomyosin. Biopolymers. doi:10.1002/bip.22853
21. Hayer-Hartl M, Bracher A, Hartl FU (2016) The GroEL-GroES chaperonin machine: a nano-cage for protein folding. Trends Biochem Sci 41(1):62–76. doi:10.1016/j.tibs.2015.07.009
22. Just WW, Peranen J (2016) Small GTPases in peroxisome dynamics. Biochim Biophys Acta 1863(5):1006–1013. doi:10.1016/j.bbamcr.2016.01.004
23. Skjaerven L, Cuellar J, Martinez A, Valpuesta JM (2015) Dynamics, flexibility, and allostery in molecular chaperonins. FEBS Lett 589(19 Pt A):2522–2532. doi:10.1016/j.febslet.2015.06.019
24. van de Willige D, Hoogenraad CC, Akhmanova A (2016) Microtubule plus-end tracking proteins in neuronal development. Cell Mol Life Sci 73(10):2053–2077. doi:10.1007/s00018-016-2168-3
25. White SH, Wimley WC (1999) Membrane protein folding and stability: physical principles. Annu Rev Biophys Biomol Struct 28:319–365. doi:10.1146/annurev.biophys.28.1.319
26. McCulloch CA, Downey GP, El-Gabalawy H (2006) Signalling platforms that modulate the inflammatory response: new targets for drug development. Nat Rev Drug Discov 5 (10):864–876. doi:10.1038/nrd2109
27. Kline KA, Falker S, Dahlberg S, Normark S, Henriques-Normark B (2009) Bacterial adhesins in host-microbe interactions. Cell Host Microbe 5(6):580–592. doi:10.1016/j.chom.2009.05.011
28. Ritter SL, Hall RA (2009) Fine-tuning of GPCR activity by receptor-interacting

proteins. Nat Rev Mol Cell Biol 10 (12):819–830. doi:10.1038/nrm2803
29. Hennessy EJ, Parker AE, O'Neill LA (2010) Targeting Toll-like receptors: emerging therapeutics? Nat Rev Drug Discov 9 (4):293–307. doi:10.1038/nrd3203
30. Venkatakrishnan AJ, Deupi X, Lebon G, Tate CG, Schertler GF, Babu MM (2013) Molecular signatures of G-protein-coupled receptors. Nature 494(7436):185–194. doi:10.1038/nature11896
31. Verkman AS, Anderson MO, Papadopoulos MC (2014) Aquaporins: important but elusive drug targets. Nat Rev Drug Discov 13 (4):259–277. doi:10.1038/nrd4226
32. Cournia Z, Allen TW, Andricioaei I, Antonny B, Baum D, Brannigan G, Buchete NV, Deckman JT, Delemotte L, Del Val C, Friedman R, Gkeka P, Hege HC, Henin J, Kasimova MA, Kolocouris A, Klein ML, Khalid S, Lemieux MJ, Lindow N, Roy M, Selent J, Tarek M, Tofoleanu F, Vanni S, Urban S, Wales DJ, Smith JC, Bondar AN (2015) Membrane protein structure, function, and dynamics: a perspective from experiments and theory. J Membr Biol 248(4):611–640. doi:10.1007/s00232-015-9802-0
33. Selkoe DJ (1999) Translating cell biology into therapeutic advances in Alzheimer's disease. Nature 399(6738 Suppl):A23–A31
34. Harte NP, Klyubin I, McCarthy EK, Min S, Garrahy SA, Xie Y, Davey GP, Boland JJ, Rowan MJ, Mok KH (2015) Amyloid oligomers and mature fibrils prepared from an innocuous protein cause diverging cellular death mechanisms. J Biol Chem 290 (47):28343–28352. doi:10.1074/jbc.M115.676072
35. Ng DP, Poulsen BE, Deber CM (2012) Membrane protein misassembly in disease. Biochim Biophys Acta 1818(4):1115–1122. doi:10.1016/j.bbamem.2011.07.046
36. Smilansky A, Dangoor L, Nakdimon I, Ben-Hail D, Mizrachi D, Shoshan-Barmatz V (2015) The voltage-dependent anion channel 1 mediates amyloid beta toxicity and represents a potential target for Alzheimer disease therapy. J Biol Chem 290(52):30670–30683. doi:10.1074/jbc.M115.691493
37. Parakh S, Atkin JD (2016) Protein folding alterations in amyotrophic lateral sclerosis. Brain Res. doi:10.1016/j.brainres.2016.04.010
38. le Maire M, Champeil P, Moller JV (2000) Interaction of membrane proteins and lipids with solubilizing detergents. Biochim Biophys Acta 1508(1-2):86–111
39. Seddon AM, Curnow P, Booth PJ (2004) Membrane proteins, lipids and detergents: not just a soap opera. Biochim Biophys Acta 1666(1-2):105–117. doi:10.1016/j.bbamem.2004.04.011
40. Zhou HX, Cross TA (2013) Influences of membrane mimetic environments on membrane protein structures. Annu Rev Biophys 42:361–392. doi:10.1146/annurev-biophys-083012-130326
41. Popot JL (2014) Folding membrane proteins in vitro: a table and some comments. Arch Biochem Biophys 564:314–326. doi:10.1016/j.abb.2014.06.029
42. Mogensen JE, Otzen DE (2005) Interactions between folding factors and bacterial outer membrane proteins. Mol Microbiol 57 (2):326–346. doi:10.1111/j.1365-2958.2005.04674.x
43. Koehler J, Woetzel N, Staritzbichler R, Sanders CR, Meiler J (2009) A unified hydrophobicity scale for multispan membrane proteins. Proteins 76(1):13–29. doi:10.1002/prot.22315
44. Foffi G, Pastore A, Piazza F, Temussi PA (2013) Macromolecular crowding: chemistry and physics meet biology (Ascona, Switzerland, 10–14 June 2012). Phys Biol 10(4):040301. doi:10.1088/1478-3975/10/4/040301
45. Otzen DE, Andersen KK (2013) Folding of outer membrane proteins. Arch Biochem Biophys 531(1-2):34–43. doi:10.1016/j.abb.2012.10.008
46. Wimley WC, Creamer TP, White SH (1996) Solvation energies of amino acid side chains and backbone in a family of host-guest pentapeptides. Biochemistry 35(16):5109–5124. doi:10.1021/bi9600153
47. Wimley WC, White SH (1996) Experimentally determined hydrophobicity scale for proteins at membrane interfaces. Nat Struct Biol 3(10):842–848
48. Killian JA, von Heijne G (2000) How proteins adapt to a membrane-water interface. Trends Biochem Sci 25(9):429–434
49. Hessa T, Kim H, Bihlmaier K, Lundin C, Boekel J, Andersson H, Nilsson I, White SH, von Heijne G (2005) Recognition of transmembrane helices by the endoplasmic reticulum translocon. Nature 433(7024):377–381. doi:10.1038/nature03216
50. Jackups R Jr, Cheng S, Liang J (2006) Sequence motifs and antimotifs in beta-barrel membrane proteins from a genome-wide analysis: the Ala-Tyr dichotomy and chaperone binding motifs. J Mol Biol 363(2):611–623. doi:10.1016/j.jmb.2006.07.095

51. Hessa T, Meindl-Beinker NM, Bernsel A, Kim H, Sato Y, Lerch-Bader M, Nilsson I, White SH, von Heijne G (2007) Molecular code for transmembrane-helix recognition by the Sec61 translocon. Nature 450 (7172):1026–1030. doi:10.1038/nature06387
52. MacCallum JL, Tieleman DP (2011) Hydrophobicity scales: a thermodynamic looking glass into lipid-protein interactions. Trends Biochem Sci 36(12):653–662. doi:10.1016/j.tibs.2011.08.003
53. Moon CP, Fleming KG (2011) Side-chain hydrophobicity scale derived from transmembrane protein folding into lipid bilayers. Proc Natl Acad Sci U S A 108(25):10174–10177. doi:10.1073/pnas.1103979108
54. Ojemalm K, Higuchi T, Jiang Y, Langel U, Nilsson I, White SH, Suga H, von Heijne G (2011) Apolar surface area determines the efficiency of translocon-mediated membrane-protein integration into the endoplasmic reticulum. Proc Natl Acad Sci U S A 108 (31):E359–E364. doi:10.1073/pnas.1100120108
55. Ojemalm K, Botelho SC, Studle C, von Heijne G (2013) Quantitative analysis of SecYEG-mediated insertion of transmembrane alpha-helices into the bacterial inner membrane. J Mol Biol 425(15):2813–2822. doi:10.1016/j.jmb.2013.04.025
56. Lin M, Gessmann D, Naveed H, Liang J (2016) Outer membrane protein folding and topology from a computational transfer free energy scale. J Am Chem Soc 138 (8):2592–2601. doi:10.1021/jacs.5b10307
57. Sosinsky GE, Nicholson BJ (2005) Structural organization of gap junction channels. Biochim Biophys Acta 1711(2):99–125. doi:10.1016/j.bbamem.2005.04.001
58. Kedrov A, Janovjak H, Sapra KT, Muller DJ (2007) Deciphering molecular interactions of native membrane proteins by single-molecule force spectroscopy. Annu Rev Biophys Biomol Struct 36:233–260. doi:10.1146/annurev.biophys.36.040306.132640
59. White SH, von Heijne G (2008) How translocons select transmembrane helices. Annu Rev Biophys 37:23–42. doi:10.1146/annurev.biophys.37.032807.125904
60. Ojemalm K, Halling KK, Nilsson I, von Heijne G (2012) Orientational preferences of neighboring helices can drive ER insertion of a marginally hydrophobic transmembrane helix. Mol Cell 45(4):529–540. doi:10.1016/j.molcel.2011.12.024
61. Chang YC, Bowie JU (2014) Measuring membrane protein stability under native conditions. Proc Natl Acad Sci U S A 111 (1):219–224. doi:10.1073/pnas.1318576111
62. Schlebach JP, Woodall NB, Bowie JU, Park C (2014) Bacteriorhodopsin folds through a poorly organized transition state. J Am Chem Soc 136(47):16574–16581. doi:10.1021/ja508359n
63. Deatherage CL, Lu Z, Kim JH, Sanders CR (2015) Notch transmembrane domain: secondary structure and topology. Biochemistry 54(23):3565–3568. doi:10.1021/acs.biochem.5b00456
64. Nilsson OB, Hedman R, Marino J, Wickles S, Bischoff L, Johansson M, Muller-Lucks A, Trovato F, Puglisi JD, O'Brien EP, Beckmann R, von Heijne G (2015) Cotranslational protein folding inside the ribosome exit tunnel. Cell Rep 12(10):1533–1540. doi:10.1016/j.celrep.2015.07.065
65. Sanders CR (2015) Perplexing new insight into the dynamics of the EmrE transporter. J Gen Physiol 146(6):441–444. doi:10.1085/jgp.201511523
66. Schlebach JP, Sanders CR (2015) The safety dance: biophysics of membrane protein folding and misfolding in a cellular context. Q Rev Biophys 48(1):1–34. doi:10.1017/S0033583514000110
67. Schlebach JP, Narayan M, Alford C, Mittendorf KF, Carter BD, Li J, Sanders CR (2015) Conformational stability and pathogenic misfolding of the integral membrane protein PMP22. J Am Chem Soc 137 (27):8758–8768. doi:10.1021/jacs.5b03743
68. Cymer F, von Heijne G, White SH (2015) Mechanisms of integral membrane protein insertion and folding. J Mol Biol 427 (5):999–1022. doi:10.1016/j.jmb.2014.09.014
69. Shanmugavadivu B, Apell HJ, Meins T, Zeth K, Kleinschmidt JH (2007) Correct folding of the beta-barrel of the human membrane protein VDAC requires a lipid bilayer. J Mol Biol 368(1):66–78. doi:10.1016/j.jmb.2007.01.066
70. Nesper J, Brosig A, Ringler P, Patel GJ, Muller SA, Kleinschmidt JH, Boos W, Diederichs K, Welte W (2008) Omp85(Tt) from Thermus thermophilus HB27: an ancestral type of the Omp85 protein family. J Bacteriol 190(13):4568–4575. doi:10.1128/JB.00369-08
71. Fairman JW, Noinaj N, Buchanan SK (2011) The structural biology of beta-barrel

membrane proteins: a summary of recent reports. Curr Opin Struct Biol 21 (4):523–531. doi:10.1016/j.sbi.2011.05.005

72. Hinnebusch BJ, Jarrett CO, Callison JA, Gardner D, Buchanan SK, Plano GV (2011) Role of the Yersinia pestis Ail protein in preventing a protective polymorphonuclear leukocyte response during bubonic plague. Infect Immun 79(12):4984–4989. doi:10.1128/IAI.05307-11
73. Andersen KK, Wang H, Otzen DE (2012) A kinetic analysis of the folding and unfolding of OmpA in urea and guanidinium chloride: single and parallel pathways. Biochemistry 51 (42):8371–8383. doi:10.1021/bi300974y
74. Noinaj N, Buchanan SK, Cornelissen CN (2012) The transferrin-iron import system from pathogenic Neisseria species. Mol Microbiol 86(2):246–257. doi:10.1111/mmi.12002
75. Noinaj N, Kuszak AJ, Gumbart JC, Lukacik P, Chang H, Easley NC, Lithgow T, Buchanan SK (2013) Structural insight into the biogenesis of beta-barrel membrane proteins. Nature 501(7467):385–390. doi:10.1038/nature12521
76. Kleinschmidt JH (2015) Folding of beta-barrel membrane proteins in lipid bilayers – unassisted and assisted folding and insertion. Biochim Biophys Acta 1848(9):1927–1943. doi:10.1016/j.bbamem.2015.05.004
77. Bakelar J, Buchanan SK, Noinaj N (2016) The structure of the beta-barrel assembly machinery complex. Science 351 (6269):180–186. doi:10.1126/science.aad3460
78. Pocanschi CL, Apell HJ, Puntervoll P, Hogh B, Jensen HB, Welte W, Kleinschmidt JH (2006) The major outer membrane protein of Fusobacterium nucleatum (FomA) folds and inserts into lipid bilayers via parallel folding pathways. J Mol Biol 355(3):548–561. doi:10.1016/j.jmb.2005.10.060
79. Schafer NP, Truong HH, Otzen DE, Lindorff-Larsen K, Wolynes PG (2016) Topological constraints and modular structure in the folding and functional motions of GlpG, an intramembrane protease. Proc Natl Acad Sci U S A 113(8):2098–2103. doi:10.1073/pnas.1524027113
80. Noinaj N, Kuszak AJ, Buchanan SK (2015) Heat modifiability of outer membrane proteins from gram-negative bacteria. Methods Mol Biol 1329:51–56. doi:10.1007/978-1-4939-2871-2_4
81. White SH (2004) The progress of membrane protein structure determination. Protein Sci 13(7):1948–1949. doi:10.1110/ps.04712004
82. Carpenter EP, Beis K, Cameron AD, Iwata S (2008) Overcoming the challenges of membrane protein crystallography. Curr Opin Struct Biol 18(5):581–586. doi:10.1016/j.sbi.2008.07.001
83. Cherezov V, Abola E, Stevens RC (2010) Recent progress in the structure determination of GPCRs, a membrane protein family with high potential as pharmaceutical targets. Methods Mol Biol 654:141–168. doi:10.1007/978-1-60761-762-4_8
84. Kozma D, Simon I, Tusnady GE (2013) PDBTM: Protein Data Bank of transmembrane proteins after 8 years. Nucleic Acids Res 41(Database issue):D524–D529. doi:10.1093/nar/gks1169
85. http://blanco.biomol.uci.edu/mpstruc/. Membrane Proteins of Known 3D Structure
86. Koebnik R, Locher KP, Van Gelder P (2000) Structure and function of bacterial outer membrane proteins: barrels in a nutshell. Mol Microbiol 37(2):239–253
87. Bartlett AI, Radford SE (2009) An expanding arsenal of experimental methods yields an explosion of insights into protein folding mechanisms. Nat Struct Mol Biol 16 (6):582–588. doi:10.1038/nsmb.1592
88. Cladera J, Rigaud JL, Villaverde J, Dunach M (1997) Liposome solubilization and membrane protein reconstitution using Chaps and Chapso. Eur J Biochem 243(3):798–804
89. Figler RA, Omote H, Nakamoto RK, Al-Shawi MK (2000) Use of chemical chaperones in the yeast Saccharomyces cerevisiae to enhance heterologous membrane protein expression: high-yield expression and purification of human P-glycoprotein. Arch Biochem Biophys 376(1):34–46. doi:10.1006/abbi.2000.1712
90. Tate CG (2001) Overexpression of mammalian integral membrane proteins for structural studies. FEBS Lett 504(3):94–98
91. Bannwarth M, Schulz GE (2003) The expression of outer membrane proteins for crystallization. Biochim Biophys Acta 1610(1):37–45
92. Sarramegna V, Talmont F, Demange P, Milon A (2003) Heterologous expression of G-protein-coupled receptors: comparison of expression systems from the standpoint of large-scale production and purification. Cell Mol Life Sci 60(8):1529–1546
93. Tate CG, Grisshammer R (1996) Heterologous expression of G-protein-coupled

receptors. Trends Biotechnol 14 (11):426–430. doi:10.1016/0167-7799(96)10059-7

94. Rai M, Padh H (2001) Expression systems for production of heterologous proteins. Curr Sci 80(9):1121–1128
95. Zerbs S, Frank AM, Collart FR (2009) Bacterial systems for production of heterologous proteins. Methods Enzymol 463:149–168. doi:10.1016/S0076-6879(09)63012-3
96. Lasken RS (2009) Genomic DNA amplification by the multiple displacement amplification (MDA) method. Biochem Soc Trans 37 (Pt 2):450–453. doi:10.1042/BST0370450
97. Bundo M, Sunaga F, Ueda J, Kasai K, Kato T, Iwamoto K (2012) A systematic evaluation of whole genome amplification of bisulfite-modified DNA. Clin Epigenetics 4(1):22. doi:10.1186/1868-7083-4-22
98. Macaulay IC, Voet T (2014) Single cell genomics: advances and future perspectives. PLoS Genet 10(1), e1004126. doi:10.1371/journal.pgen.1004126
99. Singh MI, Jain V (2013) Tagging the expressed protein with 6 histidines: rapid cloning of an amplicon with three options. PLoS One 8(5), e63922. doi:10.1371/journal.pone.0063922
100. Dubey AA, Singh MI, Jain V (2016) Rapid and robust PCR-based all-recombinant cloning methodology. PLoS One 11(3), e0152106. doi:10.1371/journal.pone.0152106
101. Bonander N, Hedfalk K, Larsson C, Mostad P, Chang C, Gustafsson L, Bill RM (2005) Design of improved membrane protein production experiments: quantitation of the host response. Protein Sci 14(7):1729–1740. doi:10.1110/ps.051435705
102. Midgett CR, Madden DR (2007) Breaking the bottleneck: eukaryotic membrane protein expression for high-resolution structural studies. J Struct Biol 160(3):265–274. doi:10.1016/j.jsb.2007.07.001
103. Wagner S, Klepsch MM, Schlegel S, Appel A, Draheim R, Tarry M, Hogbom M, van Wijk KJ, Slotboom DJ, Persson JO, de Gier JW (2008) Tuning Escherichia coli for membrane protein overexpression. Proc Natl Acad Sci U S A 105(38):14371–14376. doi:10.1073/pnas.0804090105
104. Freigassner M, Pichler H, Glieder A (2009) Tuning microbial hosts for membrane protein production. Microb Cell Fact 8:69. doi:10.1186/1475-2859-8-69
105. Jamshad M, Lin YP, Knowles TJ, Parslow RA, Harris C, Wheatley M, Poyner DR, Bill RM, Thomas OR, Overduin M, Dafforn TR (2011) Surfactant-free purification of membrane proteins with intact native membrane environment. Biochem Soc Trans 39 (3):813–818. doi:10.1042/BST0390813
106. Wang DN, Safferling M, Lemieux MJ, Griffith H, Chen Y, Li XD (2003) Practical aspects of overexpressing bacterial secondary membrane transporters for structural studies. Biochim Biophys Acta 1610(1):23–36
107. Bujard H, Gentz R, Lanzer M, Stueber D, Mueller M, Ibrahimi I, Haeuptle MT, Dobberstein B (1987) A T5 promoter-based transcription-translation system for the analysis of proteins in vitro and in vivo. Methods Enzymol 155:416–433
108. Sawasaki T, Ogasawara T, Morishita R, Endo Y (2002) A cell-free protein synthesis system for high-throughput proteomics. Proc Natl Acad Sci U S A 99(23):14652–14657. doi:10.1073/pnas.232580399
109. Yokoyama S (2003) Protein expression systems for structural genomics and proteomics. Curr Opin Chem Biol 7(1):39–43
110. Spirin AS (2004) High-throughput cell-free systems for synthesis of functionally active proteins. Trends Biotechnol 22 (10):538–545. doi:10.1016/j.tibtech.2004.08.012
111. Ishihara G, Goto M, Saeki M, Ito K, Hori T, Kigawa T, Shirouzu M, Yokoyama S (2005) Expression of G protein coupled receptors in a cell-free translational system using detergents and thioredoxin-fusion vectors. Protein Expr Purif 41(1):27–37. doi:10.1016/j.pep.2005.01.013
112. Katzen F, Chang G, Kudlicki W (2005) The past, present and future of cell-free protein synthesis. Trends Biotechnol 23(3):150–156. doi:10.1016/j.tibtech.2005.01.003
113. Endo Y, Sawasaki T (2006) Cell-free expression systems for eukaryotic protein production. Curr Opin Biotechnol 17(4):373–380. doi:10.1016/j.copbio.2006.06.009
114. Katzen F, Peterson TC, Kudlicki W (2009) Membrane protein expression: no cells required. Trends Biotechnol 27(8):455–460. doi:10.1016/j.tibtech.2009.05.005
115. Sorensen HP, Mortensen KK (2005) Advanced genetic strategies for recombinant protein expression in Escherichia coli. J Biotechnol 115(2):113–128. doi:10.1016/j.jbiotec.2004.08.004
116. Sheffield P, Garrard S, Derewenda Z (1999) Overcoming expression and purification problems of RhoGDI using a family of "parallel"

expression vectors. Protein Expr Purif 15 (1):34–39. doi:10.1006/prep.1998.1003

117. Takeshita S, Sato M, Toba M, Masahashi W, Hashimoto-Gotoh T (1987) High-copy-number and low-copy-number plasmid vectors for lacZ alpha-complementation and chloramphenicol- or kanamycin-resistance selection. Gene 61(1):63–74

118. Mandecki W, Hayden MA, Shallcross MA, Stotland E (1990) A totally synthetic plasmid for general cloning, gene expression and mutagenesis in Escherichia coli. Gene 94 (1):103–107

119. Saida F, Uzan M, Odaert B, Bontems F (2006) Expression of highly toxic genes in E. coli: special strategies and genetic tools. Curr Protein Pept Sci 7(1):47–56

120. Wagner S, Bader ML, Drew D, de Gier JW (2006) Rationalizing membrane protein over-expression. Trends Biotechnol 24(8):364–371. doi:10.1016/j.tibtech.2006.06.008

121. Maurya SR, Chaturvedi D, Mahalakshmi R (2013) Modulating lipid dynamics and membrane fluidity to drive rapid folding of a transmembrane barrel. Sci Rep 3:1989. doi:10.1038/srep01989

122. Maurya SR, Mahalakshmi R (2013) Modulation of human mitochondrial voltage-dependent anion channel 2 (hVDAC-2) structural stability by cysteine-assisted barrel-lipid interactions. J Biol Chem 288 (35):25584–25592. doi:10.1074/jbc.M113.493692

123. Maurya SR, Mahalakshmi R (2014) Influence of protein-micelle ratios and cysteine residues on the kinetic stability and unfolding rates of human mitochondrial VDAC-2. PLoS One 9 (1), e87701. doi:10.1371/journal.pone.0087701

124. Ramon A, Senorale-Pose M, Marin M (2014) Inclusion bodies: not that bad. Front Microbiol 5:56. doi:10.3389/fmicb.2014.00056

125. Middelberg AP (2002) Preparative protein refolding. Trends Biotechnol 20 (10):437–443

126. Singh SM, Panda AK (2005) Solubilization and refolding of bacterial inclusion body proteins. J Biosci Bioeng 99(4):303–310. doi:10.1263/jbb.99.303

127. Peternel S, Komel R (2010) Isolation of biologically active nanomaterial (inclusion bodies) from bacterial cells. Microb Cell Fact 9:66. doi:10.1186/1475-2859-9-66

128. Rodriguez-Carmona E, Cano-Garrido O, Seras-Franzoso J, Villaverde A, Garcia-Fruitos E (2010) Isolation of cell-free bacterial inclusion bodies. Microb Cell Fact 9:71. doi:10.1186/1475-2859-9-71

129. Thakur KG, Jaiswal RK, Shukla JK, Praveena T, Gopal B (2010) Over-expression and purification strategies for recombinant multi-protein oligomers: a case study of Mycobacterium tuberculosis sigma/anti-sigma factor protein complexes. Protein Expr Purif 74 (2):223–230. doi:10.1016/j.pep.2010.06.018

130. Gatti-Lafranconi P, Natalello A, Ami D, Doglia SM, Lotti M (2011) Concepts and tools to exploit the potential of bacterial inclusion bodies in protein science and biotechnology. FEBS J 278(14):2408–2418. doi:10.1111/j.1742-4658.2011.08163.x

131. Garcia-Fruitos E, Vazquez E, Diez-Gil C, Corchero JL, Seras-Franzoso J, Ratera I, Veciana J, Villaverde A (2012) Bacterial inclusion bodies: making gold from waste. Trends Biotechnol 30(2):65–70. doi:10.1016/j.tibtech.2011.09.003

132. Peternel S (2013) Bacterial cell disruption: a crucial step in protein production. N Biotechnol 30(2):250–254. doi:10.1016/j.nbt.2011.09.005

133. Gupta A, Iyer BR, Chaturvedi D, Maurya SR, Mahalakshmi R (2015) Thermodynamic, structural and functional properties of membrane protein inclusion bodies are analogous to purified counterparts: case study from bacteria and humans. RSC Adv 5(2):1227–1234. doi:10.1039/c4ra11207e

134. Mohanty AK, Wiener MC (2004) Membrane protein expression and production: effects of polyhistidine tag length and position. Protein Expr Purif 33(2):311–325

135. Liu W, Kamensky Y, Kakkar R, Foley E, Kulmacz RJ, Palmer G (2005) Purification and characterization of bovine adrenal cytochrome b561 expressed in insect and yeast cell systems. Protein Expr Purif 40 (2):429–439. doi:10.1016/j.pep.2004.12.027

136. Smith SM (2011) Strategies for the purification of membrane proteins. Methods Mol Biol 681:485–496. doi:10.1007/978-1-60761-913-0_29

137. Colombini M (1989) Voltage gating in the mitochondrial channel, VDAC. J Membr Biol 111(2):103–111

138. Colombini M (2012) VDAC structure, selectivity, and dynamics. Biochim Biophys Acta 1818(6):1457–1465. doi:10.1016/j.bbamem.2011.12.026

139. McCommis KS, Baines CP (2012) The role of VDAC in cell death: friend or foe? Biochim

Biophys Acta 1818(6):1444–1450. doi:10.1016/j.bbamem.2011.10.025

140. Messina A, Reina S, Guarino F, De Pinto V (2012) VDAC isoforms in mammals. Biochim Biophys Acta 1818(6):1466–1476. doi:10.1016/j.bbamem.2011.10.005
141. Raghavan A, Sheiko T, Graham BH, Craigen WJ (2012) Voltage-dependant anion channels: novel insights into isoform function through genetic models. Biochim Biophys Acta 1818(6):1477–1485. doi:10.1016/j.bbamem.2011.10.019
142. Shoshan-Barmatz V, Ben-Hail D (2012) VDAC, a multi-functional mitochondrial protein as a pharmacological target. Mitochondrion 12(1):24–34. doi:10.1016/j.mito.2011.04.001
143. De Pinto V, Reina S, Gupta A, Messina A, Mahalakshmi R (2016) Role of cysteines in mammalian VDAC isoforms' function. Biochim Biophys Acta. doi:10.1016/j.bbabio.2016.02.020
144. Diane Johnson HL (1967) Isolation of liver or kidney mitochondria. Methods Enzymol 10:94–96
145. Linden M, Gellerfors P, Nelson BD (1982) Purification of a protein having pore forming activity from the rat liver mitochondrial outer membrane. Biochem J 208(1):77–82
146. Palmieri F, De Pinto V (1989) Purification and properties of the voltage-dependent anion channel of the outer mitochondrial membrane. J Bioenerg Biomembr 21 (4):417–425
147. Hiller S, Abramson J, Mannella C, Wagner G, Zeth K (2010) The 3D structures of VDAC represent a native conformation. Trends Biochem Sci 35(9):514–521. doi:10.1016/j.tibs.2010.03.005
148. Arora A, Rinehart D, Szabo G, Tamm LK (2000) Refolded outer membrane protein A of Escherichia coli forms ion channels with two conductance states in planar lipid bilayers. J Biol Chem 275(3):1594–1600
149. Hingorani KS, Gierasch LM (2014) Comparing protein folding in vitro and in vivo: foldability meets the fitness challenge. Curr Opin Struct Biol 24:81–90. doi:10.1016/j.sbi.2013.11.007
150. Reina S, Checchetto V, Saletti R, Gupta A, Chaturvedi D, Guardiani C, Guarino F, Scorciapino MA, Magri A, Foti S, Ceccarelli M, Messina AA, Mahalakshmi R, Szabo I, De Pinto V (2016) VDAC3 as a sensor of oxidative state of the intermembrane space of mitochondria: the putative role of cysteine residue modifications. Oncotarget 7(3):2249–2268. doi:10.18632/oncotarget.6850
151. Chaturvedi D, Mahalakshmi R (2013) Methionine mutations of outer membrane protein X influence structural stability and beta-barrel unfolding. PLoS One 8(11), e79351. doi:10.1371/journal.pone.0079351
152. Gupta A, Zadafiya P, Mahalakshmi R (2014) Differential contribution of tryptophans to the folding and stability of the attachment invasion locus transmembrane beta-barrel from Yersinia pestis. Sci Rep 4:6508. doi:10.1038/srep06508
153. Iyer BR, Mahalakshmi R (2015) Residue-dependent thermodynamic cost and barrel plasticity balances activity in the PhoPQ-activated enzyme PagP of Salmonella typhimurium. Biochemistry 54(37):5712–5722. doi:10.1021/acs.biochem.5b00543
154. Hartl FU, Hayer-Hartl M (2009) Converging concepts of protein folding in vitro and in vivo. Nat Struct Mol Biol 16(6):574–581. doi:10.1038/nsmb.1591
155. Gessmann D, Chung YH, Danoff EJ, Plummer AM, Sandlin CW, Zaccai NR, Fleming KG (2014) Outer membrane beta-barrel protein folding is physically controlled by periplasmic lipid head groups and BamA. Proc Natl Acad Sci U S A 111(16):5878–5883. doi:10.1073/pnas.1322473111
156. McMorran LM, Brockwell DJ, Radford SE (2014) Mechanistic studies of the biogenesis and folding of outer membrane proteins in vitro and in vivo: what have we learned to date? Arch Biochem Biophys 564:265–280. doi:10.1016/j.abb.2014.02.011
157. Sandlin CW, Zaccai NR, Fleming KG (2015) Skp trimer formation is insensitive to salts in the physiological range. Biochemistry 54 (48):7059–7062. doi:10.1021/acs.biochem.5b00806
158. Thoma J, Burmann BM, Hiller S, Muller DJ (2015) Impact of holdase chaperones Skp and SurA on the folding of beta-barrel outer-membrane proteins. Nat Struct Mol Biol 22 (10):795–802. doi:10.1038/nsmb.3087
159. Rudolph R, Lilie H (1996) In vitro folding of inclusion body proteins. FASEB J 10 (1):49–56
160. Li M, Su ZG, Janson JC (2004) In vitro protein refolding by chromatographic procedures. Protein Expr Purif 33(1):1–10. doi:10.1016/j.pep.2003.08.023
161. Huysmans GH, Radford SE, Brockwell DJ, Baldwin SA (2007) The N-terminal helix is a post-assembly clamp in the bacterial outer membrane protein PagP. J Mol Biol 373

(3):529–540. doi:10.1016/j.jmb.2007.07.072

162. Stanley AM, Fleming KG (2008) The process of folding proteins into membranes: challenges and progress. Arch Biochem Biophys 469(1):46–66. doi:10.1016/j.abb.2007.09.024
163. Alibolandi M, Mirzahoseini H (2011) Chemical assistance in refolding of bacterial inclusion bodies. Biochem Res Int 2011:631607. doi:10.1155/2011/631607
164. Rath A, Deber CM (2012) Protein structure in membrane domains. Annu Rev Biophys 41:135–155. doi:10.1146/annurev-biophys-050511-102310
165. Serebryany E, Zhu GA, Yan EC (2012) Artificial membrane-like environments for in vitro studies of purified G-protein coupled receptors. Biochim Biophys Acta 1818 (2):225–233. doi:10.1016/j.bbamem.2011.07.047
166. Dekker N, Merck K, Tommassen J, Verheij HM (1995) In vitro folding of Escherichia coli outer-membrane phospholipase A. Eur J Biochem 232(1):214–219
167. Surrey T, Schmid A, Jahnig F (1996) Folding and membrane insertion of the trimeric beta-barrel protein OmpF. Biochemistry 35(7):2283–2288. doi:10.1021/bi951216u
168. Vogt J, Schulz GE (1999) The structure of the outer membrane protein OmpX from Escherichia coli reveals possible mechanisms of virulence. Structure 7(10):1301–1309
169. Albrecht R, Zeth K, Soding J, Lupas A, Linke D (2006) Expression, crystallization and preliminary X-ray crystallographic studies of the outer membrane protein OmpW from Escherichia coli. Acta Crystallogr Sect F Struct Biol Cryst Commun 62(Pt 4):415–418. doi:10.1107/S1744309106010190
170. Anbazhagan V, Qu J, Kleinschmidt JH, Marsh D (2008) Incorporation of outer membrane protein OmpG in lipid membranes: protein-lipid interactions and beta-barrel orientation. Biochemistry 47 (23):6189–6198. doi:10.1021/bi800203g
171. Huysmans GH, Baldwin SA, Brockwell DJ, Radford SE (2010) The transition state for folding of an outer membrane protein. Proc Natl Acad Sci U S A 107(9):4099–4104. doi:10.1073/pnas.0911904107
172. Mager F, Gessmann D, Nussberger S, Zeth K (2011) Functional refolding and characterization of two Tom40 isoforms from human mitochondria. J Membr Biol 242(1):11–21. doi:10.1007/s00232-011-9372-8
173. Yamashita S, Lukacik P, Barnard TJ, Noinaj N, Felek S, Tsang TM, Krukonis ES, Hinnebusch BJ, Buchanan SK (2011) Structural insights into Ail-mediated adhesion in Yersinia pestis. Structure 19(11):1672–1682. doi:10.1016/j.str.2011.08.010
174. Barnard TJ, Gumbart J, Peterson JH, Noinaj N, Easley NC, Dautin N, Kuszak AJ, Tajkhorshid E, Bernstein HD, Buchanan SK (2012) Molecular basis for the activation of a catalytic asparagine residue in a self-cleaving bacterial autotransporter. J Mol Biol 415(1):128–142. doi:10.1016/j.jmb.2011.10.049
175. Colombini M (2009) The published 3D structure of the VDAC channel: native or not? Trends Biochem Sci 34(8):382–389. doi:10.1016/j.tibs.2009.05.001
176. Colombini M, Mannella CA (2012) VDAC, the early days. Biochim Biophys Acta 1818 (6):1438–1443. doi:10.1016/j.bbamem.2011.11.014
177. Maurya SR, Mahalakshmi R (2015) N-helix and cysteines inter-regulate human mitochondrial VDAC-2 function and biochemistry. J Biol Chem 290(51):30240–30252. doi:10.1074/jbc.M115.693978
178. Arndt C, Koristka S, Bartsch H, Bachmann M (2012) Native polyacrylamide gels. Methods Mol Biol 869:49–53. doi:10.1007/978-1-61779-821-4_5
179. Wittig I, Braun HP, Schagger H (2006) Blue native PAGE. Nat Protoc 1(1):418–428. doi:10.1038/nprot.2006.62
180. Rath A, Cunningham F, Deber CM (2013) Acrylamide concentration determines the direction and magnitude of helical membrane protein gel shifts. Proc Natl Acad Sci U S A 110(39):15668–15673. doi:10.1073/pnas.1311305110
181. Rath A, Glibowicka M, Nadeau VG, Chen G, Deber CM (2009) Detergent binding explains anomalous SDS-PAGE migration of membrane proteins. Proc Natl Acad Sci U S A 106(6):1760–1765. doi:10.1073/pnas.0813167106
182. Dunker AK, Kenyon AJ (1976) Mobility of sodium dodecyl sulphate – protein complexes. Biochem J 153(2):191–197
183. Burgess NK, Dao TP, Stanley AM, Fleming KG (2008) Beta-barrel proteins that reside in the Escherichia coli outer membrane in vivo demonstrate varied folding behavior in vitro. J Biol Chem 283(39):26748–26758. doi:10.1074/jbc.M802754200

184. Kleinschmidt JH, Tamm LK (1996) Folding intermediates of a beta-barrel membrane protein. Kinetic evidence for a multi-step membrane insertion mechanism. Biochemistry 35 (40):12993–13000. doi:10.1021/bi961478b
185. Tamm LK, Hong H, Liang B (2004) Folding and assembly of beta-barrel membrane proteins. Biochim Biophys Acta 1666(1-2):250–263. doi:10.1016/j.bbamem.2004.06.011
186. Bulieris PV, Behrens S, Holst O, Kleinschmidt JH (2003) Folding and insertion of the outer membrane protein OmpA is assisted by the chaperone Skp and by lipopolysaccharide. J Biol Chem 278(11):9092–9099. doi:10.1074/jbc.M211177200
187. Kleinschmidt JH (2006) Folding kinetics of the outer membrane proteins OmpA and FomA into phospholipid bilayers. Chem Phys Lipids 141(1-2):30–47. doi:10.1016/j.chemphyslip.2006.02.004
188. Roman EA, Gonzalez Flecha FL (2014) Kinetics and thermodynamics of membrane protein folding. Biomolecules 4 (1):354–373. doi:10.3390/biom4010354
189. Schagger H, von Jagow G (1987) Tricine-sodium dodecyl sulfate-polyacrylamide gel electrophoresis for the separation of proteins in the range from 1 to 100 kDa. Anal Biochem 166(2):368–379
190. McMorran LM, Bartlett AI, Huysmans GH, Radford SE, Brockwell DJ (2013) Dissecting the effects of periplasmic chaperones on the in vitro folding of the outer membrane protein PagP. J Mol Biol 425(17):3178–3191. doi:10.1016/j.jmb.2013.06.017
191. Pocanschi CL, Popot JL, Kleinschmidt JH (2013) Folding and stability of outer membrane protein A (OmpA) from Escherichia coli in an amphipathic polymer, amphipol A8-35. Eur Biophys J 42(2-3):103–118. doi:10.1007/s00249-013-0887-z
192. Dewald AH, Hodges JC, Columbus L (2011) Physical determinants of beta-barrel membrane protein folding in lipid vesicles. Biophys J 100(9):2131–2140. doi:10.1016/j.bpj.2011.03.025
193. Khan MA, Neale C, Michaux C, Pomes R, Prive GG, Woody RW, Bishop RE (2007) Gauging a hydrocarbon ruler by an intrinsic exciton probe. Biochemistry 46 (15):4565–4579. doi:10.1021/bi602526k
194. Mogensen JE, Tapadar D, Schmidt MA, Otzen DE (2005) Barriers to folding of the transmembrane domain of the Escherichia coli autotransporter adhesin involved in diffuse adherence. Biochemistry 44 (11):4533–4545. doi:10.1021/bi0475121
195. Surrey T, Jahnig F (1995) Kinetics of folding and membrane insertion of a beta-barrel membrane protein. J Biol Chem 270 (47):28199–28203
196. Schlebach JP, Kim MS, Joh NH, Bowie JU, Park C (2011) Probing membrane protein unfolding with pulse proteolysis. J Mol Biol 406(4):545–551. doi:10.1016/j.jmb.2010.12.018
197. Okada J, Koga Y, Takano K, Kanaya S (2012) Slow unfolding pathway of hyperthermophilic Tk-RNase H2 examined by pulse proteolysis using the stable protease Tk-subtilisin. Biochemistry 51(45):9178–9191. doi:10.1021/bi300973n
198. Missiakas D, Betton JM, Chaffotte A, Minard P, Yon JM (1992) Kinetic studies of the refolding of yeast phosphoglycerate kinase: comparison with the isolated engineered domains. Protein Sci 1(11):1485–1493. doi:10.1002/pro.5560011110
199. Lang K, Schmid FX (1986) Use of a trypsin-pulse method to study the refolding pathway of ribonuclease. Eur J Biochem 159 (2):275–281
200. Park C, Marqusee S (2005) Pulse proteolysis: a simple method for quantitative determination of protein stability and ligand binding. Nat Methods 2(3):207–212. doi:10.1038/nmeth740
201. Li HM, Moore T, Keegstra K (1991) Targeting of proteins to the outer envelope membrane uses a different pathway than transport into chloroplasts. Plant Cell 3(7):709–717
202. Na YR, Park C (2009) Investigating protein unfolding kinetics by pulse proteolysis. Protein Sci 18(2):268–276. doi:10.1002/pro.29
203. Chang Y, Park C (2009) Mapping transient partial unfolding by protein engineering and native-state proteolysis. J Mol Biol 393 (2):543–556. doi:10.1016/j.jmb.2009.08.006
204. Kim MS, Song J, Park C (2009) Determining protein stability in cell lysates by pulse proteolysis and Western blotting. Protein Sci 18 (5):1051–1059. doi:10.1002/pro.115
205. Guo R, Gaffney K, Yang Z, Kim M, Sungsuwan S, Huang X, Hubbell WL, Hong H (2016) Steric trapping reveals a cooperativity network in the intramembrane protease GlpG. Nat Chem Biol 12(5):353–360. doi:10.1038/nchembio.2048
206. Kelly SM, Price NC (2000) The use of circular dichroism in the investigation of protein

structure and function. Curr Protein Pept Sci 1(4):349–384

207. Wallace BA, Lees JG, Orry AJ, Lobley A, Janes RW (2003) Analyses of circular dichroism spectra of membrane proteins. Protein Sci 12(4):875–884. doi:10.1110/ps.0229603
208. Sreerama N, Woody RW (2004) On the analysis of membrane protein circular dichroism spectra. Protein Sci 13(1):100–112. doi:10.1110/ps.03258404
209. Khan MA, Moktar J, Mott PJ, Bishop RE (2010) A thiolate anion buried within the hydrocarbon ruler perturbs PagP lipid acyl chain selection. Biochemistry 49 (11):2368–2379. doi:10.1021/bi901669q
210. Zoldak G, Zubrik A, Musatov A, Stupak M, Sedlak E (2004) Irreversible thermal denaturation of glucose oxidase from Aspergillus niger is the transition to the denatured state with residual structure. J Biol Chem 279 (46):47601–47609. doi:10.1074/jbc.M406883200
211. Allen B, Blum M, Cunningham A, Tu GC, Hofmann T (1990) A ligand-induced, temperature-dependent conformational change in penicillopepsin. Evidence from nonlinear Arrhenius plots and from circular dichroism studies. J Biol Chem 265 (9):5060–5065
212. Chen Y, Mao H, Zhang X, Gong Y, Zhao N (1999) Thermal conformational changes of bovine fibrinogen by differential scanning calorimetry and circular dichroism. Int J Biol Macromol 26(2-3):129–134
213. Ahn VE, Lo EI, Engel CK, Chen L, Hwang PM, Kay LE, Bishop RE, Prive GG (2004) A hydrocarbon ruler measures palmitate in the enzymatic acylation of endotoxin. EMBO J 23(15):2931–2941. doi:10.1038/sj.emboj.7600320
214. Hwang PM, Bishop RE, Kay LE (2004) The integral membrane enzyme PagP alternates between two dynamically distinct states. Proc Natl Acad Sci U S A 101 (26):9618–9623. doi:10.1073/pnas.0402324101
215. Moon CP, Zaccai NR, Fleming PJ, Gessmann D, Fleming KG (2013) Membrane protein thermodynamic stability may serve as the energy sink for sorting in the periplasm. Proc Natl Acad Sci U S A 110 (11):4285–4290. doi:10.1073/pnas.1212527110
216. Huysmans GH, Radford SE, Baldwin SA, Brockwell DJ (2012) Malleability of the folding mechanism of the outer membrane protein PagP: parallel pathways and the effect of membrane elasticity. J Mol Biol 416 (3):453–464. doi:10.1016/j.jmb.2011.12.039
217. Chaturvedi D, Mahalakshmi R (2014) Juxtamembrane tryptophans have distinct roles in defining the OmpX barrel-micelle boundary and facilitating protein-micelle association. FEBS Lett 588(23):4464–4471. doi:10.1016/j.febslet.2014.10.017
218. Kleinschmidt JH, Tamm LK (1999) Time-resolved distance determination by tryptophan fluorescence quenching: probing intermediates in membrane protein folding. Biochemistry 38(16):4996–5005. doi:10.1021/bi9824644
219. Pattnaik BR, Ghosh S, Rajeswari MR (1997) Selective excitation of tryptophans in OmpF: a fluorescence emission study. Biochem Mol Biol Int 42(1):173–181
220. Lovelle M, Mach T, Mahendran KR, Weingart H, Winterhalter M, Gameiro P (2011) Interaction of cephalosporins with outer membrane channels of Escherichia coli. Revealing binding by fluorescence quenching and ion conductance fluctuations. Phys Chem Chem Phys 13(4):1521–1530. doi:10.1039/c0cp00969e
221. Moon CP, Fleming KG (2011) Using tryptophan fluorescence to measure the stability of membrane proteins folded in liposomes. Methods Enzymol 492:189–211. doi:10.1016/B978-0-12-381268-1.00018-5
222. Schulz GE (2002) The structure of bacterial outer membrane proteins. Biochim Biophys Acta 1565(2):308–317
223. Lakowicz JR (2006) Principles of fluorescence spectroscopy, 3rd edn. Springer, London
224. Kang G, Lopez-Pena I, Oklejas V, Gary CS, Cao W, Kim JE (2012) Forster resonance energy transfer as a probe of membrane protein folding. Biochim Biophys Acta 1818 (2):154–161. doi:10.1016/j.bbamem.2011.08.029
225. Lella M, Kamilla S, Jain V, Mahalakshmi R (2016) Molecular mechanism of holin transmembrane domain I in pore formation and bacterial cell death. ACS Chem Biol 11 (4):910–920. doi:10.1021/acschembio.5b00875
226. Giepmans BN, Adams SR, Ellisman MH, Tsien RY (2006) The fluorescent toolbox for assessing protein location and function. Science 312(5771):217–224. doi:10.1126/science.1124618
227. O'Hare HM, Johnsson K, Gautier A (2007) Chemical probes shed light on protein

function. Curr Opin Struct Biol 17 (4):488–494. doi:10.1016/j.sbi.2007.07.005

228. Knopp MM, Lobmann K, Elder DP, Rades T, Holm R (2015) Recent advances and potential applications of modulated differential scanning calorimetry (mDSC) in drug development. Eur J Pharm Sci. doi:10.1016/j.ejps.2015.12.024
229. Olsen CM, Shikiya R, Ganugula R, Reiling-Steffensmeier C, Khutsishvili I, Johnson SE, Marky LA (2016) Application of differential scanning calorimetry to measure the differential binding of ions, water and protons in the unfolding of DNA molecules. Biochim Biophys Acta 1860(5):990–998. doi:10.1016/j.bbagen.2015.10.002
230. Spink CH (2015) The deconvolution of differential scanning calorimetry unfolding transitions. Methods 76:78–86. doi:10.1016/j.ymeth.2014.12.001
231. Kim NA, Jin JH, Kim KH, Lim DG, Cheong H, Kim YH, Ju W, Kim SC, Jeong SH (2016) Investigation of early and advanced stages in ovarian cancer using human plasma by differential scanning calorimetry and mass spectrometry. Arch Pharm Res. doi:10.1007/s12272-016-0722-z
232. Brandes N, Welzel PB, Werner C, Kroh LW (2006) Adsorption-induced conformational changes of proteins onto ceramic particles: differential scanning calorimetry and FTIR analysis. J Colloid Interface Sci 299 (1):56–69. doi:10.1016/j.jcis.2006.01.065
233. Vermeer AW, Norde W (2000) The thermal stability of immunoglobulin: unfolding and aggregation of a multi-domain protein. Biophys J 78(1):394–404. doi:10.1016/S0006-3495(00)76602-1
234. Lepock JR, Frey HE, Inniss WE (1990) Thermal analysis of bacteria by differential scanning calorimetry: relationship of protein denaturation in situ to maximum growth temperature. Biochim Biophys Acta 1055 (1):19–26
235. Lepock JR, Frey HE, Rodahl AM, Kruuv J (1988) Thermal analysis of CHL V79 cells using differential scanning calorimetry: implications for hyperthermic cell killing and the heat shock response. J Cell Physiol 137 (1):14–24. doi:10.1002/jcp.1041370103
236. Joanne P, Galanth C, Goasdoue N, Nicolas P, Sagan S, Lavielle S, Chassaing G, El Amri C, Alves ID (2009) Lipid reorganization induced by membrane-active peptides probed using differential scanning calorimetry. Biochim Biophys Acta 1788(9):1772–1781. doi:10.1016/j.bbamem.2009.05.001
237. Cinelli S, Onori G, Zuzzi S, Bordi F, Cametti C, Sennato S, Diociaiuti M (2007) Properties of mixed DOTAP-DPPC bilayer membranes as reported by differential scanning calorimetry and dynamic light scattering measurements. J Phys Chem B 111 (33):10032–10039. doi:10.1021/jp071722g
238. Mabrey S, Sturtevant JM (1976) Investigation of phase transitions of lipids and lipid mixtures by sensitivity differential scanning calorimetry. Proc Natl Acad Sci U S A 73 (11):3862–3866
239. Ortiz A, Aranda FJ, Gomez-Fernandez JC (1987) A differential scanning calorimetry study of the interaction of alpha-tocopherol with mixtures of phospholipids. Biochim Biophys Acta 898(2):214–222
240. Sanchez-Ruiz JM (1992) Theoretical analysis of Lumry-Eyring models in differential scanning calorimetry. Biophys J 61(4):921–935. doi:10.1016/S0006-3495(92)81899-4
241. Chapman D, Urbina J (1974) Biomembrane phase transitions. Studies of lipid-water systems using differential scanning calorimetry. J Biol Chem 249(8):2512–2521
242. Sanchez-Ruiz JM, Lopez-Lacomba JL, Cortijo M, Mateo PL (1988) Differential scanning calorimetry of the irreversible thermal denaturation of thermolysin. Biochemistry 27(5):1648–1652
243. Sanchez-Ruiz JM (1995) Differential scanning calorimetry of proteins. In: Proteins: structure, function, and engineering. Series subcellular biochemistry, vol 24, pp 133–176
244. Beatriz Ibarra-Molero JMS-R (2006) Differential scanning calorimetry of proteins: an overview and some recent developments. In: Advanced techniques in biophysics. Springer series in biophysics, vol 10, pp 27–48
245. Jelesarov I, Bosshard HR (1999) Isothermal titration calorimetry and differential scanning calorimetry as complementary tools to investigate the energetics of biomolecular recognition. J Mol Recognit 12(1):3–18. doi:10.1002/(SICI)1099-1352(199901/02)12:1<3::AID-JMR441>3.0.CO;2-6
246. Brandts JF, Lin LN (1990) Study of strong to ultratight protein interactions using differential scanning calorimetry. Biochemistry 29 (29):6927–6940
247. Blagoev B, Kratchmarova I, Ong SE, Nielsen M, Foster LJ, Mann M (2003) A proteomics strategy to elucidate functional protein-protein interactions applied to EGF signaling.

Nat Biotechnol 21(3):315–318. doi:10.1038/nbt790

248. Wu CC, Yates JR 3rd (2003) The application of mass spectrometry to membrane proteomics. Nat Biotechnol 21(3):262–267. doi:10.1038/nbt0303-262
249. Eichacker LA, Granvogl B, Mirus O, Muller BC, Miess C, Schleiff E (2004) Hiding behind hydrophobicity. Transmembrane segments in mass spectrometry. J Biol Chem 279 (49):50915–50922. doi:10.1074/jbc.M405875200
250. Savas JN, Stein BD, Wu CC, Yates JR 3rd (2011) Mass spectrometry accelerates membrane protein analysis. Trends Biochem Sci 36 (7):388–396. doi:10.1016/j.tibs.2011.04.005
251. Cravatt BF, Simon GM, Yates JR 3rd (2007) The biological impact of mass-spectrometry-based proteomics. Nature 450 (7172):991–1000. doi:10.1038/nature06525
252. Barrera NP, Isaacson SC, Zhou M, Bavro VN, Welch A, Schaedler TA, Seeger MA, Miguel RN, Korkhov VM, van Veen HW, Venter H, Walmsley AR, Tate CG, Robinson CV (2009) Mass spectrometry of membrane transporters reveals subunit stoichiometry and interactions. Nat Methods 6(8):585–587. doi:10.1038/nmeth.1347
253. Sharon M (2010) How far can we go with structural mass spectrometry of protein complexes? J Am Soc Mass Spectrom 21 (4):487–500. doi:10.1016/j.jasms.2009.12.017
254. Kirshenbaum N, Michaelevski I, Sharon M (2010) Analyzing large protein complexes by structural mass spectrometry. J Vis Exp (40). doi:10.3791/1954
255. Leney AC, McMorran LM, Radford SE, Ashcroft AE (2012) Amphipathic polymers enable the study of functional membrane proteins in the gas phase. Anal Chem 84 (22):9841–9847. doi:10.1021/ac302223s
256. Schey KL, Grey AC, Nicklay JJ (2013) Mass spectrometry of membrane proteins: a focus on aquaporins. Biochemistry 52 (22):3807–3817. doi:10.1021/bi301604j
257. Laganowsky A, Reading E, Hopper JT, Robinson CV (2013) Mass spectrometry of intact membrane protein complexes. Nat Protoc 8(4):639–651. doi:10.1038/nprot.2013.024
258. Borysik AJ, Robinson CV (2012) Formation and dissociation processes of gas-phase detergent micelles. Langmuir 28(18):7160–7167. doi:10.1021/la3002866
259. Barrera NP, Robinson CV (2011) Advances in the mass spectrometry of membrane proteins: from individual proteins to intact complexes. Annu Rev Biochem 80:247–271. doi:10.1146/annurev-biochem-062309-093307
260. Landreh M, Robinson CV (2015) A new window into the molecular physiology of membrane proteins. J Physiol 593(2):355–362. doi:10.1113/jphysiol.2014.283150
261. Barrera NP, Zhou M, Robinson CV (2013) The role of lipids in defining membrane protein interactions: insights from mass spectrometry. Trends Cell Biol 23(1):1–8. doi:10.1016/j.tcb.2012.08.007
262. Plesniak LA, Mahalakshmi R, Rypien C, Yang Y, Racic J, Marassi FM (2011) Expression, refolding, and initial structural characterization of the Y. pestis Ail outer membrane protein in lipids. Biochim Biophys Acta 1808 (1):482–489. doi:10.1016/j.bbamem.2010.09.017
263. Speers AE, Blackler AR, Wu CC (2007) Shotgun analysis of integral membrane proteins facilitated by elevated temperature. Anal Chem 79(12):4613–4620. doi:10.1021/ac0700225
264. Shevchenko A, Tomas H, Havlis J, Olsen JV, Mann M (2006) In-gel digestion for mass spectrometric characterization of proteins and proteomes. Nat Protoc 1 (6):2856–2860. doi:10.1038/nprot.2006.468
265. Okazaki M, Kurabayashi K, Asanuma M, Saito Y, Dodo K, Sodeoka M (2015) VDAC3 gating is activated by suppression of disulfide-bond formation between the N-terminal region and the bottom of the pore. Biochim Biophys Acta 1848(12):3188–3196. doi:10.1016/j.bbamem.2015.09.017
266. Hamill OP, Marty A, Neher E, Sakmann B, Sigworth FJ (1981) Improved patch-clamp techniques for high-resolution current recording from cells and cell-free membrane patches. Pflugers Arch 391(2):85–100
267. Dietrich D, Clusmann H, Kral T (2002) Improved hybrid clamp: resolution of tail currents following single action potentials. J Neurosci Methods 116(1):55–63
268. Pottosin I, Dobrovinskaya O (2015) Ion channels in native chloroplast membranes: challenges and potential for direct patch-clamp studies. Front Physiol 6:396. doi:10.3389/fphys.2015.00396
269. Gutsmann T, Heimburg T, Keyser U, Mahendran KR, Winterhalter M (2015) Protein reconstitution into freestanding planar lipid

membranes for electrophysiological characterization. Nat Protoc 10(1):188–198. doi:10.1038/nprot.2015.003

270. Ries RS, Choi H, Blunck R, Bezanilla F, Heath JR (2004) Black lipid membranes: visualizing the structure, dynamics, and substrate dependence of membranes. J Phys Chem B 108(41):16040–16049. doi:10.1021/jp048098h
271. Basle A, Iyer R, Delcour AH (2004) Subconductance states in OmpF gating. Biochim Biophys Acta 1664(1):100–107. doi:10.1016/j.bbamem.2004.04.009
272. Kreir M, Farre C, Beckler M, George M, Fertig N (2008) Rapid screening of membrane protein activity: electrophysiological analysis of OmpF reconstituted in proteoliposomes. Lab Chip 8(4):587–595. doi:10.1039/b713982a
273. Basle A, Qutub R, Mehrazin M, Wibbenmeyer J, Delcour AH (2004) Deletions of single extracellular loops affect pH sensitivity, but not voltage dependence, of the Escherichia coli porin OmpF. Proteing Eng Des Sel 17 (9):665–672. doi:10.1093/protein/gzh078
274. Zakharian E, Reusch RN (2003) Outer membrane protein A of Escherichia coli forms temperature-sensitive channels in planar lipid bilayers. FEBS Lett 555(2):229–235
275. Arnold T, Poynor M, Nussberger S, Lupas AN, Linke D (2007) Gene duplication of the eight-stranded beta-barrel OmpX produces a functional pore: a scenario for the evolution of transmembrane beta-barrels. J Mol Biol 366(4):1174–1184. doi:10.1016/j.jmb.2006.12.029
276. Kuszak AJ, Jacobs D, Gurnev PA, Shiota T, Louis JM, Lithgow T, Bezrukov SM, Rostovtseva TK, Buchanan SK (2015) Evidence of distinct channel conformations and substrate binding affinities for the mitochondrial outer membrane protein translocase pore Tom40. J Biol Chem 290 (43):26204–26217. doi:10.1074/jbc.M115.642173
277. Noskov SY, Rostovtseva TK, Chamberlin AC, Teijido O, Jiang W, Bezrukov SM (2016) Current state of theoretical and experimental studies of the voltage-dependent anion channel (VDAC). Biochim Biophys Acta. doi:10.1016/j.bbamem.2016.02.026
278. Hong M, Zhang Y, Hu F (2012) Membrane protein structure and dynamics from NMR spectroscopy. Annu Rev Phys Chem 63:1–24. doi:10.1146/annurev-physchem-032511-143731
279. Wang S, Ladizhansky V (2014) Recent advances in magic angle spinning solid state NMR of membrane proteins. Prog Nucl Magn Reson Spectrosc 82:1–26. doi:10.1016/j.pnmrs.2014.07.001
280. Das N, Murray DT, Cross TA (2013) Lipid bilayer preparations of membrane proteins for oriented and magic-angle spinning solid-state NMR samples. Nat Protoc 8 (11):2256–2270. doi:10.1038/nprot.2013.129
281. Bechinger B, Resende JM, Aisenbrey C (2011) The structural and topological analysis of membrane-associated polypeptides by oriented solid-state NMR spectroscopy: established concepts and novel developments. Biophys Chem 153(2-3):115–125. doi:10.1016/j.bpc.2010.11.002
282. Gong XM, Franzin CM, Thai K, Yu J, Marassi FM (2007) Nuclear magnetic resonance structural studies of membrane proteins in micelles and bilayers. Methods Mol Biol 400:515–529. doi:10.1007/978-1-59745-519-0_35
283. Hwang PM, Choy WY, Lo EI, Chen L, Forman-Kay JD, Raetz CR, Prive GG, Bishop RE, Kay LE (2002) Solution structure and dynamics of the outer membrane enzyme PagP by NMR. Proc Natl Acad Sci U S A 99 (21):13560–13565. doi:10.1073/pnas.212344499
284. Fernandez C, Hilty C, Wider G, Guntert P, Wuthrich K (2004) NMR structure of the integral membrane protein OmpX. J Mol Biol 336(5):1211–1221. doi:10.1016/j.jmb.2003.09.014
285. Arora A, Abildgaard F, Bushweller JH, Tamm LK (2001) Structure of outer membrane protein A transmembrane domain by NMR spectroscopy. Nat Struct Biol 8(4):334–338. doi:10.1038/86214
286. Hiller S, Garces RG, Malia TJ, Orekhov VY, Colombini M, Wagner G (2008) Solution structure of the integral human membrane protein VDAC-1 in detergent micelles. Science 321(5893):1206–1210. doi:10.1126/science.1161302
287. Raschle T, Rios Flores P, Opitz C, Muller DJ, Hiller S (2016) Monitoring backbone hydrogen bond formation in beta-barrel membrane protein folding. Angew Chem Int Ed. doi:10.1002/anie.201509910
288. Dufrene YF (2006) Atomic force microscopy of membrane proteins separating two aqueous compartments. Nat Methods 3 (12):973–975. doi:10.1038/nmeth1206-973

289. Scheuring S, Goncalves RP, Prima V, Sturgis JN (2006) The photosynthetic apparatus of Rhodopseudomonas palustris: structures and organization. J Mol Biol 358(1):83–96. doi:10.1016/j.jmb.2006.01.085
290. Drews G, Peters J, Dierstein R (1983) Molecular-organization and biosynthesis of pigment-protein complexes of Rhodopseudomonas capsulata. Ann Microbiol 134B (1):151–158
291. Sturgis JN, Tucker JD, Olsen JD, Hunter CN, Niederman RA (2009) Atomic force microscopy studies of native photosynthetic membranes. Biochemistry 48 (17):3679–3698. doi:10.1021/bi900045x
292. Goncalves RP, Agnus G, Sens P, Houssin C, Bartenlian B, Scheuring S (2006) Two-chamber AFM: probing membrane proteins separating two aqueous compartments. Nat Methods 3(12):1007–1012. doi:10.1038/nmeth965
293. Muller DJ, Engel A (2002) Conformations, flexibility, and interactions observed on individual membrane proteins by atomic force microscopy. Methods Cell Biol 68:257–299
294. Ge L, Villinger S, Mari SA, Giller K, Griesinger C, Becker S, Muller DJ, Zweckstetter M (2016) Molecular plasticity of the human voltage-dependent anion channel embedded into a membrane. Structure 24(4):585–594. doi:10.1016/j.str.2016.02.012
295. Hane F, Drolle E, Gaikwad R, Faught E, Leonenko Z (2011) Amyloid-beta aggregation on model lipid membranes: an atomic force microscopy study. J Alzheimers Dis 26 (3):485–494. doi:10.3233/JAD-2011-102112
296. Adamcik J, Jung JM, Flakowski J, De Los Rios P, Dietler G, Mezzenga R (2010) Understanding amyloid aggregation by statistical analysis of atomic force microscopy images. Nat Nanotechnol 5(6):423–428. doi:10.1038/nnano.2010.59
297. Stolz M, Stoffler D, Aebi U, Goldsbury C (2000) Monitoring biomolecular interactions by time-lapse atomic force microscopy. J Struct Biol 131(3):171–180. doi:10.1006/jsbi.2000.4301
298. Modler AJ, Gast K, Lutsch G, Damaschun G (2003) Assembly of amyloid protofibrils via critical oligomers—a novel pathway of amyloid formation. J Mol Biol 325(1):135–148
299. Garcia-Saez AJ, Schwille P (2007) Single molecule techniques for the study of membrane proteins. Appl Microbiol Biotechnol 76 (2):257–266. doi:10.1007/s00253-007-1007-8
300. Zhang S, Andreasen M, Nielsen JT, Liu L, Nielsen EH, Song J, Ji G, Sun F, Skrydstrup T, Besenbacher F, Nielsen NC, Otzen DE, Dong M (2013) Coexistence of ribbon and helical fibrils originating from hIAPP(20-29) revealed by quantitative nanomechanical atomic force microscopy. Proc Natl Acad Sci U S A 110(8):2798–2803. doi:10.1073/pnas.1209955110
301. Zhou XE, Gao X, Barty A, Kang Y, He Y, Liu W, Ishchenko A, White TA, Yefanov O, Han GW, Xu Q, de Waal PW, Suino-Powell KM, Boutet S, Williams GJ, Wang M, Li D, Caffrey M, Chapman HN, Spence JC, Fromme P, Weierstall U, Stevens RC, Cherezov V, Melcher K, Xu HE (2016) X-ray laser diffraction for structure determination of the rhodopsin-arrestin complex. Sci Data 3:160021. doi:10.1038/sdata.2016.21
302. Panneels V, Wu W, Tsai CJ, Nogly P, Rheinberger J, Jaeger K, Cicchetti G, Gati C, Kick LM, Sala L, Capitani G, Milne C, Padeste C, Pedrini B, Li XD, Standfuss J, Abela R, Schertler G (2015) Time-resolved structural studies with serial crystallography: a new light on retinal proteins. Struct Dyn 2(4):041718. doi:10.1063/1.4922774
303. Kumar RP, Ranaghan MJ, Ganjei AY, Oprian DD (2015) Crystal structure of recoverin with calcium ions bound to both functional EF hands. Biochemistry 54(49):7222–7228. doi:10.1021/acs.biochem.5b01160
304. Strynadka NC, James MN (1989) Crystal structures of the helix-loop-helix calcium-binding proteins. Annu Rev Biochem 58:951–998. doi:10.1146/annurev.bi.58.070189.004511
305. Szczepek M, Beyriere F, Hofmann KP, Elgeti M, Kazmin R, Rose A, Bartl FJ, von Stetten D, Heck M, Sommer ME, Hildebrand PW, Scheerer P (2014) Crystal structure of a common GPCR-binding interface for G protein and arrestin. Nat Commun 5:4801. doi:10.1038/ncomms5801
306. Sommer ME, Elgeti M, Hildebrand PW, Szczepek M, Hofmann KP, Scheerer P (2015) Structure-based biophysical analysis of the interaction of rhodopsin with G protein and arrestin. Methods Enzymol 556:563–608. doi:10.1016/bs.mie.2014.12.014
307. Schredelseker J, Paz A, Lopez CJ, Altenbach C, Leung CS, Drexler MK, Chen JN, Hubbell WL, Abramson J (2014) High resolution structure and double electron-electron resonance of the zebrafish voltage-dependent anion channel 2 reveal an oligomeric

population. J Biol Chem 289 (18):12566–12577. doi:10.1074/jbc. M113.497438

308. Ujwal R, Cascio D, Colletier JP, Faham S, Zhang J, Toro L, Ping P, Abramson J (2008) The crystal structure of mouse VDAC1 at 2.3 A resolution reveals mechanistic insights into metabolite gating. Proc Natl Acad Sci U S A 105(46):17742–17747. doi:10.1073/pnas.0809634105
309. Cuesta-Seijo JA, Neale C, Khan MA, Moktar J, Tran CD, Bishop RE, Pomes R, Prive GG (2010) PagP crystallized from SDS/cosolvent reveals the route for phospholipid access to the hydrocarbon ruler. Structure 18 (9):1210–1219. doi:10.1016/j.str.2010.06.014
310. Khan MA, Bishop RE (2009) Molecular mechanism for lateral lipid diffusion between the outer membrane external leaflet and a beta-barrel hydrocarbon ruler. Biochemistry 48 (41):9745–9756. doi:10.1021/bi9013566
311. Dalebroux ZD, Matamouros S, Whittington D, Bishop RE, Miller SI (2014) PhoPQ regulates acidic glycerophospholipid content of the Salmonella Typhimurium outer membrane. Proc Natl Acad Sci U S A 111 (5):1963–1968. doi:10.1073/pnas.1316901111
312. Song L, Hobaugh MR, Shustak C, Cheley S, Bayley H, Gouaux JE (1996) Structure of staphylococcal alpha-hemolysin, a heptameric transmembrane pore. Science 274 (5294):1859–1866
313. Iacovache I, Bischofberger M, van der Goot FG (2010) Structure and assembly of pore-forming proteins. Curr Opin Struct Biol 20 (2):241–246. doi:10.1016/j.sbi.2010.01.013
314. Tilley SJ, Orlova EV, Gilbert RJ, Andrew PW, Saibil HR (2005) Structural basis of pore formation by the bacterial toxin pneumolysin. Cell 121(2):247–256. doi:10.1016/j.cell.2005.02.033
315. Unno H, Goda S, Hatakeyama T (2014) Hemolytic lectin CEL-III heptamerizes via a large structural transition from alpha-helices to a beta-barrel during the transmembrane pore formation process. J Biol Chem 289 (18):12805–12812. doi:10.1074/jbc.M113.541896
316. Noinaj N, Kuszak AJ, Balusek C, Gumbart JC, Buchanan SK (2014) Lateral opening and exit pore formation are required for BamA function. Structure 22 (7):1055–1062. doi:10.1016/j.str.2014.05.008
317. Jansen KB, Baker SL, Sousa MC (2015) Crystal structure of BamB bound to a periplasmic domain fragment of BamA, the central component of the beta-barrel assembly machine. J Biol Chem 290(4):2126–2136. doi:10.1074/jbc.M114.584524
318. Leyton DL, Belousoff MJ, Lithgow T (2015) The beta-barrel assembly machinery complex. Methods Mol Biol 1329:1–16. doi:10.1007/978-1-4939-2871-2_1
319. Gu Y, Li H, Dong H, Zeng Y, Zhang Z, Paterson NG, Stansfeld PJ, Wang Z, Zhang Y, Wang W, Dong C (2016) Structural basis of outer membrane protein insertion by the BAM complex. Nature 531(7592):64–69. doi:10.1038/nature17199
320. Han L, Zheng J, Wang Y, Yang X, Liu Y, Sun C, Cao B, Zhou H, Ni D, Lou J, Zhao Y, Huang Y (2016) Structure of the BAM complex and its implications for biogenesis of outer-membrane proteins. Nat Struct Mol Biol 23(3):192–196. doi:10.1038/nsmb.3181
321. Bergal HT, Hopkins AH, Metzner SI, Sousa MC (2016) The structure of a BamA-BamD fusion illuminates the architecture of the beta-barrel assembly machine core. Structure 24 (2):243–251. doi:10.1016/j.str.2015.10.030
322. Fleming KG (2015) A combined kinetic push and thermodynamic pull as driving forces for outer membrane protein sorting and folding in bacteria. Philos Trans R Soc Lond B Biol Sci 370(1679). doi:10.1098/rstb.2015.0026
323. Plummer AM, Fleming KG (2015) BamA alone accelerates outer membrane protein folding in vitro through a catalytic mechanism. Biochemistry 54(39):6009–6011. doi:10.1021/acs.biochem.5b00950
324. Danoff EJ, Fleming KG (2015) Membrane defects accelerate outer membrane beta-barrel protein folding. Biochemistry 54 (2):97–99. doi:10.1021/bi501443p
325. Chan NC, Lithgow T (2008) The peripheral membrane subunits of the SAM complex function codependently in mitochondrial outer membrane biogenesis. Mol Biol Cell 19(1):126–136. doi:10.1091/mbc.E07-08-0796

Probing the Dynamics of Plasma Membrane Glutamate Transporters in Real Time by Total Internal Fluorescence Reflection Microscopy

Eliana S. Di Cairano, Paola Marciani, Stefania Moretti, and Carla Perego

Abstract

Plasma membrane solute carriers (SLCs) mediate the transport of a variety of substrates across the plasma membranes. There is a great interest in understanding how they work and how their density at the plasma membrane is dynamically regulated, given their involvement in important physiological processes ranging from neurotransmission to nutrient and drug absorption. Genetically encoded fluorescent fusion proteins of transporters and total internal reflection fluorescence microscopy (TIRFM) provide the ideal methodological approaches to follow their dynamics in living cells. Owing to its ability to selectively excite a very thin fluorescent section (~100 nm) immediately above the glass cover on which cells are grown, TIRFM is becoming the technique of election for measuring in a quantitative manner events occurring at or near the plasma membrane with high temporal and spatial resolution. Here we provide a method for using TIRF microscopy to study the basal and regulated trafficking of the excitatory amino acid transporter 3 (EAAT3/EAAC1), a glutamate transporter of the solute carrier family 1 (SLC1). A detailed protocol of how carrying out image recording under TIRF microscopy and how to extract quantitative data on transporter density at the plasma membrane is presented. The method and the applied analyses can be extended to other plasma membrane proteins (solute and ion transporters, channels, and receptors) in different cellular contexts.

Keywords Cell signal transduction, Constitutive and regulated trafficking, Endocytosis, Exocytosis, Glutamate transporters, Total internal reflection fluorescence microscopy

1 Introduction

Solute carrier (SLC) membrane transport proteins mediate the transport of solute and ions across biological membranes. The family comprises more than 400 members, many of which are targets of important drugs or are associated with human genetic diseases. They are emerging as key molecules involved in the control of important physiological processes ranging from synaptic transmission to nutrient or drug absorption and waste removal [1, 2].

Arun K. Shukla (ed.), *Chemical and Synthetic Approaches in Membrane Biology*, Springer Protocols Handbooks, (2017) 117–139, DOI 10.1007/8623_2016_9, Published online: 19 April 2017

In order to be effective, SLCs must be located at the plasma membrane. Emerging data indicate they are not simply synthesized and targeted to the final destination of action (the plasma membrane), but their surface density is controlled by constitutive cycling between the plasma membrane and a recycling compartment, with the proportion at the plasma membrane depending on the rate of plasma membrane transporter delivery or removal (Fig. 1a) [3]. Although apparently expensive from an energetic point of view, this mechanism ensures the rapid adaptation of the cell to modifications of the extracellular environment or to metabolic demand.

Elucidating the mechanisms that control their density in specific plasma membrane sub-domains is essential not only for a better understanding of the physiological processes they regulate but also for defining general rules in plasma membrane protein dynamics and for designing appropriate pharmacological approaches to treat neurodegenerative, neuropsychiatric, and metabolic diseases.

Comprehensive analysis of these phenomena requires quantitative methodological approaches with a time resolution within the minute or even second range, corresponding to the lifetime of neurotransmission, and a spatial resolution close to nanometers, in the proximity of the plasma membrane.

In the past, several techniques have been used, but none of them was sufficiently comprehensive.

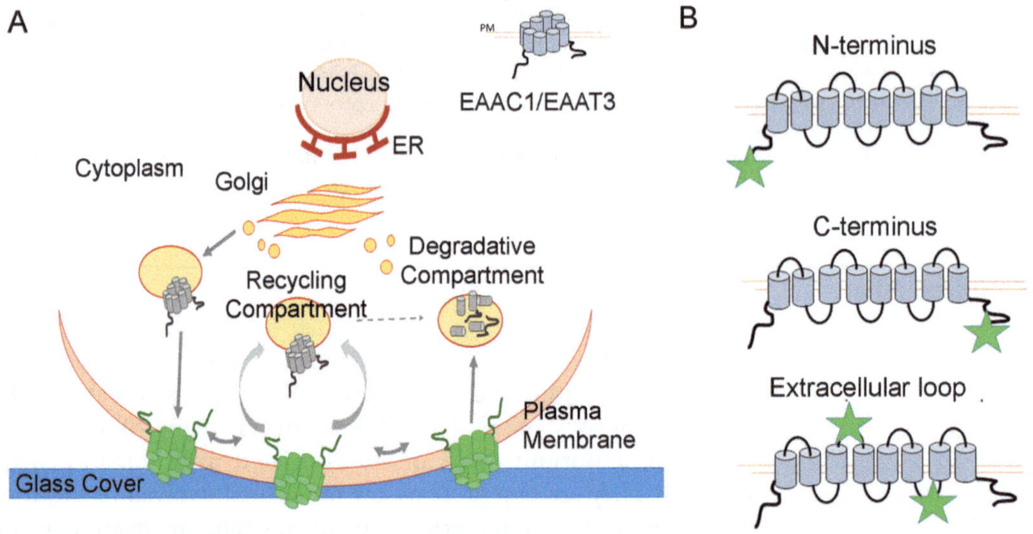

Fig. 1 Glutamate transporter density at the plasma membrane. (**A**) Mechanisms controlling glutamate transporter density at the plasma membrane. In addition to canonical synthetic and degradative pathways, the transporter density at the plasma membrane is maintained by constitutive cycling between the plasma membrane and a recycling compartment. (**B**) Schematic representation of the membrane topology of mammalian glutamate transporters of the SLC1A family. The potential positions of the tag (*green star*) are depicted

Traditionally, the transporter density at the plasma membrane has been measured by functional or biochemical studies [4, 5]. Different from receptor–ligand interaction, substrates are released inside the cell by transporters; therefore, in order to label the proteins, the reaction must be performed at 0°C to minimize substrate uptake. However, the dynamics of the event will be lost.

Biochemical approaches such as biotinylation give excellent quantitative information [6, 7]. In these experiments, an exogenous biotin molecule is covalently linked to primary amine groups of lysine residues exposed to extracellular compartment. Thanks to the high affinity of streptavidin for biotin, biotinylated molecules of interest can be isolated and analyzed by western blotting techniques [3, 8, 9].

Unfortunately, transporters are really complex integral membrane proteins; their extracellular loops (at least those of the SLC1 and SLC6 gene family) are very small and often contain glycosylation sites which mask possible biotinylation residues, thus making the biotinylation process inefficient [10, 11]. Furthermore, high expression of the proteins of interest is necessary to have a good yield, and this may perturb the normal targeting of transporters to the plasma membrane. Finally, the temporal resolution is within some minutes, too long for some physiological events.

Alternative biochemical approaches consider labeled ligands or antibodies directed against extracellular epitopes of transporters, but high-quality antibodies are not available for many transporters [12].

On the other hand, genetically encoded fluorescent transporters coupled with epifluorescence microscopy provide important descriptive information about the expression and localization of proteins and the dynamic nature of signal transduction-mediated events [13]. However, events cannot be described quantitatively because the spatial resolution of the signal, particularly that presents at the plasma membrane, is confused with signals from outside the area of interest.

Confocal microscopy restricts the penetration field of the laser beam to <900 nm; the spatial resolution is greatly improved, but the temporal resolution of the events is restricted by the scan process.

Total internal reflection fluorescence microscopy (TIRFM), a specific application of the fluorescence microscopy technique, overcomes these limitations. In TIRFM, a laser beam illuminates the sample with an incident angle greater than the critical angle. In this way, the incident light is completely reflected by the glass coverslips; an evanescent wave is generated at the glass–sample interface and penetrates the biological sample, because of the lower reflection index. As the evanescent wave decades exponentially, it illuminates only fluorophores located in a narrow (<100 nm) section at the cell–coverslip interface, where plasma membrane transporters reside and work. The high signal-to-noise ratio of image generated by the evanescent wave greatly enhances the spatial resolution of the signal, particularly in the *z*-axis (nanometer resolution) [14].

The chip-based image acquisition strategy provides the temporal resolution necessary to follow highly dynamic events. Finally, the relatively low-powered evanescent wave illumination minimizes fluorophores photobleaching and allows long-lasting recording.

This technique has been widely used to characterize the molecular mechanisms responsible for the delivery of the GLUT4 glucose transporter to the plasma membrane in response to insulin, in adipocytes and muscle cells [15]. We have exploited this procedure to study the dynamics of neurotransmitter transporters in neuronal and epithelial cells [3, 16].

The proposal of this paper is to provide a method to perform single-cell, time-lapse TIRF microscopy useful to follow the glutamate transporter dynamics at the plasma membrane under basal or stimulated conditions. A detailed protocol to analyze images and to extract quantitative information is also provided. A critical analysis of techniques available to tag transporters for time-lapse microscopy experiments and the problems linked to the evaluation of the transporter density are reported in the Notes section. The method has been settled up to follow glutamate transporter dynamics at the plasma membrane; however, the approach to the image analysis and quantification can be extended to other plasma membrane proteins such as solute and ion transporters, channels, and receptors, in different cellular contexts.

2 Materials

2.1 Cell Culture and Transfection

2.1.1 Equipment

Standard laboratory equipment, e.g., biohazard hoods, CO_2 incubator, inverted microscope, microfuge, vortex mixer, water bath (37°C), standard pipettes and micropipettes, and standard sterile plastic culture supplies (3.5-cm Petri dishes, polypropylene tubes, micro tubes).

2.1.2 COS-7 Cell Growing and Transfection Reagents

- *COS-7* (available at ATCC® CRL-1651™) is a fibroblast-like cell line, derived from *Cercopithecus aethiops* kidney [17]. The cells are grown adherent in Dulbecco's Modified Eagle Medium (DMEM) and can be easily transfected with several transfection reagents. COS-7 cells are maintained in culture in 75 cc flasks, at 37°C in a 5% CO_2 incubator. Subculture once a week or when they have covered 80–90% of the surface area.
- *Culture medium*: DMEM with low glucose, 10% fetal bovine serum (FBS), penicillin (100 U/mL), streptomycin (100 μg/mL), and L-glutamine (2 mM). Filter the solution and conserve at 4°C. Prior to use, warm up at 37°C.

- *Phosphate buffer saline (PBS) solution*: 150 mM NaCl, 24 mM phosphate buffer, and pH 7.4. Filter the solution.
- *Trypsin solution*: 0.25% (w/v) trypsin in 0.53 mM ethylenediaminetetraacetic acid (EDTA) solution.
- *Polyethylenimine (PEI) solution*: 10% polyethylenimine (25 kDa linear), 150 mM NaCl solution, pH 7.8.

  Prepare 100 mL of a 150 mM NaCl solution. Adjust the pH to 5.5 with 0.01 N HCl. Add 10% PEI and adjust again the pH of the solution to 7.8 with 0.01 N HCl. Filter and fractionate the solution in suitable volumes and store at 4°C up to 6 months. PEI cannot be frozen.
- *EGFP-EAAC1 plasmid*: high-quality, intact plasmid DNA at 0.3–1 μg/μL concentration in Tris HCl buffer. The pC1-EGFP-EAAC1 plasmid has been used. The construct has been generated by in-frame fusion protein of the rabbit glutamate transporter EAAC1 with a red-shifted variant of wild-type green fluorescent protein (EGFP) which has been optimized for brighter fluorescence and higher expression in mammalian cells (excitation maximum = 488 nm; emission maximum = 507 nm) (pEGFP-C1 vector, Clontech). The correct delivery to the plasma membrane of the tagged protein has been tested by fluorescence and biotinylation experiments [3]. A scheme of the constructs is shown in Fig. 1b.

2.1.3 Cell Plating Materials and Reagents

- *Glass covers*: Thermo Scientific circled cover glasses, 24-mm diameter, 0.17 ± 0.005-mm thickness, and 1.5255 ± 0.00015 refractive index.
- *MilliQ water*: >18 MOhm-cm, 0.22 μm filtered.
- *Absolute ethanol*
- *6 N KOH solution*: solution must be freshly prepared and filtered through 0.22-μm filters (insoluble potassium carbonate may form with time in the bottle).
- Glass slide-strainer containers (Wash-N-Dry™ Coverslip Rack, Electron Microscopy Sciences).

2.2 TIRFM Recording

2.2.1 Laser TIRF Microscopy Equipment

The procedure requires a motorized, software-assisted, inverted microscope with both wide field (epifluorescence) and TIRF illumination modules. The TIRFM is achieved through the object configuration. The fluorescence microscope has to be capable of time-lapse imaging and must be equipped with a fast shutter and a sensitive monochromatic camera to capture low-light levels. The image acquisition software must record each frames. Major microscope components are:

- Inverted microscope with high numerical aperture lens (NA 1.45 Alpha Plan-Fluar) 100× oil, immersion objective.

- High sensitive, cooled CCD camera.
- Motorized filter turret or filter wheels.
- High-speed, software-controlled shutter, for a fast control of laser illumination.
- Laser 488- or 514-nm lines. Green fluorescence is excited using the 488-nm laser line.
- Filter sets: a customized filter set achieves better signal-to-noise ratio. For the EGFP signal, the band pass 488-/10-nm excitation and 525-/50-nm emission filters are employed.
- Microscope software to acquire, process, and analyze images over various time points.

In this experiment, images are acquired using a Zeiss Axiovert inverted microscope equipped with a Retiga SRV Fast 1394 CCD camera. TIRF is achieved through an objective-based approach and use of a multi-line (458/488/514 nm) 100-mW argon-ion laser as a light source. The 488-nm line laser light necessary to excite GFP is filtered through a band pass 488-/10-nm excitation filter mounted on the filter wheel. The linearly polarized laser light is introduced into the beam path with a monomode fiber, via the TIRF slider. The wide field illumination is obtained through a conventional mercury short-arc lamp HBO white light. A polarization-maintaining double prism in the slider ensures the combination of TIRF illumination and white light [18]. The Image ProPlus software, version 6.3 (Media Cybernetics), is used to capture images. A scheme of the microscope setup is shown in Fig. 2a.

2.2.2 Imaging Solutions

- *Krebs' (KRH) solution*: 125 mM NaCl, 5 mM KCl, 1.2 mM $MgSO_4$, 1.2 mM KH_2PO_4, 25 mM 4-(2-hydroxyethyl)piperazine-1-ethanesulfonic acid (HEPES) (buffered to pH 7.4), 2 mM $CaCl_2$, and 6 mM glucose. Warm the solution at 37°C.
- *2× glutamate KRH solution*: Add 10 mM L-glutamate to the KRH solution. Warm the solution at 37°C. This solution must be mixed 1:1 to KRH solution to obtain a 1× working glutamate solution (final glutamate concentration 5 mM).
- *2× HIP-A KRH solution*: Make 50 nM 3-hydroxy-4,5,6,6a-tetrahydro-3aH-pyrrolo[3,4-d]isoxazole-4-carboxylic acid (HIP-A) in KRH solution. HIP-A is a high-affinity glutamate transporter blocker [19]. Warm the solution at 37°C. This solution must be mixed to an equal volume of the incubation solution to obtain a 1× working HIP-A solution (final HIP-A concentration 25 nM).

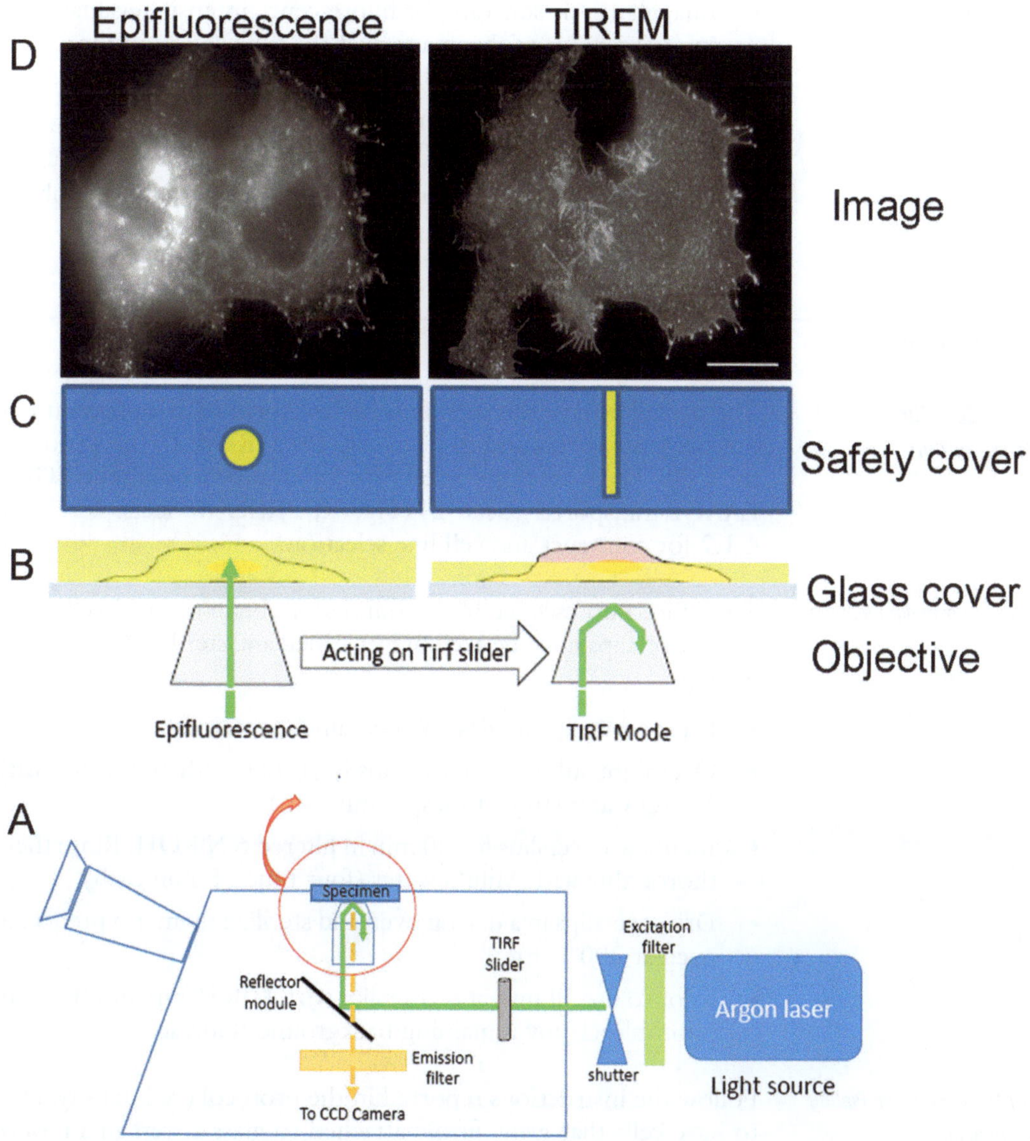

Fig. 2 Reaching the TIRF microscope configuration. (**A**) Schematic view of the TIRF microscope setup. The inverted Zeiss Axiovert microscope, the objective, the specimen, the laser beam (*green line*), the reflector module, and the TIRF slider controlling the incidence angle are indicated. Cells are imaged using a 100× oil immersion objective. Digital images are captured on a cooled Retiga SRV CCD camera. The reflector module, the excitation (488/10 nm, *green rectangle*), and the emission (band pass 525/50 nm, *yellow rectangle*) filters are indicated. (**B**) Cartoon showing the objective, the position of the excitation beam (*green line*), the glass cover, and the sample. *Left*, the excitation beam travels directly through the coverslip–sample interface. The sample is excited in epifluorescence mode. *Right*, the excitation beam forms an incident angle greater than the critical angle, the light is completely reflected back into the objective lens, and an evanescent field propagates in the sample (100 nm). (**C**) Cartoon showing the sample safety cover and the position of the excitation beam (*yellow circle*) that emerges out of the objective, under epifluorescence (*left*), and TIRF illumination (*right*). (**D**) Epifluorescence (*left*) and TIRFM (*right*) images of EGFP-EAAC1 in a live Cos-7 cell. Scale bar: 10 μm

2.3 Data Analysis Software

- Image analysis software for fluorescence intensity evaluation in specific regions of the image over the time course of the experiment (Image ProPlus 6.3 Software, Media Cybernetics).
- A spreadsheet application for data analysis, processing, and storage (Excel, Microsoft).
- Statistical analysis software (GraphPad Prism 4.00, GraphPad Software, Inc).

3 Methods

3.1 Cell Culture and Transfection

The overall goal of this section is to describe the basic procedure to prepare glass covers and to plate cells for TIRFM. In this protocol, the COS-7 cells are transfected with a plasmid encoding the EGFP-EAAC1 transporter (Sect. 2) (Fig. 1b). Refer to Sects. 4.1.1 and 4.1.2 for construct and cell line selection.

3.1.1 Coverslip Cleaning

To reduce fluorescence background and obtain optimal cell adhesion, coverslips must be accurately cleaned and sterilized, before cell plating.

- Place coverslips in glass slide-strainer containers.
- Overnight submerge coverslips in absolute ethanol. Rinse with MilliQ water (three times, 5 min each).
- Incubate coverslips for 20 min in filtered 6 N KOH. Rinse them thoroughly with MilliQ water (four times, 5 min each).
- Dry coverslips in a drying oven and sterilize them in a preheated oven at 200°C for 3 h.
- Prior to use, illuminate coverslips with HBO lamp for 30 min to photobleach any remaining background fluorescence.

3.1.2 COS-7 Cell Plating for Imaging

Follow the instructions reported in the protocol (cell density, etc.) to have cells that grow firmly attached to glass covers in a monolayer, which is crucial for TIRFM.

- The day before transfection, place each coverslip in a 3.5-cm Petri dish, add 1 mL of complete culture medium, and keep at 37°C, in a 5% CO_2 incubator.
- Trypsinize cells. Remove complete growth medium and wash the cells with 3 mL of PBS.
- Treat cells with trypsin solution (2 mL for a 75-cc flasks), incubate for 5 min at 37°C, 5% CO_2, and detach cells using pipette.
- Inactivate trypsin by adding 2 mL of DMEM without serum and collect cells by centrifugation at 300 × *g* for 5 min.

- Remove the supernatant, add 1 mL of complete medium to the pellet, and disperse cells into a single-cell suspension, by pipetting up and down the solution. Count cells.
- For optimal cell growth and efficient transfection, plate 3×10^5 cells/well (3.5 cm in diameter). Incubate at 37°C in a 5% CO_2 incubator overnight.

3.1.3 COS-7 Transfection by Polyethylenimine (PEI)

We here report a basic protocol for plasmid transfection with PEI. This reagent is no toxic and cost-effective. Thirty percent transfection efficiency is expected following exactly the outlined procedure [20].

Twenty-four hours after plating, remove the medium and refresh with 1.5 mL of complete medium. Keep the cells at 37°C, in a 5% CO_2 incubator.

- Under the laminar flow biosafety hood, in a sterile 1.5-mL microtube, add 3 µg of plasmid DNA to 100 µL of PEI solution (amount for 3.5-cm Petri dish).
- Mix for 10 s and then incubate the DNA/PEI mixture for 30 min at room temperature.
- Carefully add the DNA/PEI mixture to the Petri dish containing coverslips with adherent cells and gently shake to distribute the reagent in the Petri dish.
- Return the cells to the 5% CO_2 incubator, at 37°C.
- The next day, substitute the transfection mixture with 1.5 mL of fresh complete medium and keep the cells in the 5% CO_2 incubator.
- Perform the TIRFM experiment 48 h after transfection. This expression time ensures the efficient targeting of transporter proteins to the plasma membrane.

The protocol outlined here uses a 3:1 ratio of PEI to DNA (w/w). This ratio is optimal for the majority of plasmid we have tested and for several cell lines. However, for different plasmids or cell lines, media, PEI:DNA ratio AND expression time should be optimized to maximize protein expression. We routinely screen ratios between 1:1 and 5:1.

3.2 TIRFM Configuration, Experiment Settings, and Recording

The main objectives of this section are to describe the procedure to reach the correct TIRFM configuration and to show how to perform time-lapse experiments.

TIRFM varies greatly in instrument setup, the described setup employs laser beam as incident light and adopts through the object configuration [18]. Refer to Sect. 4.2 for TIRFM configurations and for photobleaching.

3.2.1 Reaching TIRF Illumination (Fig. 2)

- Turn on the lasers 30 min before starting the experiment, as it needs to warm up and stabilize.
- Insert the glass cover with transfected cells in the appropriate imaging chamber, assemble the chamber, and add 500 μL of KRH solution.
- Place the chamber on the microscope stage and coat the specimen with the safety cover (Fig. 2c).
- In epifluorescence mode, look through the eyepiece and move the focus knob until the image comes into focus. Select the transfected cells to be recorded. The ideal cells should be flat, located in the center of the recording chamber (to avoid the generation of interference fringes on the specimen), and their fluorescent signal should be recorded with a low exposure time.
- Under software control, switch to the laser illumination *in live* mode and achieve the TIRF configuration.
- Last-generation TIRF microscopes have fully motorized systems with automatic laser incident angle adjustment and focus adjustment for TIRFM. Alternatively, you can reach the TIRF configuration simply by checking the position of laser beam on the sample safety cover (Fig. 2c).

  (a) When the laser beam is positioned in the center of the objective lens (Fig. 2b, left), a spot is visible in the center of the TIRF safety cover (Fig. 2c, left), and the cell is imaged in the epifluorescence mode (several focus planes are visible) (Fig. 2d, left).

  (b) To reach the critical angle, turn the "angle adjustment" screw on the TIRF slider and move the spot in the *Y* direction (forward or backward) (Fig. 2c, left). When you reach the critical angle (Fig. 2b, right), the spot disappears, and a thin, focused line emerges in the middle of the safety cover (Fig. 2c, right).

  (c) To fine-tune the TIRF angle, observe the cell image on the video. At this point, an epifluorescence-like image of the cell is still visible. Gently, move the screw until a flat, highly contrasted image of the cell appears (only one optical plane is visible: the plasma membrane in contact with the coverslip) (Fig. 2d, right). You have reached the TIRF configuration.

3.2.2 Setting Up of Experimental Recording Conditions

Before starting, set up the experimental recording conditions using the software-assisted menu.

- *Set up the laser power.* Capture the images using low laser power and make use, whenever possible, of attenuated filters.

- *Set up the exposition time.* Capture the images using low exposure time and high gain. Appropriate exposure times are between 50 and 100 msec.
- *Set up the sampling frequency.* Appropriate to the phenomenon (in the example, every 15 s).
- *Set up the experiment duration.* Time scale ranging from few minutes to 1 h.

Adjust the sampling frequency and the total recording time according to the phenomenon you are interested in and keep these parameters constant within the same experiment.

Refer to Sect. 4.2 for comments on photobleaching.

3.2.3 Sample Imaging

Envisage at least two different experimental conditions:

(a) Control (or resting) condition: the solution where cells are normally kept (KRH solution)

(b) Stimulated condition: resting solution plus the stimulus

As an example, we monitor the effect of glutamate on transporter density. To evaluate whether the effect is mediated by the transporter activity, in a subset of "stimulated conditions," we also test the effect of HIP-A, a selective inhibitor of EAATs transporters [19].

Under TIRF illumination add the solution of interest and start recording the time-sequential images.

- *Resting condition.* Add 500 μL of KRH solution to the existing 500 μL, to reach the final volume of 1 mL in the recording chamber. Start recording the time-lapse experiment in TIRFM mode. Save the time-sequential images as "resting condition."
- *Stimulated condition.* Focus on the same cell, remove 500 μL of KRH solution, and substitute with 500 μL of 2× glutamate KRH solution (or 2× HIP-A). Record and save the time-sequential images as "glutamate-stimulated condition" (or "HIP-A stimulated condition").
- Stick to the same settings during experiments (laser power, time exposure, frame number).

To obtain the *bleaching rate*, register a sequence of frames in the cell where no stimulation has been applied, using the identical setting conditions.

Refer to Sects. 4.1 and 4.2 for cell variability and recording conditions.

3.3 Data Collection and Analysis

The main objective of this section is to show how to extract quantitative information from the image sequence, how to correct data for photobleaching, and how to interpret the experimental results.

To analyze images automatically, macros have been developed in the lab, based on existing functions of the image analysis software [21]. The majority of image analysis software offers this possibility. Useful macros can be also found online, freely available (http://wiki.cmci.info/downloads/bleach_corrector).

3.3.1 Fluorescence Intensity Quantification

Open the experimental (resting or stimulated) time-sequential images to quantify the fluorescence intensity in area of interest (AOI) of the image, over the course of the sequence.

Go to the macro menu and set capture "multiple AOIs." Choose one of the selection tools (circle, square, etc.) in the menu to create an AOI in a relevant region of the cell. Create several AOIs within the same image, for simultaneous processing of data.

Go to the macro menu, and choose measurements. Select "mean intensity" to evaluate the mean fluorescence intensity within the selected AOIs.

Experimental AOI: *To calculate the density of transporters at the plasma membrane, place AOIs in regions of interest of the cell surface (i.e., filopodia, lamellipodia, plasma membrane) (Fig. 3B, white square). Consider two to three AOIs for each cell.*

Background AOI: *To calculate the background of the image, place three AOIs, identical to experimental AOI, in a region of the image without cells/fluorescence (Fig. 3A, white square).*

If you have created a macro, with the AOIs selected, automatically the software calculates the average fluorescence intensity of each AOI, over the course of the movie. Otherwise, scroll the movies and in each frame select count/measure intensity.

Collect and export data to a spreadsheet program for further analysis.

3.3.2 Photobleaching Evaluation (Fig. 3C)

Similarly, open the "Bleaching" file and place AOIs in a region of the image without cells (to create *Background Bleaching AOI*) or over the cell surface (to create *Bleaching AOI*).

Employ these data to evaluate the photobleaching rate.

First open the file "Background Bleaching AOI" data in a spreadsheet program and average the values.

To evaluate the photobleaching rate, open the file "Bleaching AOI" in a spreadsheet program. From each time point, subtract the averaged background value from all the frames (Fig. 3Ca).

Normalize the fluorescence intensity values in each frame to the initial intensity value ($F0$) ($F/F0$) (Fig. 3Cb).

Fit background-subtracted normalized bleaching data with an exponential curve. Many different types of software display the equation on the chart and automatically assign the values to the variables in the equation (Fig. 3Cc). Use the experimental variables to correct fluorescence data.

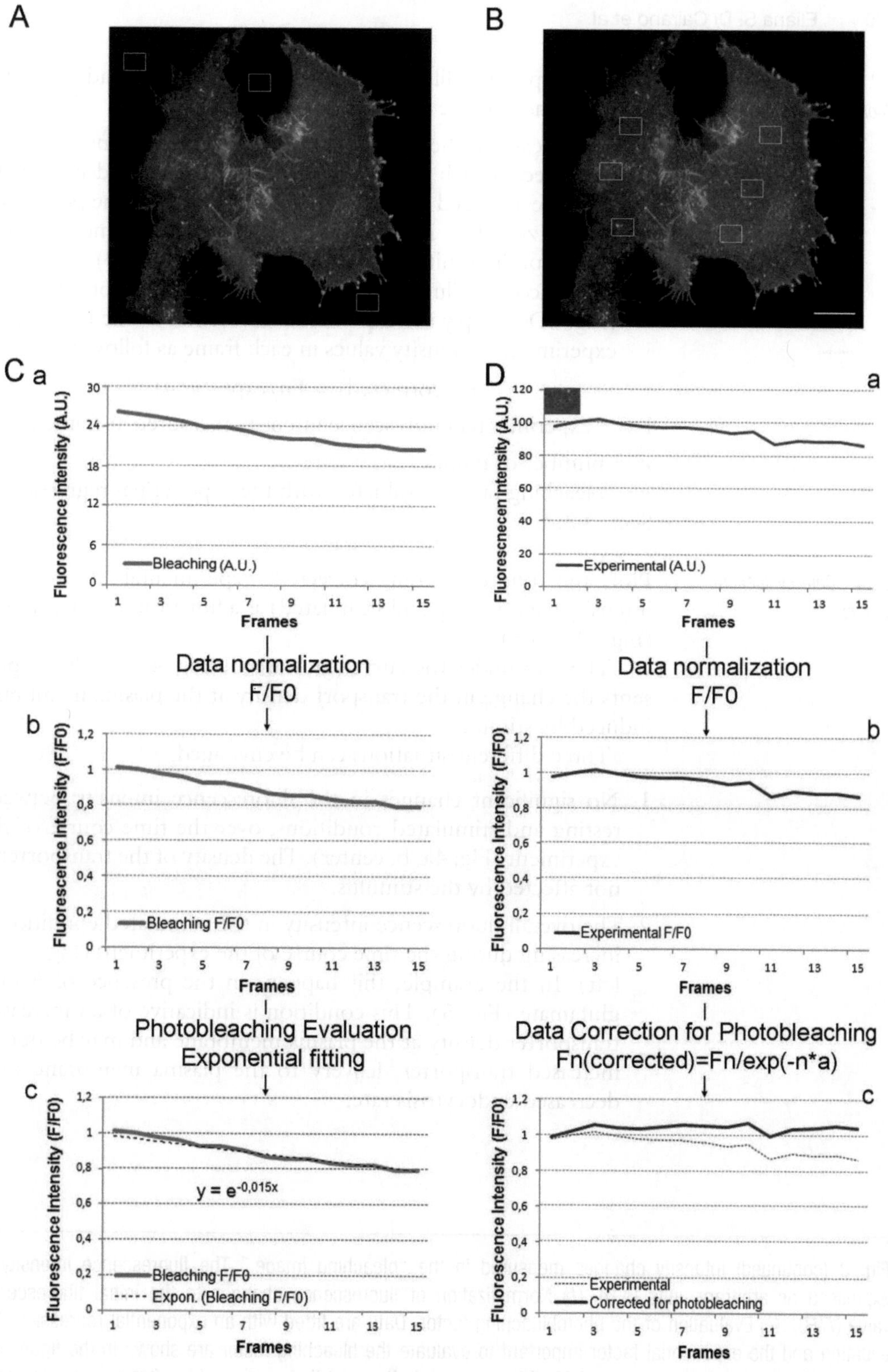

Fig. 3 Workflow of data processing. (**A**, **B**) TIRFM images of EGFP-EAAC1 in a live Cos-7 cell. The *white squares* indicate representative "background AOI" in (**A**) or an "experimental AOI" in (**B**). Scale bar 10 μm. (**C**) Workflow for the evaluation of the photobleaching rate. From top to bottom: (*a*) time course of fluorescence

3.3.3 Data Correction for Photobleaching (Fig. 3D)

- First, open the file "Background AOI" in a spreadsheet program and average the values.
- Then, export the time course of the fluorescence changes measured in each "Experimental AOI" to a spreadsheet. Subtract the averaged background value from each time point, and normalize the intensity value in each frame to the corresponding initial fluorescence intensity ($F/F0$) (Fig. 3Da, b). To correct fluorescence intensity data for photobleaching (Fig. 3Dc), apply the exponential correction to the normalized experimental intensity values in each frame as follows:

$$\text{Fn(corrected)} = \text{Fn}/\exp(-n * a)$$

Fn = experimental fluorescence intensity measured at frame n

n = number of frames
a = bleaching factor (evaluated with the exponential regression in Sect. 3.3.2)

3.3.4 Data Evaluation (Fig. 4)

Plot the photobleaching-corrected experimental fluorescence intensity data (resting and stimulated) as a function of time/frames (Figs. 4 and 5).

The area under the curve (gray area in Fig. 4b) (AUC) represents the change in the transport density at the plasma membrane induced by stimulus.

Three different situations can be envisaged:

1. No significant changes in the fluorescence intensity between resting and stimulated conditions, over the time course of the experiment (Fig. 4a, b, center). The density of the transporter is not affected by the stimulus.
2. The overall fluorescence intensity in the stimulated condition is increasing during the time course of the experiment (Fig. 4a, b, left). In the example, this happens in the presence of 5 mM glutamate (Fig. 5). This condition is indicative of an increased transporter density at the plasma membrane and may be due to increased transporter delivery to the plasma membrane or a decreased endocytosis rate.

Fig. 3 (continued) intensity changes measured in the "bleaching image." The fluorescence intensity is expressed as arbitrary units (A.U). (*b*) Normalization of fluorescence changes to the initial fluorescence value ($F/F0$). (*c*) Evaluation of the photobleaching factor. Data are fitted with an exponential regression; the equation and the exponential factor important to evaluate the bleaching factor are shown in the figure. (**D**) Proposed workflow for "experimental AOI." From top to bottom: (*a*) time course of the fluorescence intensity changes measured in the "experimental image"; (*b*) data normalization to the initial fluorescence value ($F/F0$); (*c*) data correction for photobleaching (*gray continuous line*)

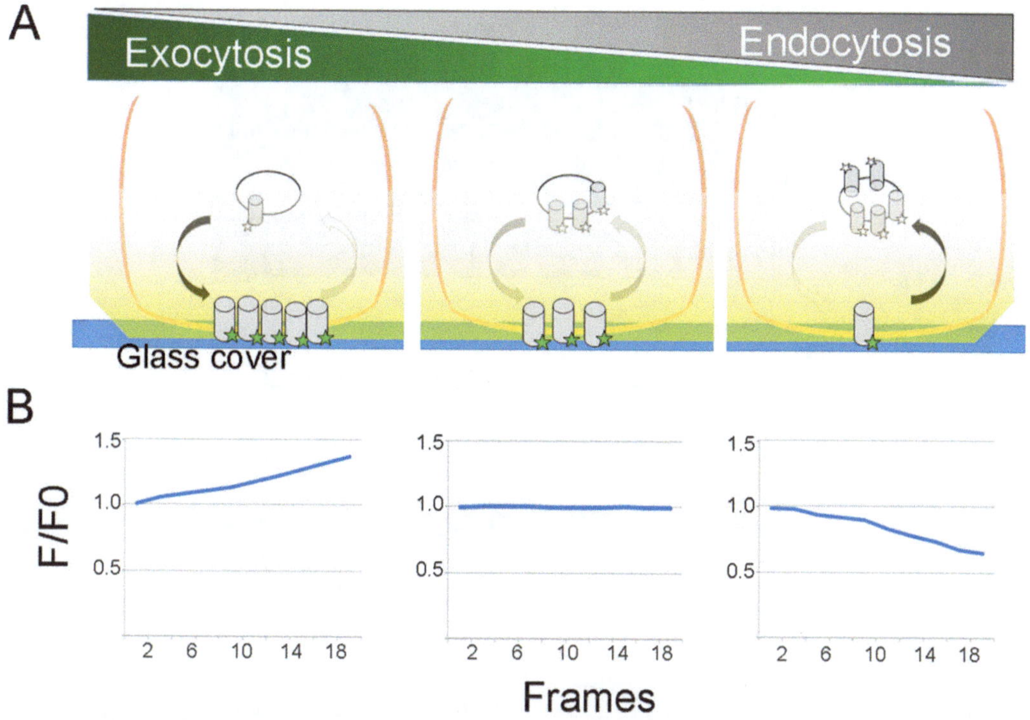

Fig. 4 Modification of transporter density at the plasma membrane evoked by stimulus application. (**A**) The transporter density at the plasma membrane is the result of a fine balance between exocytic (*green line*) and endocytic (*gray line*) processes. Under stimulated conditions, the transporter density at the plasma membrane increases if the exocytosis prevails over endocytosis (*left*), remains unchanged if endocytic and exocytic processes are balanced (*center*), or decreases if endocytosis prevails over exocytosis (*right*). (**B**) Expected modification of the normalized fluorescence intensity (*blue line*) over the time in case of increased (*left*), unchanged (*center*), or decreased (*right*) transporter density at the plasma membrane under stimulated conditions

3. The fluorescence intensity in the stimulated condition decreases over the time course of the experiment if compared to the resting condition (Fig. 4a, b, right). In the example, this corresponds to HIP-A, the EAAT1 inhibitor (Fig. 5). This situation is indicative of a decreased transporter density at the plasma membrane and may be due to increased endocytosis or decreased delivery.

The copresence of drugs blocking the exocytosis or endocytosis pathway in the stimulated condition may be helpful in understanding the mechanism of action of the stimuli (exocytosis versus endocytosis).

4 Notes

The method allows the selective visualization of transporters density at the plasma membrane based on the change in the fluorescence intensity signal. The ability of TIRFM to perform a really

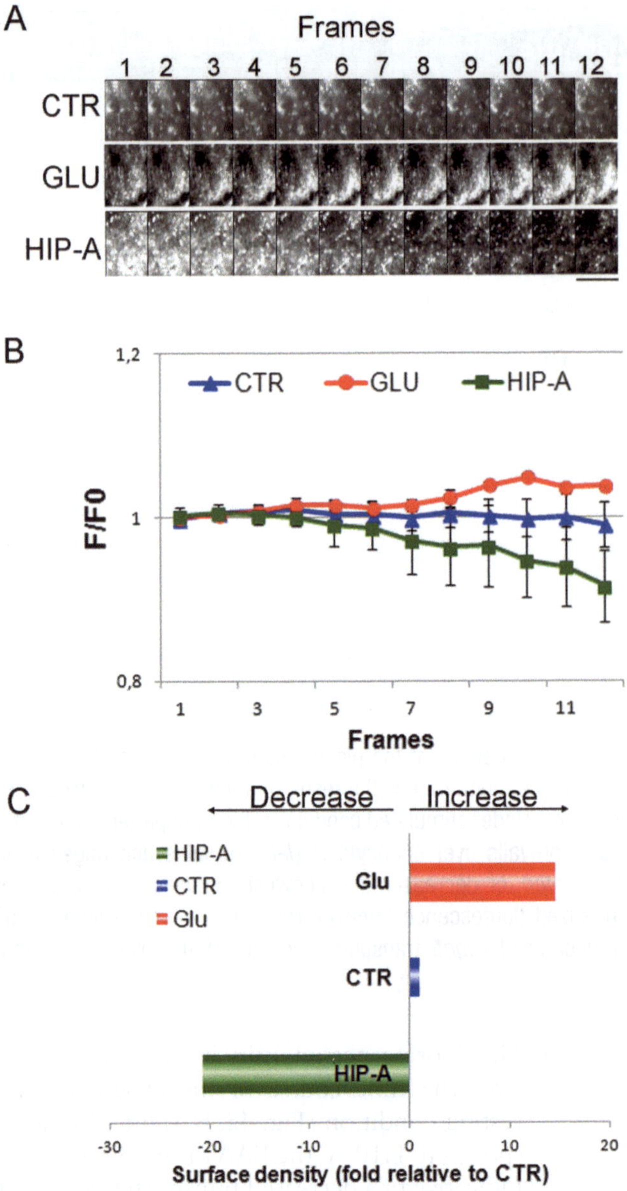

Fig. 5 An example, glutamate-induced increase of EAAC1 density at the plasma membrane. (**A**) Time-lapse imaging of EGFP-EAAC1 constructs in a live Cos-7 cell. Cells are recorded under resting (CTR) and stimulated conditions (5 mM glutamate [Glu] or 25 nM HIP-A application). Cells are sampled every 15 s, for a total of 3 min. Scale bar: 4 μm. (**B**) Normalized fluorescence intensity profiles (*F*/*F0*) showing the amount of proteins detected at the cell surface by TIRFM under resting conditions (*blue*, *triangles*), 5 mM glutamate (*red*, *circles*), or 25 nM HIP-A (*green*, *squares*). $P < 0.001$ Ctr vs Glu; $p < 0.01$ Ctr vs HIP-A; $p < 0.00005$ Glu vs HIP-A. (**C**) Histograms representing mean changes (Total AUC) in the fluorescence intensity under control (*blue bar*) or stimulated (glutamate *red bar*, HIP-A *green bar*) conditions

narrow optical sectioning at the level of the cover glass–sample interface ensures the production of surface-selective images. Three points can be particularly challenging and deserve particular attention: the generation of the probe (the intrinsically fluorescent protein of interest), the correct image capture, and the conversion of images information in quantitative data.

4.1 Selection of Fluorescent Probes and Cell Line for Transfection and Imaging

4.1.1 Creation of Genetically Encoded Fluorescent Recombinant Transporters

The ability to evaluate the transporters density at the plasma membrane in real time, in living cells, relies on the possibility to mark them with intrinsically fluorescent tag or a fluorescent exogenous ligand. An ideal tag should allow the visualization of the protein of interest without perturbing its folding, targeting, and functional activity. Furthermore, its emitted fluorescence should be sufficiently bright and stable to be detected as long as the experiment length, without damaging the cell.

Tag Selection: *Probably, the exponential development of live cell microscopy techniques has been fueled by the introduction of genetically encoded fluorescent probes. The prototype is the GFP protein that exhibits bright green fluorescence when exposed to light in the blue to ultraviolet range [22]. Many different mutants of GFP have been engineered, and several spectral variants are now commercially available [23]. They can be fused to protein of interest and can be simultaneously transfected in cells to monitor in vivo protein–protein interactions and signal transduction pathways [24]. pH-sensitive GFP variants have been produced in the last years and have been used as sensors to monitor pH-sensitive processes. Given that glutamate transport by EAATs carriers is characterized by the transport of a net* H^+ *ion inside the cell for each cycle, an interesting application of these probes is the study of intracellular pH changes associated with glutamate transport activity [25].*

However, GFP is a rather huge protein (238 amino acid residues, 26.8 kDa), and its fusion, sometimes, can perturb the folding and trafficking of the protein of interest. Furthermore, GFP proteins have a spontaneous tendency to oligomerize, and this may alter the interaction of transporters with accessory proteins. Finally, the photo-physical properties of naturally fluorescent proteins are not as good as organic dyes. They can blink, they are not very bright, and some are not photostable (see below).

To overcome these problems, other classes of protein-based fusion tags have been developed in the last years. An interesting class of tags consists in "self-labeling" enzymes that covalently attach a fluorescent ligand to one of its own amino acid residues. These enzymes are called Halo tags and SNAP/CLIP and are fused in frame with the protein of interest. When expressed, the proteins are not innately fluorescent but become fluorescent when the cells are exposed to cell-permeable fluorescent ligands [26]. Tag strategies for protein localization can be found at https://www.neb.com/applications/cellular-analysis/protein-localization. Other interesting tags are those consisting in short peptide sequences enriched in cysteine residues that can specifically react with small membrane-permeable biarsenical dyes (FlAsH and ReAsH). The optimal tag sequence is "CCPGCC," where C is cysteine, G glycine, and P proline. The first advantage of this class of tags is that they are small; thus, the introduced sequences have less opportunity to disrupt the overall fold and function of the labeled protein. The second advantage of these probes is the easy of pulse-labeling procedures. For some of them, labeling can be restricted in both time and space, and sequential labeling schemes can be employed to target different pools of proteins [27].

Tag Positions: *The position of the tag can be crucial. Based on transporter structure, three positions can be envisaged: N-terminus, C-terminus, and intracellular or extracellular loops of the selected protein.*

In our experience, tags (small amino acid sequences or complex proteins) work well when fused to the N-end of transporters [3, 16]. To maintain the correct folding and trafficking of the recombinant protein, it may be useful to add a flexible linker between the transporter and the GFP. The standard flexible linkers consist of glycines interspersed with serines or threonines (e.g., GGGSGGGS) [28].

The C-terminal end may be more critical. Indeed, several transporters carry at the very end of their COOH-terminal domain, target motives for interaction with PDZ proteins [3, 29]. The presence of a tag in this region may mask the recognition motif and abolish interaction with PDZ protein, thus profoundly affecting the transporter function and/or trafficking.

Although potentially interesting from the experimental point of view, a tag insertion in the extracellular loops is a risky choice. In our experience, even the presence of the small c-myc tag (EQKLISEEDL sequence) in the second extracellular loop of the glutamate transporter causes the transporter retention in the endoplasmic reticulum, probably for folding problems.

Independently of the tag selection and position, we recommend testing the activity and the targeting of the generated recombinant protein with functional and/or biochemical experiments [3].

4.1.2 Cell Line Selection and Transfection

TIRFM allows the collection of information regarding processes occurring at the plasma membrane or in an environment of 100–200 nm under the plasma membrane, which is a very narrow strip in living cells.

Therefore, the ideal cell line must grow adherent to the cell substrate and easily transfected. Its cell surface must be sufficiently flat and stable over the time in order to produce good images for quantitative analysis. To improve cell adhesion, it is possible to coat the glass cover with matrix adhesion proteins like collagen, gelatin, or polylysine. Mind that the sub-domain localization and the activity of channels and transporters may be modified by their interaction with cell-substrate adhesion molecules [30].

Cell transfection does not generally represent a problem because 30–50% transfection efficiency is sufficient to easily perform single-cell analysis by imaging techniques. A wide variety of gene delivery techniques to introduce plasmid DNA is commercially available and can be tested and optimized. In general, avoid transfection system and/or conditions which cause excessive protein overexpression, since they can alter the transporter targeting to the plasma membrane and its dynamics. We found particularly efficient the PEI-mediated transfection [20]. This method does not require serum withdrawal, and the reagent is not toxic; therefore, the cell morphology is preserved, and cells retain large adhesive surfaces with the substrate. Usually the PEI reagent does not result in excessive overexpression of recombinant proteins, is cost-effective, and applicable to a broad range of cell lines.

4.2 TIRFM Recording and Data Analysis

This is probably the most critical part. TIRFM enables the analysis of individual molecular events through detection of changes in the fluorescent signal derived from tagged proteins that move in or out the evanescent field. However, several confounding factors can modify this signal over the time, without necessarily implying modification of the transporter density. Generally, they can be separated in those that cause modification of cell surface domain in focus in the TIRF zone and those which modify the emitted fluorescent light of the fluorophore.

4.2.1 Modification of the Cell Morphology

The evanescent wave generated by the TIRF illumination decays exponentially in the sample; therefore, also subtle modifications in the cell morphology like ruffle and filopodia generation, and cell movement, can modify the intensity of the fluorescent signal. Some of the modifications are caused by changes in temperature and cell medium volumes.

Temperature: *Temperature changes due to the experimental environment, and medium addition and intense laser illumination may modify the focus, thus changing the region under TIRF illumination during recording. As the evanescent wave decays exponentially, this in turn may modify the intensity of the fluorescent signal. Cell morphology and transporter trafficking as well are extremely sensitive to temperature.*

To avoid this problem, the sample needs to be thermostated to 37°C. Chamber incubators are ideal to guarantee a constant temperature of all devices inside the incubator and the specimen but also a large part of the microscope, thus preserving the experimental conditions and minimizing the focus drift resulting from temperature gradient. In the absence of a temperature control system, we recommend to prewarm the imaging chamber, all added solutions, and pipette tips used for addition and to avoid long-term imaging experiments.

Volume Sample: *Given that the cell reacts to mechanical stress, modification of the total volume in the imaging chamber may change the strip under TIRF illumination. To solve this artifact, try to keep constant the volume of the medium in the imaging chamber. Avoid complete removal of the cell medium and change the medium preferably using a perfusion system, possibly connected with a vacuum pump, to reduce shearing stress and noise.*

4.2.2 Modification of the Emitted Fluorescent Light

pH Changes: *Fluorescent molecules are particularly sensitive to pH modification. Drug application, transporter activity, and cell metabolism may alter the pH of the medium over the course of the experiments, thus modifying the fluorescent signal intensity. This may be particularly relevant in the tiny volume between the cell and the surface of the coverslip where TIRF illumination takes place.*

To avoid this problem, solutions must always be buffered. Furthermore, it is useful to monitor possible pH modification induced by the presence of the drug/treatment, in preliminary experiments.

Photobleaching: *is probably the most relevant change of fluorescence intensity occurring in the sample. It is a progressive modification of the fluorescence signal due to photon-induced chemical damage and covalent modification of the fluorophore. Normally, it causes a permanent loss of the fluorescence emission intensity and the progressive fading of the fluorescent signal [31].*

Controlling photobleaching is critical in order to capture satisfactory images that can be analyzed and quantified. There are several possible interventions.

1. The best strategy is to avoid photoeffects during the recording section. In a constant environment, photobleaching depends on the intensity of the light source, the time, and the cycle of exposure to the excitation source. Therefore, the experimental settings should be accurately optimized.

   Illumination intensity. Photobleaching is proportional to the illumination intensity. To reduce photobleaching, use attenuated and shuttered illumination. Attenuate the illumination intensity until image looks grainy and baseline recordings are stable.

   Exposure time. The same applies for the exposure time. The longer the exposure time, the higher the photobleaching rate. Reduce the time until image looks grainy. Optimal exposure times are between 50 and 100 ms.

   The recording settings in resting and stimulated condition must be identical. Mind the overall expectation of your experiment. If you envisage an increase in the transporter density (e.g., increased fluorescence), keep the exposure time (or the illumination intensity) in the lower part of the detector range.

If data analysis is not excessively affected by the spatial resolution of the signal, bin the camera (2 × 2 binning can be a good compromise).

Number of frame capture. The maximum interval between two subsequent frames is dictated by how fast is the phenomenon. High temporal resolution may exert photoeffects, and with low temporal resolution, the effect of the stimulus may be lost. In preliminary experiments, evaluate the time scale of the phenomenon you are interested in (how long it lasts, whether the compound to be tested needs a preincubation time or acts immediately, etc.).

2. Selection of a different fluorescent probe. As above reported, genetically encoded fluorescent reporter proteins are not particularly bright and can photobleach. In time-lapse experiments requiring high sample frequency, it may be strategic to make use of "self-labeling" enzyme tags whose fluorescence relies on incubation with organic dyes which are more stable and can be chemically modified [26].
3. If despite controlling recording conditions, the baseline is not stable and photobleaching still occurs, it is possible to estimate the effect mathematically and correct data. In many cases, photobleaching follows a simple exponential decay function [18, 31], which makes its assessment and its correction easier by performing control recordings (Sect. 3). Several correction formulas are available; in the paper, a mono-exponential equation has been used to fit the photobleaching, but also linear regression can be used. Given the cell-to-cell variability in the change in baseline fluorescence over time (see below), before the beginning of the experiment, evaluate the bleaching rate in the same cell you want to record from.

4.2.3 Sample Variability

The presented protocol relies on single-cell analysis. In our experience, there is a great variability from cell to cell, in particular if you are studying rapid phenomena, the effect of signal transduction pathways, or the basal trafficking of plasma membrane proteins. To reduce the variability, when possible, record always the same cell under resting and stimulated conditions. To be sure that the effects you are observing rely on the specific treatment you are studying, replicate the experiments using a different sequence (first stimulated then resting conditions). This applies only to drugs/treatments whose effects are reversible, as in this case.

References

1. César-Razquin A, Snijder B, Frappier-Brinton T, Isserlin R, Gyimesi G, Bai X, Reithmeier RA, Hepworth D, Hediger MA, Edwards AM, Superti-Furga GA (2015) A call for systematic research on solute carriers. Cell 162:478–487
2. Lin L, Yee SW, Kim RB, Giacomini KM (2015) SLC transporters as therapeutic targets: emerging opportunities. Nat Rev Drug Discov 14:543–560
3. D'Amico A, Soragna A, Di Cairano E, Panzeri N, Anzai N, Vellea Sacchi F, Perego C (2010) The surface density of the glutamate transporter EAAC1 is controlled by interactions with PDZK1 and AP2 adaptor complexes. Traffic 11:1455–1470. doi:10.1111/j.1600-0854.2010.01110.x
4. Ahn J, Pietrini G, Muth TR, Caplan MJ (1998) Expression of neurotransmitter transport systems in polarized cells. Methods Enzymol 296:370–388
5. Qian Y, Galli A, Ramamoorthy S, Risso S, DeFelice LJ, Blakely RD (1997) Protein kinase C activation regulates human serotonin transporters in HEK-293 cells via altered cell surface expression. J Neurosci 17:45–57
6. Gabriel L, Stevens Z, Melikian H (2009) Measuring plasma membrane protein endocytic rates by reversible biotinylation. J Vis Exp 34: e1669. doi:10.3791/1669
7. Sargiacomo M, Lisanti M, Graeve L, Le Bivic A, Rodriguez-Boulan E (1989) Integral and peripheral protein composition of the apical and basolateral membrane domains in MDCK cells. J Membr Biol 107:277–286
8. Perego C, Bulbarelli A, Longhi R, Caimi M, Villa A, Caplan MJ, Pietrini G (1997) Sorting of two polytopic proteins, the gamma-aminobutyric acid and betaine transporters, in polarized epithelial cells. J Biol Chem 272:6584–6592
9. Perego C, Vanoni C, Villa A, Longhi R, Kaech SM, Fröhli E, Hajnal A, Kim SK, Pietrini G (1999) PDZ-mediated interactions retain the epithelial GABA transporter on the basolateral surface of polarized epithelial cells. EMBO J 18:2384–2393
10. Kanai Y, Hediger MA (2004) The glutamate/neutral amino acid transporter family SLC1: molecular, physiological and pharmacological aspects. Pflugers Arch 447:469–479
11. Kristensen AS, Andersen J, Jørgensen TN, Sørensen L, Eriksen J, Loland CJ, Strømgaard K, Gether U (2011) SLC6 neurotransmitter transporters: structure, function, and regulation. Pharmacol Rev 63:585–640
12. Kumar V, Rahbek-Clemmensen T, Billesbølle CB, Jorgensen TN, Gether U, Newman AH (2014) Novel and high affinity fluorescent ligands for the serotonin transporter based on (s)-citalopram. ACS Med Chem Lett 5:696–699
13. Stephens DJ, Allan VJ (2003) Light microscopy techniques for live cell imaging. Science 300:82–86
14. Axelrod D (2008) Total internal reflection fluorescence microscopy. Methods Cell Biol 89:169–221. doi:10.1016/S0091-679X(08)00607-9
15. Burchfield JG, Lu J, Fazakerley DJ, Tan SX, Ng Y, Mele K, Buckley MJ, Han W, Hughes WE, James DE (2013) Novel systems for dynamically assessing insulin action in live cells reveals heterogeneity in the insulin response. Traffic 14:259–273. doi:10.1111/tra.12035, Epub 2013 Jan 18
16. Perego C, Di Cairano ES, Ballabio M, Magnaghi V (2012) Neurosteroid allopregnanolone regulates EAAC1-mediated glutamate uptake and triggers actin changes in Schwann cells. J Cell Physiol 227:1740–1751. doi:10.1002/jcp.22898
17. Gluzman Y (1981) SV40-transformed simian cells support the replication of early SV40 mutants. Cell 23:175–182
18. Daniele F, Di Cairano ES, Moretti S, Piccoli G, Perego C (2015) TIRFM and pH-sensitive GFP-probes to evaluate neurotransmitter vesicle dynamics in SH-SY5Y neuroblastoma cells: cell imaging and data analysis. J Vis Exp 95: e52267. doi:10.3791/52267.
19. Funicello M, Conti P, De Amici M, De Micheli C, Mennini T, Gobbi M (2004) Dissociation of [3H]L-glutamate uptake from L-glutamate-induced [3H]D-aspartate release by 3-hydroxy-4,5,6,6a-tetrahydro-3aH-pyrrolo[3,4-d]isoxazole-4-carboxylic acid and 3-hydroxy-4,5,6,6a-tetrahydro-3aH-pyrrolo[3,4-d]isoxazole-6-carboxylic acid, two conformationally constrained aspartate and glutamate analogs. Mol Pharmacol 66:522–552
20. Neuberg P, Kichler A (2014) Recent developments in nucleic acid delivery with polyethylenimines. Adv Genet 88:263–288
21. Treccani G, Musazzi L, Perego C, Milanese M, Nava N, Bonifacino T, Lamanna J, Malgaroli A, Drago F, Racagni G, Nyengaard JR, Wegener G, Bonanno G, Popoli M (2014) Stress and corticosterone increase the readily releasable pool of glutamate vesicles in synaptic terminals

of prefrontal and frontal cortex. Mol Psychiatry 19:433–443

22. Tsien RY (1998) The green fluorescent protein. Annu Rev Biochem 67:509–544. doi:10.1146/annurev.biochem.67.1.509
23. Shaner NC, Steinbach PA, Tsien RY (2005) A guide to choosing fluorescent proteins. Nat Methods 2:905–909
24. Sorkina T, Doolen S, Galperin E, Zahniser NR, Sorkin A (2003) Oligomerization of dopamine transporters visualized in living cells by fluorescence resonance energy transfer microscopy. J Biol Chem 278:28274–28283
25. Watts SD, Torres-Salazar D, Divito CB, Amara SG (2014) Cysteine transport through excitatory amino acid transporter 3 (EAAT3). PLoS One 9, e109245. doi:10.1371/journal.pone.0109245, eCollection 2014
26. Crivat G, Taraska JW (2012) Imaging proteins inside cells with fluorescent tags. Trends Biotechnol 30:8–16. doi:10.1016/j.tibtech.2011.08.002, Epub 2011 Sep 15
27. Lai YT, Chang YY, Hu L, Yang Y, Chao A, Du ZY, Tanner JA, Chye ML, Qian C, Ng KM, Li H, Sun H (2015) Rapid labeling of intracellular His-tagged proteins in living cells. Proc Natl Acad Sci U S A 112:2948–2953
28. Huston JS, Levinson D, Mudgett-Hunter M, Tai MS, Novotný J, Margolies MN, Ridge RJ, Bruccoleri RE, Haber E, Crea R (1988) Protein engineering of antibody binding sites: recovery of specific activity in an anti-digoxin single-chain Fv analogue produced in *Escherichia coli*. Proc Natl Acad Sci USA 85:5879–5883
29. Perego C, Vanoni C, Bossi M, Massari S, Basudev H, Longhi R, Pietrini G (2000) The GLT-1 and GLAST glutamate transporters are expressed on morphologically distinct astrocytes and regulated by neuronal activity in primary hippocampal cocultures. J Neurochem 75:1076–1084
30. Zuidema JM, Hyzinski-García MC, Van Vlasselaer K, Zaccor NW, Plopper GE, Mongin AA, Gilbert RJ (2014) Enhanced GLT-1 mediated glutamate uptake and migration of primary astrocytes directed by fibronectin-coated electrospun poly-L-lactic acid fibers. Biomaterials 35:1439–1449
31. Diaspro A, Chirico G, Usai C, Ramoino P, Dobrucki J (2006) Photobleaching. In: Pawley JB (ed) Handbook of biological confocal microscopy. Springer, New York. doi:10.1007/978-0-387-45524-2_39

Fluorescence-Based Screening of Membrane Proteins for Structural Studies

Jyoti Kumari, Surbhi Dhingra, and Janesh Kumar

Abstract

Structure determination of membrane proteins remains a technically challenging problem till date. Optimization of expression, purification, and stability of membrane proteins is still a bottleneck for their structure determination by either X-ray crystallography or single-particle cryo-electron microscopy. Their expression levels are low; hydrophobic transmembrane domains require detergents to extract them from cell membrane. Besides this, they are often flexible and unstable and must be stabilized before subjecting them to structure determination trials. Traditional methods of screening either require large amounts of material or are limited in their use due to the presence of detergents, which hinder with or cause high background during these evaluations.

Here, we describe two fluorescence-based methods that have enabled rapid, efficient, and economical screening of membrane proteins for construct optimization and evaluation of suitable stability conditions essential for successful structure determination. Both the techniques require only nanogram to microgram quantities of material and can be adapted to perform medium- to high-throughput screening.

Keywords Crystallization, Fluorescence, Membrane proteins, Screening, Structural biology

1 Introduction

Membrane proteins play essential roles in a variety of cellular processes and are involved in detecting and transmitting external signals, material, and information into and out of the cytoplasmic and nuclear milieu. Not surprisingly, they are critical components of signal transduction across the membranes including transport of substrates, permeation of ions, and communication of external stimuli. It is estimated that approximately 25–30% of all the proteins encoded by the human genome are membrane proteins [1] and are targets of around 60% of all the available drugs in the market [2].

Despite being immensely important, understanding the functions and the molecular mechanisms of their action and regulation are hampered by a lack of knowledge about their three-dimensional

Arun K. Shukla (ed.), *Chemical and Synthetic Approaches in Membrane Biology*, Springer Protocols Handbooks, (2017) 141–155, DOI 10.1007/8623_2016_1, © Springer Science+Business Media LLC 2016, Published online: 25 September 2016

structures, more so from eukaryotic sources. Out of ~121953 structures (August 2016) in the Protein Data Bank (PDB), only 2014 are of membrane proteins (~1.6%), and out of it, only 638 (~0.52%) are for unique membrane protein structures (http://blanco.biomol.uci.edu/mpstruc/). Above data clearly suggests that membrane protein structural biology is still a largely unconquered area. They are challenging targets for structural studies due to multiple reasons. Except for a few membrane proteins that are in high abundance and can be purified from source, most of these targets need to be overexpressed in heterologous expression systems to produce milligram quantities of well-behaved, stable, and pure protein suitable for structural studies. The overexpression requires extensive construct engineering and optimization which is a long and tedious process. The other challenging frontier is to find suitable conditions where the target membrane protein is stable and homogenous after extraction, solubilization, and purification from membranes. This requires screening of numerous buffers, detergents, lipids, ligands, antibodies, or a combination of these and is again a time- and resource-intensive endeavor.

A number of tools have been developed over the years to counter and overcome these difficulties associated with structure-function analysis of membrane proteins. Fluorescence-based methods, especially the use of a fluorescent fusion protein, are the most popular and successful as it allows monitoring of target proteins in cells, whole animals, or crude cell lysates without having to purify them [3]. The ability of fluorescent proteins to fold when tagged on either N- or C-terminus of target proteins has made it possible to directly study the biology of proteins in vivo or monitor their behavior in vitro [4]. A C-terminal fusion of green fluorescent protein (GFP) is reported to be an indicator of correctly folded target protein as GFP will only fold into fluorescent form if the target protein upstream is correctly folded [5, 6]. Besides, the sensitivity of fluorescent techniques allows detection of nanomolar quantities of tagged proteins. Development of a synthetic codon-optimized GFP construct which is more stable and ~45-fold brighter in comparison to the wild-type protein also prevented GFP-induced dimerization and aggregation of fusion proteins [7]. *Escherichia coli* that is still one of the preferred expression systems for membrane proteins of either prokaryotic or eukaryotic origin suffers from a major drawback associated with aggregation or inclusion body formation. However, GFP-based screening has facilitated overexpression, detergent selection, and purification of membrane proteins from bacterial system [8, 9]. It was also demonstrated that GFP fusion has made it possible to distinguish between membrane-integrated protein from that in the inclusion bodies [10]. GFP fusion proteins have also been used to track membrane protein production in yeast [11], insect cells [12], and mammalian cells [13] demonstrating its utility.

Further, evaluation of stability profiles of membrane proteins in different solution conditions via traditional methods like circular dichroism (CD) [14, 15], differential scanning calorimetry (DSC) [16, 17], or absorption spectroscopy [18] has been challenging and normally requires large quantities of purified protein. Highly sensitive fluorescent assays developed for analysis of membrane protein stability are critical for successful crystallization and structure determination via X-ray crystallography or cryo-EM.

Here, we review two of the most widely used techniques toward optimization of membrane protein overexpression, purification, and stability: (1) fluorescence-detection size-exclusion chromatography (FSEC) where size-exclusion chromatography can be used to investigate the state of fusion protein in crude extracts [19], formation of multiprotein complexes [20], and determination of thermal stability (FSEC-TS) [21] and (2) microscale fluorescent thermal stability assay for high-throughput stability evaluation and optimization of integral membrane proteins [22].

2 Fluorescence-Detection Size-Exclusion Chromatography (FSEC)

Membrane protein expressed as fusion with fluorescent proteins has been used for topology analysis of bacterial inner membrane proteome [23, 24] and as an indicator of overexpression by measuring fluorescence of whole cells or non-detergent-solubilized membranes [8, 10, 25]. In 2006, Gouaux Lab introduced fluorescence-detection size-exclusion chromatography (FSEC) for rapid screening of membrane protein expression levels, stability, and monodispersity [19]. This method is an adaptation and extension of the fluorescent-based concept by coupling size-exclusion chromatography with online fluorescence detection. Since then it has been widely used and has facilitated structure determination of multiple membrane proteins. Construct design for FSEC includes the target protein covalently fused with enhanced green fluorescent protein (EGFP) [7] either at N- or C-termini for screening along with an affinity tag (e.g., His-tag) for purification. In between the protein and the two tags, a proteolytic cleavage site is inserted for their removal (Fig. 1a).

FSEC involves two steps of screening before going onto large-scale purification of the membrane protein. The first step is to observe subcellular localization of the fluorescent protein. Constructs can be narrowed down based on epifluorescence or batch fluorescence measurements of the fusion protein from a small number of cells. The second step of screening involves size-exclusion chromatography of detergent-solubilized whole cells or membranes and monitoring of their SEC profiles via fluorescence spectroscopy. Cells are harvested, lysed in different detergents, and cleared of cell debris by ultracentrifugation, and the crude cell lysate

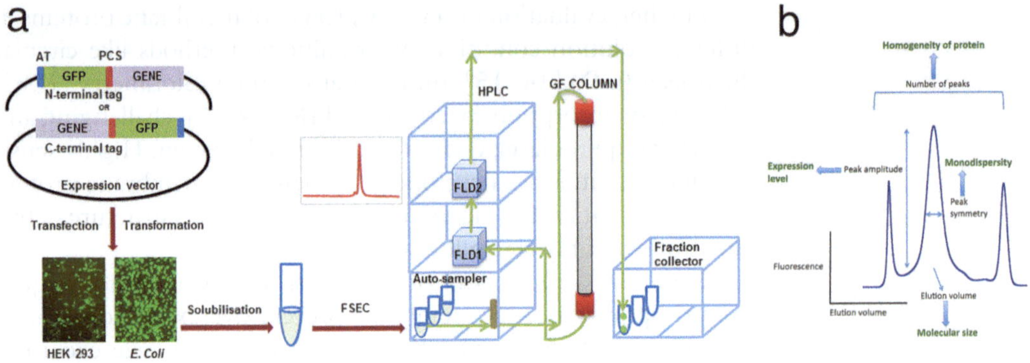

Fig. 1 Schematic representation of membrane protein screening using FSEC. (**a**) Design of fusion constructs with EGFP either on the N- or C-terminus of the target protein; AT stands for affinity tag, GFP for green fluorescent protein (can be replaced with any other fluorescent protein if desired), PCS for protease cleavage site, and GENE for target membrane protein. Protein expression can be initially observed by epifluorescence followed by detergent solubilization of cells and loading onto a SEC column connected to FSEC setup. The autosampler loads the sample, and the eluate is passed through fluorescence detectors (FLDs) that are set at appropriate excitation and emission wavelengths for fluorophore detection. (**b**) Shows a hypothetical elution profile of a GFP fusion protein detected by GFP fluorescence and the typical information this profile can generate

is resolved on a gel filtration column, and the eluate is monitored by fluorescence detectors. Careful analysis of the FSEC profile for peak amplitude and area, number of peaks, peak symmetry, and retention volume gives information about the level of expression, protein solubility, homogeneity, stability of the expressed protein, and an estimate of molecular weight (Fig. 1b). A symmetric Gaussian peak would convey that protein is homogenous or monodisperse and stable in solution. For asymmetric peaks, Gaussian peak fitting may be carried out to estimate if a peak consists of more than one species. An advantage of FSEC method is that besides its utility in construct screening and optimization, it can also be used to monitor the large-scale expression and purification of selected constructs. Even after proteolytic removal of the fluorescent tag, the intrinsic tryptophan fluorescence of the protein may be used to further track the purification and for optimization of suitable buffer, detergent, and additive conditions. So, without using up a lot of purified protein, one can still screen a multitude of conditions to optimize stability and monodispersity of membrane protein targets.

FSEC-Based Thermostability Test (FSEC-TS): FSEC-TS is an adaptation of FSEC that allows for evaluation of thermal stability of the membrane protein sample [21]. Protein stability along with homogeneity is an important parameter for X-ray crystallography or cryo-EM studies and can be improved by optimizing appropriate conditions. This may include screening additives such as detergents, lipids, or mixture of both and ligands, ions, etc. If done using traditional methods, this screening is not only labor intensive but requires large amount of purified protein as well which is difficult for membrane proteins because of their limited expression. In FSEC-TS, thermal stability is defined in terms of fluorescence of tagged protein or tryptophan fluorescence of purified protein [21]. Sample protein aliquots are incubated over a range of temperatures in a thermocycler, and their stability is analyzed by monitoring FSEC profile allowing for evaluation and optimization of thermal stability using only nanogram to microgram amount of protein.

Multicolor Fluorescence Size-Exclusion Chromatography (MC-FSEC): MC-FSEC is yet another adaptation of the FSEC technique that allows rapid assessment of assembly of protein complex, stability, stoichiometry, and pre-crystallization screening [20, 26]. This technique requires an FSEC setup that is equipped with multiple detectors for analyzing different fluorophores at the same time. It enables tracking of individual subunits during expression, solubilization, and purification steps and optimization of protein complex stability.

2.1 Materials

(1) Fluorescence tag fusion constructs – tag can be on either side of the protein (C′ or N′) and has to be empirically determined where it is tolerated. However, a C-terminal tag is preferred.

(2) Expression system – for prokaryotes, *E. coli* is found to be a more robust expression system, and for eukaryotes, HEK293 or Sf9 cells are more suitable for membrane protein overexpression.

(3) An appropriate size-exclusion column for gel filtration.

(4) FSEC instrument – SEC column connected in line with FPLC/HPLC connected to fluorescent detectors. An ideal FSEC setup would consist of a HPLC pump, a thermostatted autosampler equipped with a cooling/heating module, and at least two fluorescence detectors to simultaneously look at two different fluorophores. Further, a UV detector, a refractive index (RI) detector, and a fraction collector may also be added if needed. One can in principle connect any FPLC system with fluorescence detector for FSEC experiments. However, using an HPLC pump/setup enables the use of shorter HPLC size-exclusion columns reducing runtimes without compromising on the peak resolution. Also for high-throughput experimentation, an autosampler is desirable as it enables automatic injection of multiple samples in required sequence without continuous monitoring and user intervention.

2.2 Method

Construct optimization – Depending on bioinformatic analysis of protein sequence, different N- and C-termini deletion constructs are made. Primers are designed keeping thrombin sites or any other proteolytic site in between fusion tag and protein of interest. PCR-amplified products are cloned into suitable expression vector. It may be useful to adopt a high-throughput cloning method to enable engineering of multiple constructs.

Expression test – Transformation/transfection or infection into suitable expression system. Expression may be confirmed by detecting epifluorescence or batch fluorescence 16–72 h post transfection. For FSEC experiments, cells are harvested and lysed, and membrane proteins are solubilized in detergent for 1 h at 4°C. One may either use whole-cell lysates or do a membrane fractionation for this step. Insoluble cell debris are removed using high-speed centrifugation, and supernatant is loaded onto a suitable equilibrated SEC column, e.g., Superose 6 (GE), Bio SEC-5 (Agilent), Shodex semimicro KW404-4F, TSK columns, etc. The column eluate is monitored via fluorescence detectors (Fig. 1a). One may estimate the expression level of proteins by using a standard curve derived from known concentrations and fluorescence yields of purified fluorescent proteins.

Figure 2a, b shows optimization of two different synaptic receptors. C-terminal EGFP-tagged receptor constructs transiently expressed in HEK293 cells for 96 h were solubilized in a buffer containing the nonionic detergent *n*-dodecyl-β-D-Maltoside (DDM), and the resulting supernatant was analyzed by FSEC. The EGFP-fluorescence peak associated with the mutant 1 and mutant 2 constructs of protein X shows better expression, and the peak observed was nearly symmetric, whereas the native protein has poor expression and low levels of functional oligomer with majority of protein eluting at position corresponding to monomer. Similarly, for receptor protein Y, only mutant 1 has a fluorescent peak corresponding to functional oligomer, while native protein or mutant 2 does not.

For membrane biochemistry, the choice of detergent has an enormous impact on protein behavior in solution, during purification and crystallization. FSEC can be used to determine the degree of monodispersity and stability of the target protein in various detergents/lipids or mixtures of both either from whole-cell lysates or from small amount of purified protein. We show a representative example of detergent screening via FSEC with a bacterial integral membrane protein. Approximately 1 μg of purified protein in DDM (*n*-dodecyl-β-D-Maltoside) was mixed with 10× CMC (critical micelle concentration) of six different detergents, DDM, DM (*n*-decyl-β-D-Maltoside), nDM (*n*-nonyl-β-D-Maltoside), MNG (maltose neopentyl glycol), β-OG (*n*-octyl-β-D-Glucoside), and DDM/CHS (*n*-dodecyl-β-D-Maltoside/cholesteryl hemisuccinate), incubated on ice for 1 h and centrifuged, and supernatant was subjected to FSEC analysis by using a column equilibrated in DDM. It has been reported that the adverse effects of a destabilizing detergent are not rescued by a mild and typically non-denaturing detergent such as DDM in the mobile phase of the SEC column [19, 21]. We also observe the same; as shown in

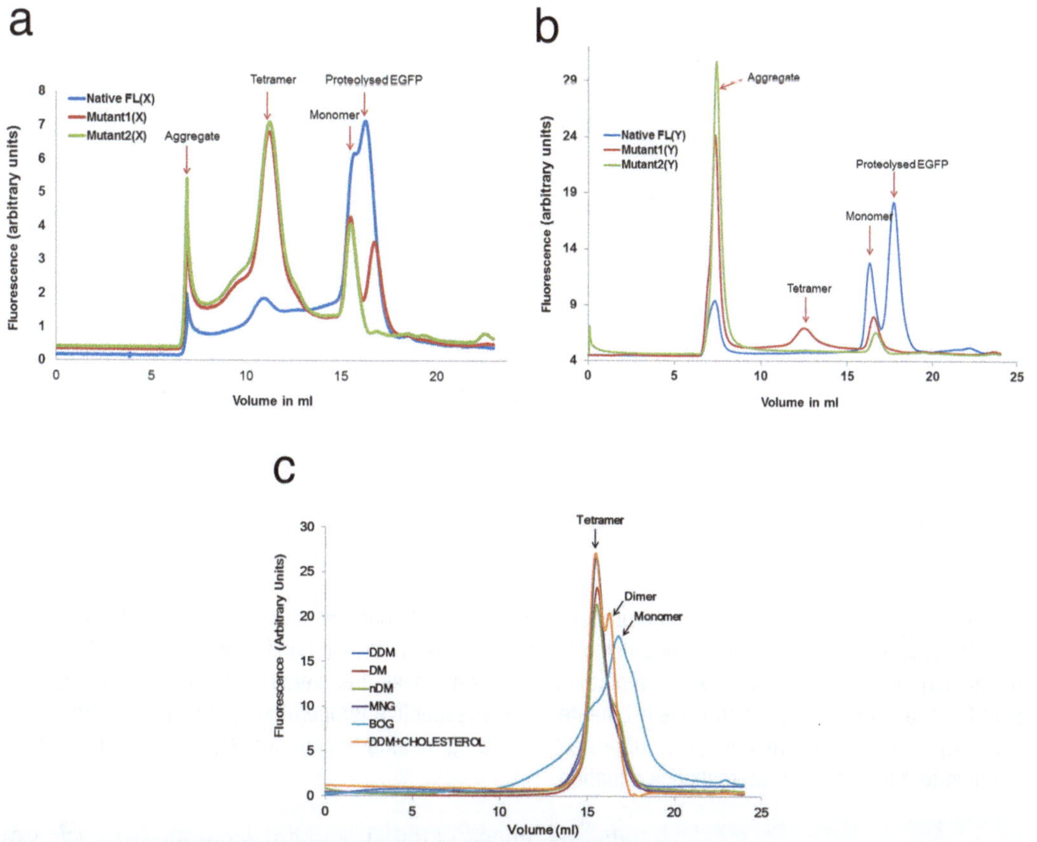

Fig. 2 Example of construct optimization using FSEC. FSEC traces from (**a**) synaptic membrane protein X and (**b**) synaptic membrane protein Y fused with a C-terminal EGFP. The *arrows* indicate the estimated elution position of the aggregated protein in void volume, functional receptor oligomer (tetramer), monomeric receptor subunit, and free EGFP, respectively. (**c**) Shows detergent screening by FSEC via monitoring tryptophan fluorescence of a purified bacterial membrane protein. *Arrows* indicate functional oligomeric state and lower oligomeric states of destabilized protein, respectively

Fig. 2c, protein profile looks symmetrical and the functional tetrameric state is most stable in DDM and MNG detergents. However, the peak height is reduced, and the tetramer is destabilized in DM, nDM, β-OG, and even in DDM/CHS suggesting that for this protein DDM and MNG are more suitable.

FSEC-TS – Equal aliquots (~1 μg) of either whole-cell lysates or purified protein are taken in thin-walled PCR tubes, incubated in a thermocycler for 10 min over a range of temperatures (usually 4–80°C), and centrifuged to remove precipitated material followed by loading of the supernatant onto a SEC column attached to FSEC setup. The fluorophore signals, e.g., GFP in case of whole-cell lysates or tryptophan fluorescence in case of purified protein, are monitored. Peak height or peak area corresponding to functional protein oligomer is plotted against temperature to generate a

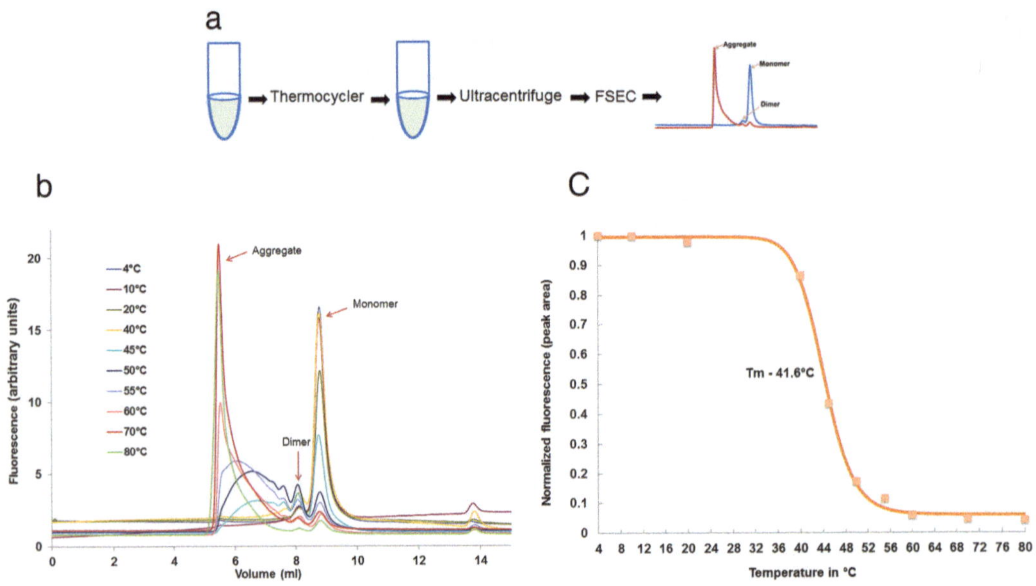

Fig. 3 Evaluation of thermostability of a protein via FSEC-TS. (**a**) Schematic representations of FSEC-TS work flow. Cell lysates/purified sample heated using thermocycler is centrifuged and subjected to FSEC analysis. The melting temperature is calculated via plotting peak height/peak area versus temperature. (**b**) Representative FSEC profiles from purified protein Z heated at the respective temperatures, detected by Trp fluorescence. (**c**) Representative melting curve of protein Z showing melting temperature (T_m) determined by fitting the curve to a sigmoidal dose–response equation

thermostability profile from which a melting temperature (T_m) for the protein can be calculated (Fig. 3a).

As shown in Fig. 3b, tryptophan fluorescence peak profiles from purified protein Z incubated between 4 and 40°C were sharp and symmetrical, and the protein recovery was high, indicating that the protein remained stable and monodisperse at these temperatures. In contrast, samples incubated between temperatures 45 and 80°C had a significant decrease in the peak height due to protein denaturation, aggregation, formation of higher oligomers, and precipitation after heating. The peak area of the functional monomeric peak when plotted against temperature shows a melting curve; from the fluorescence signal intensity at the respective temperatures (Fig. 3c), the calculated T_m for this protein is 41.6°C.

FSEC-TS can also be used to evaluate and optimize effects of different detergents, lipids, and additives like ions, ligands, etc. on the thermostability of target membrane protein [21]. Once a T_m has been determined, rather than plotting complete melting curve, protein aliquots can be incubated at the calculated T_m for 10 min in presence of a panel of additives, detergents, lipids, etc. to assess their effects and to screen for the ones that enhance thermostability.

The FSEC profiles could be normalized to control sample at 4°C without any additive.

MC-FSEC or multicolor FSEC can be used to investigate assembly and stability of heteromeric membrane protein complexes. Various subunits of the complex should be tagged with monomeric fluorescent proteins that have sufficiently separate spectral profiles to discriminate them from each other, e.g., yellow (YFP) and cyan (CFP, mCerulean), green (GFP, EGFP), and red (RFP, mRuby, etc.). However, due to differences in the fluorescence quantum yield and relative brightness of the various fluorescent tags used, stoichiometry assessments are not straightforward via MC-FSEC. One may estimate the expression levels of proteins and their stoichiometry by using a standard curve derived from known concentrations and fluorescence yields of recombinant fluorescent proteins. The fluorescence detectors should also be adjusted accordingly to reliably plot and assess expression levels of individual subunits in a complex.

2.3 Notes

Merits – FSEC is both a time- and resource-efficient method for mid- to high-throughput screening of membrane proteins for structure determination [27]. Expression can be evaluated and optimized at small scale from crude cell lysates. A number of different deletion constructs can be tested rapidly, and maximally fluorescent construct can be chosen for further optimization or large-scale expression. Also, different orthologues of same gene can be tested at the same time to screen for most promising candidate [28]. Many detergents can be screened to check maximum extraction and solubility of the protein from cell lysate. Homogeneity and stability of sample can be monitored at each and every step of purification. Protein stability can be tested using FSEC-TS that allows for screening of various stabilizing conditions efficiently with either unpurified or purified protein. Further, MC-FSEC can be used prior to pre-crystallization screening, to evaluate affinity between different subunits, stability of heteromeric complex to facilitate its crystallization, or structure determination by single-particle cryo-EM.

Demerits – Fusion of a fluorescence tag to the membrane protein can alter its normal biophysical as well as biochemical property in solution [29]. It has been reported that sometimes N- or C-termini of proteins are not accessible for fusion of big tags such as GFP, and hence construct screening may be required to determine if the tag can be placed at any of the intra- or extracellular loops. At times the proteolytic removal of the GFP tag might be difficult and may require playing with the linker length between target protein and the GFP. In some cases, removal of GFP may also significantly reduce the yield and affect the homogeneity of sample along with its biophysical and biochemical properties. Further, GFP fusion can also alter the function of the target protein, its cellular localization, and processing. To eliminate the use of fluorescent fusion, a technique based on a fluorescent multivalent NTA probe that interacts with histidine tag on recombinant protein has been proposed [29]. However, in our hands we get high background probably due to nonspecific labeling of histidine-rich proteins present in cell lysates.

3 Microscale Fluorescent Thermal Stability Assay (CPM Method)

3.1 Introduction

Membrane protein expression for structural studies is a challenging, time-consuming, and laborious task as a large part of it is taken up by construct optimization for getting stable protein for crystallization and biophysical studies. Traditional methods used to check the thermostability of integral membrane proteins require a large amount of material and are not suited for high-throughput analysis [14–18]. Fluorescence-based methods have become popular as they precisely indicate the stability of protein in question with high sensitivity. Reporter dyes like SYPRO Orange [30] were used earlier and had several shortcomings, one being the high background in presence of detergents typically present in membrane protein purifications. In 2008, Raymond Stevens and group introduced a new protocol using thiol-specific fluorescent dye *N*-(4-[7-diethylamino-4-methyl-3-coumarinyl]phenyl)maleimide, CPM [22], which would be described here. In proteins, cysteines are generally embedded in the interior of protein and hence can be used as a sensor for checking their stability. CPM is a lipophilic dye that forms an adduct with cysteines via thioether bond. This involvement of maleimide group in covalent bond releases the coumarin group from maleimide quenching effect allowing it to fluoresce. This fluorescence increases with increase in temperature as more protein unfolds and cysteines get exposed (Fig. 4a). CPM dye works efficiently in buffers within the pH range of 6–8. Below pH 6, the rate of formation of CPM-thiol adducts decreases, and above pH 8, the chemical selectivity for thiol group is decreased.

3.2 Materials

(1) Purified protein sample: 10–20 μg per assay.

(2) CPM dye: Master stock is made at 4 mg/ml in DMSO, stored in −80°C, and working stock is made by diluting it 40-fold in assay buffer.

(3) Heat blocks or thermocycler.

(4) Cuvettes and spectrofluorometer or 96-well plates and fluorescence plate reader.

3.3 Method

Membrane protein sample is diluted in the appropriate buffer solution containing detergent (optimum pH 6.0–8.0) to a final volume of 200 μl. Five microliters of diluted dye solution

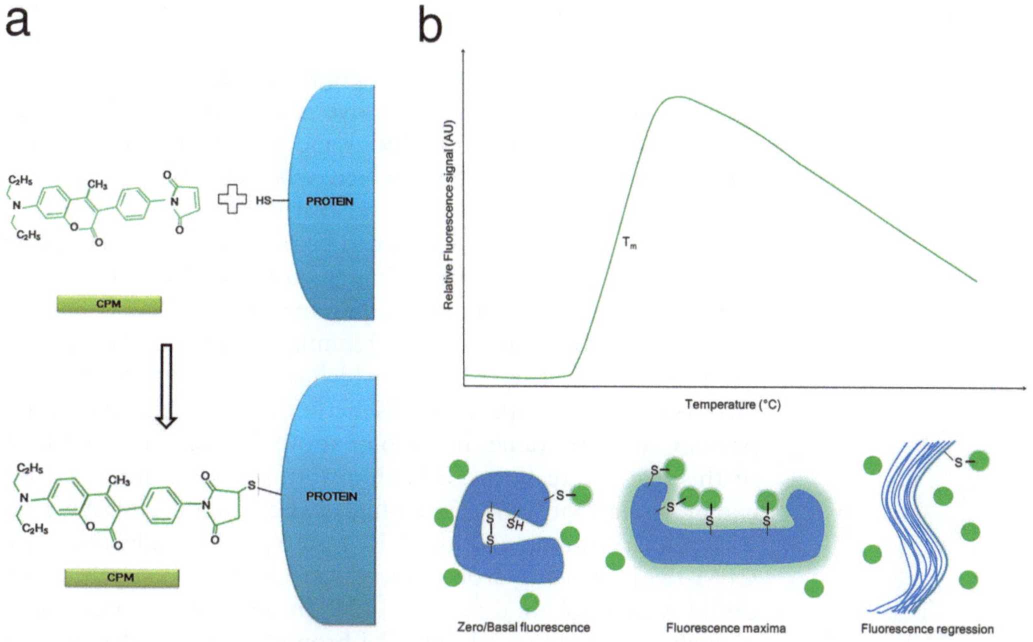

Fig. 4 (**a**) Schematic of reaction between CPM *N*-(4-[7-diethylamino-4-methyl-3-coumarinyl]phenyl)maleimide and protein. The reaction takes place between the maleimide group of the dye and cysteine of protein via a thioether bond. (**b**) Graphic illustration showing various stages of fluorescence-based thermostability profile of a hypothetical membrane protein

(appropriate amount of dye to be added should be determined for each protein) is then added to the protein and incubated for 30 min on ice in dark (to prevent photobleaching). The mixture is then slowly heated from low to higher temperatures (e.g., 25–100°C) in a thermocycler. Fluorescence measurements are taken at regular intervals in a spectrofluorometer (excitation, 387 nm; emission, 463 nm) and are plotted against temperature. Melting temperature (T_m) is obtained by fitting to a Boltzmann sigmoidal function. Simultaneously, buffer-only and dye-only controls should be run to get a higher signal-to-noise ratio. The same method can be adapted for 96-well plates to get a high-throughput data using various parameters like different detergents, ionic strength conditions, buffers, additives, etc.

The dye shows basal level of fluorescence when the protein is intact at lower temperatures as no cysteines or very few are available on the surface for interaction. As the temperature rises, the protein begins to unfold exposing the cysteines present in the interior of protein structure. The dye now interacts with these exposed cysteines, and the fluorescence increases to reach a fluorescence maxima. With higher temperatures, beyond T_m, aggregation of protein and dissociation of dye may lead to a decrease in fluorescence signal (Fig. 4b).

3.4 Notes

Advantages of this method include high sensitivity as it requires small quantities (~10 μg/reaction) of the protein sample, thus making it a method of choice. It gives reproducible results under a defined set of conditions. The dye is compatible with multiple detergents and their concentrations and gives high signal-to-noise ratio for membrane proteins which was not possible by earlier techniques. This method gives relative stability value of protein under various conditions which can be used for construct optimization and crystallization trials. It is an economical as well as a time-saving method with minimum requirements [22].

However, there are few shortcomings of this method as well. To begin with, unlike FSEC-TS which can work with both unpurified lysates and pure protein, this method is suited only for purified protein. Also, pH values below 6 or above 8 hamper the usefulness of this dye (decrease in CPM-thiol formation <6 and decreased chemical selectivity >8). Under high-pH conditions, it can interact with primary amines and give false results; additionally, the rate of CPM-thiol adduct hydrolysis increases, and it can compete significantly with thiol modification [22]. In addition, fluorescence is quenched in presence of salts like bromide, and the dye interacts with lipid bilayers used for protein stability and hence should be used with proper controls [31]. Further, reducing agents like dithiothreitol/β-mercaptoethanol as well as compounds like imidazole (used for purification of His-tagged proteins) react with CPM resulting in higher background signal [31]. Hence, dialysis of protein against a suitable buffer should be carried out before proceeding with this method. Also, in some cases protein-CPM interactions might affect stability of the protein and result in melting point transitions. Proteins containing cysteine residues in disordered regions give a higher fluorescence signal in the lower plateau of melting curve, while absence of cysteine residues in the protein prevents the use of this method, and alternative strategies have to be used like engineering cysteine residues. However, a recent article by Wang et al. suggests that cysteine residue is not essential for CPM thermostability assay, and the dye might be binding to hydrophobic areas exposed upon unfolding of protein [32]. This method cannot differentiate between oligomers and monomers and it is difficult to detect if the oligomer is falling apart. Hence, this protocol is more suited for studying monomeric proteins.

4 Discussion

Structural investigations of membrane proteins usually suffer glitches at each step starting from protein expression in heterologous system, its purification, to obtaining a homogenous and stable protein suitable for crystallization trials or single-particle cryo-EM. Earlier, protein stability and monodispersity were assessed using techniques like CD, DLS, and UV/Vis spectroscopy [33], which required high amounts of proteins as well as are not high throughput.

Although each of these methods have their shortcomings, the two techniques described here have played a chief role in solving the problems in construct optimizations for obtaining stable proteins in solution. FSEC of membrane proteins [19] gives an idea on the solubility and proper folding of the protein in crude cell lysates and has helped in reducing time for construct optimizations as well as other parameters to keep the protein stably in solution post purification. Similarly, CPM method [22] has been helpful in studying many GPCR proteins [34, 35] among others and setting up their successful pre-crystallization trials based on thermostability that has been directly proportional to increase in crystallization success rate. Other newer techniques being developed that have a lot of promise in understanding high-resolution structure and dynamics of membrane proteins are in-cell NMR [36, 37] and in-cell fluorescence [38] that even enable investigating the protein in its natural surroundings and are techniques to watch out for in the future.

Acknowledgments

The authors acknowledge financial assistance from the University Grants Commission (UGC), New Delhi (J.K.), and the Department of Biotechnology, New Delhi. This work was supported by the Wellcome Trust/DBT India Alliance intermediate fellowship to Janesh Kumar.

References

1. Krogh A, Larsson B, von Heijne G, Sonnhammer EL (2001) Predicting transmembrane protein topology with a hidden Markov model: application to complete genomes. J Mol Biol 305:567–580. doi:10.1006/jmbi.2000.4315
2. Overington JP, Al-Lazikani B, Hopkins AL (2006) How many drug targets are there? Nat Rev Drug Discov 5:993–996. doi:10.1038/nrd2199
3. Carpenter EP, Beis K, Cameron AD, Iwata S (2008) Overcoming the challenges of membrane protein crystallography. Curr Opin Struct Biol 18:581–586. doi:10.1016/j.sbi.2008.07.001
4. Cha HJ, Dalal NG, Bentley WE (2004) In vivo monitoring of intracellular expression of human interleukin-2 using green fluorescent protein fusion partner in Pichia pastoris. Biotechnol Lett 26:1157–1162. doi:10.1023/B:BILE.0000035489.09417.d4
5. Geertsma ER, Groeneveld M, Slotboom D-J, Poolman B (2008) Quality control of overexpressed membrane proteins. Proc Natl Acad Sci U S A 105:5722–5727. doi:10.1073/pnas.0802190105
6. Pédelacq J-D, Piltch E, Liong EC et al (2002) Engineering soluble proteins for structural genomics. Nat Biotechnol 20:927–932. doi:10.1038/nbt732
7. Crameri A, Whitehorn EA, Tate E, Stemmer WP (1996) Improved green fluorescent protein by molecular evolution using DNA shuffling. Nat Biotechnol 14:315–319. doi:10.1038/nbt0396-315
8. Drew DE, von Heijne G, Nordlund P, de Gier JW (2001) Green fluorescent protein as an indicator to monitor membrane protein overexpression in *Escherichia coli*. FEBS Lett 507:220–224
9. Hammon J, Palanivelu DV, Chen J et al (2009) A green fluorescent protein screen for

identification of well-expressed membrane proteins from a cohort of extremophilic organisms. Protein Sci Publ Protein Soc 18:121–133. doi:10.1002/pro.18

10. Drew D, Slotboom D-J, Friso G et al (2005) A scalable, GFP-based pipeline for membrane protein overexpression screening and purification. Protein Sci Publ Protein Soc 14:2011–2017. doi:10.1110/ps.051466205
11. Drew D, Newstead S, Sonoda Y, Kim H, von Heijne G, Iwata S (2008) GFP-based optimization scheme for the overexpression and purification of eukaryotic membrane proteins in *Saccharomyces cerevisiae*. Nat Protoc 3(5):784–798. doi:10.1038/nprot.2008.44
12. Hu N-J, Rada H, Rahman N et al (2015) GFP-based expression screening of membrane proteins in insect cells using the baculovirus system. Methods Mol Biol Clifton NJ 1261:197–209. doi:10.1007/978-1-4939-2230-7_11
13. Goehring A, Lee C-H, Wang KH et al (2014) Screening and large-scale expression of membrane proteins in mammalian cells for structural studies. Nat Protoc 9:2574–2585. doi:10.1038/nprot.2014.173
14. Galka JJ, Baturin SJ, Manley DM et al (2008) Stability of the glycerol facilitator in detergent solutions. Biochemistry (Mosc) 47:3513–3524. doi:10.1021/bi7021409
15. Lau FW, Bowie JU (1997) A method for assessing the stability of a membrane protein. Biochemistry (Mosc) 36:5884–5892. doi:10.1021/bi963095j
16. Grinberg AV, Gevondyan NM, Grinberg NV, Grinberg VY (2001) The thermal unfolding and domain structure of Na+/K+-exchanging ATPase. A scanning calorimetry study. Eur J Biochem FEBS 268:5027–5036
17. Maneri LR, Low PS (1988) Structural stability of the erythrocyte anion transporter, band 3, in different lipid environments. A differential scanning calorimetric study. J Biol Chem 263:16170–16178
18. Hughes AV, Rees P, Heathcote P, Jones MR (2006) Kinetic analysis of the thermal stability of the photosynthetic reaction center from *Rhodobacter sphaeroides*. Biophys J 90:4155–4166. doi:10.1529/biophysj.105.070029
19. Kawate T, Gouaux E (2006) Fluorescence-detection size-exclusion chromatography for precrystallization screening of integral membrane proteins. Struct Lond Engl 1993 14:673–681. doi:10.1016/j.str.2006.01.013
20. Parcej D, Guntrum R, Schmidt S et al (2013) Multicolour fluorescence-detection size-exclusion chromatography for structural genomics of membrane multiprotein complexes. PLoS One 8, e67112. doi:10.1371/journal.pone.0067112
21. Hattori M, Hibbs RE, Gouaux E (2012) A fluorescence-detection size-exclusion chromatography-based thermostability assay for membrane protein precrystallization screening. Struct Lond Engl 1993 20:1293–1299. doi:10.1016/j.str.2012.06.009
22. Alexandrov AI, Mileni M, Chien EYT et al (2008) Microscale fluorescent thermal stability assay for membrane proteins. Struct Lond Engl 1993 16:351–359. doi:10.1016/j.str.2008.02.004
23. Daley DO, Rapp M, Granseth E et al (2005) Global topology analysis of the *Escherichia coli* inner membrane proteome. Science 308:1321–1323. doi:10.1126/science.1109730
24. Drew D, Sjöstrand D, Nilsson J et al (2002) Rapid topology mapping of *Escherichia coli* inner-membrane proteins by prediction and PhoA/GFP fusion analysis. Proc Natl Acad Sci U S A 99:2690–2695. doi:10.1073/pnas.052018199
25. Drew D, Lerch M, Kunji E et al (2006) Optimization of membrane protein overexpression and purification using GFP fusions. Nat Methods 3:303–313. doi:10.1038/nmeth0406-303
26. Trowitzsch S, Tampé R (2015) Multicolor fluorescence-based screening toward structural analysis of multiprotein membrane complexes. Methods Enzymol 557:3–26. doi:10.1016/bs.mie.2014.11.043
27. Hsieh JM, Besserer GM, Madej MG et al (2010) Bridging the gap: a GFP-based strategy for overexpression and purification of membrane proteins with intra and extracellular C-termini. Protein Sci Publ Protein Soc 19:868–880. doi:10.1002/pro.365
28. Shaffer PL, Goehring A, Shankaranarayanan A, Gouaux E (2009) Structure and mechanism of a Na+independent amino acid transporter. Science 325:1010–1014. doi:10.1126/science.1176088
29. Backmark AE, Olivier N, Snijder A et al (2013) Fluorescent probe for high-throughput screening of membrane protein expression. Protein Sci Publ Protein Soc 22:1124–1132. doi:10.1002/pro.2297
30. Yeh AP, McMillan A, Stowell MHB (2006) Rapid and simple protein-stability screens: application to membrane proteins. Acta Crystallogr D Biol Crystallogr 62:451–457. doi:10.1107/S0907444906005233

31. Liu W, Hanson MA, Stevens RC, Cherezov V (2010) LCP-Tm: an assay to measure and understand stability of membrane proteins in a membrane environment. Biophys J 98:1539–1548. doi:10.1016/j.bpj.2009.12.4296
32. Wang Z, Ye C, Zhang X, Wei Y (2015) Cysteine residue is not essential for CPM protein thermal-stability assay. Anal Bioanal Chem 407:3683–3691. doi:10.1007/s00216-015-8587-4
33. Goñi FM, Alonso A (2000) Spectroscopic techniques in the study of membrane solubilization, reconstitution and permeabilization by detergents. Biochim Biophys Acta BBA - Biomembr 1508:51–68. doi:10.1016/S0304-4157(00)00011-3
34. Chun E, Thompson AA, Liu W et al (2012) Fusion partner toolchest for the stabilization and crystallization of G protein-coupled receptors. Struct Lond Engl 1993 20:967–976. doi:10.1016/j.str.2012.04.010
35. Xu F, Wu H, Katritch V et al (2011) Structure of an agonist-bound human A2A adenosine receptor. Science 332:322–327. doi:10.1126/science.1202793
36. Li C, Liu M (2013) Protein dynamics in living cells studied by in-cell NMR spectroscopy. FEBS Lett 587:1008–1011. doi:10.1016/j.febslet.2012.12.023
37. Selenko P, Wagner G (2007) Looking into live cells with in-cell NMR spectroscopy. J Struct Biol 158:244–253. doi:10.1016/j.jsb.2007.04.001
38. Borra R, Dong D, Elnagar AY et al (2012) In-cell fluorescence activation and labeling of proteins mediated by FRET-quenched split inteins. J Am Chem Soc 134:6344–6353. doi:10.1021/ja300209u

Single-Molecule Localization Super-Resolution Microscopy of Synaptic Proteins

Francisco J. Barrantes

Abstract

Recent years have witnessed huge progress in the field of light microscopy with the development and implementation of new approaches leading to dramatic improvements in the spatial and temporal resolution of this form of imaging, most particularly in its biological applications. The limitations in spatial resolution imposed by the diffraction of light have been circumvented by resorting to different strategies, which are briefly outlined in the Introduction. These protocols are intended to provide practical guidelines for the imaging of synaptic proteins using one such strategy, namely, single-molecule stochastic localization super-resolution microscopy.

The protocols use neuronal cells from the hippocampus of rodent embryos as the experimental paradigm and outline the steps for obtaining dissociated neurons and establishing primary cultures for in vitro studies. The techniques can be adapted to the culture of neurons from other brain regions. Procedures for handling fixed and live specimens are described, as well as the use of extrinsic fluorescent probes and fluorescent proteins, mounting media, examples of hardware configurations, software for image analysis, and some hints for the implementation of minimalist approaches to single-molecule localization nanoscopy.

Keywords Nanoscopy, Neuronal cell culture, Sample preparation, Single-molecule imaging, Staining, Super-resolution microscopy

1 Introduction

Optical ("light," "wide-field," "far-field") microscopy is undoubtedly the most popular technique for imaging cells and tissues, and the successful combinations of appropriate probes (organic dyes, fluorescent proteins, and inorganic nanoparticles, e.g., quantum dots), new light sources (solid-state lasers, light-emitting diodes), and improved detectors (CCD cameras, avalanche photodiodes) have made fluorescence microscopy the method of choice, essentially because of the unparalleled selectivity and sensitivity achieved in biological applications, pervading practically all realms of biology. Moreover, recent years have witnessed a revolution in fluorescence microscopy: using conventional lenses and visible light it has been possible to circumvent the century-old limit dictated by the

Arun K. Shukla (ed.), *Chemical and Synthetic Approaches in Membrane Biology*, Springer Protocols Handbooks, (2017) 157–198, DOI 10.1007/8623_2016_10, © Springer Science+Business Media LLC 2016, Published online: 28 October 2016 diffraction barrier, a fundamental law formulated by the German physicist Ernst Abbe in 1873. This limitation in the resolution of optical imaging instruments imposed by the diffraction of light was solved by independent and complementary methodologies which were acknowledged by the 2014 Nobel Prize in Chemistry, jointly awarded to Stefan W. Hell, William E. Moerner, and Eric Betzig. These advances were materialized in the techniques of stimulated emission depletion (STED) microscopy, conceived [1, 2] and implemented [3] by Stefan Hell and his group in Germany, and single-molecule localization microscopies, the methodologies developed by Betzig, Moerner, and Zhuang and their groups in the USA [4–11].

A third approach is structured illumination microscopy (SIM), which excites the sample with a series of spatially structured patterns in a wide-field configuration and resolves otherwise inaccessible high-resolution information in the observed image in the form of Moiré fringes. Multiple images of the emitted fluorescence are recorded in the high-frequency domain at lower spatial frequencies and computationally separated to extract low- and high-frequency data. This results in about a twofold gain in resolution in the x-, y-, and z-axes [12, 13]. A main feature of SIM is that unlike a confocal microscope, it utilizes full amplitude of the emitted fluorescence, thus producing brighter images which require relatively shorter times to acquire. It is a relatively fast super-resolution technique, currently available commercially (Zeiss Elyra, Nikon N-SIM, GE Healthcare DeltaVision).

In the STED "deterministic" point-by-point targeted nanoscopy mode, the physical tools employed to interrogate fluorescent molecules are (1) visible light and (2) regular lenses, much like the manner used in a conventional diffraction-limited wide-field laser scanning microscope. It is interesting to note that the crucial improvement in spatial resolution afforded by STED microscopy also relies on a purely physical effect: reducing the effective volume from which fluorescence emission is collected. How is this accomplished? In simple terms, the specimen is scanned with two concentrically focused laser beams: the central, interrogating laser beam is surrounded by a red-shifted doughnut-shaped laser beam depleting the emission of fluorescent molecules at the periphery of the former. An effective point-spread function (PSF) is created in which fluorescence is confined to the immediate vicinity of the central intensity zero; the volume of this region is no longer limited by diffraction. The resolution of the STED microscope can be adjusted by varying the intensity of the depletion peripheral laser beam.

The concept of targeted deterministic switching in the vicinity of an intensity zero originally resulting in the development of STED microscopy has been generalized to include any reversible and saturable optical transition in the fluorescence mode (e.g., in the so-called reversible saturable optical linear fluorescence transition ("RESOLFT") microscopy technique [14, 15]), but interestingly, this concept is not limited to fluorescence: any other on–off spectroscopic transition in the state of a molecule (e.g., absorption) can also be exploited; the “off” state need not be dark [3, 16]. Commercial versions of STED microscopes are available (Leica TCS STED, Abberior Instruments in various models ranging from the one-color pulsed STED, the easy3D STED, and RESCue 3D STED to the correlative STED nanoscope combined with atomic force microscopy module). In the case of the nervous system, STED has been successfully applied to the study of synaptic vesicles, dendritic spines, and other subcellular aspects of the synapse (reviewed in [17–19]). It was also the technique employed for the first imaging of supramolecular aggregates of a neurotransmitter receptor [20], live-cell imaging of dendritic spines [21], actin dynamics in synapses in brain slices [22], and most dramatically, the imaging of dendritic spines in the intact brain of a living mouse [23], among other applications. A new technique, known as mirror-enhanced, axial narrowing, super-resolution (MEANS) microscopy, is a spin-off of STED nanoscopy in which cells to be studied are grown on custom-made tiny mirrors instead of transparent glass coverslips [24]. The MEANS technique is claimed to improve axial resolution sixfold and lateral resolution twofold.

In contrast, in single-molecule methods such as photoactivated localization microscopy (PALM) [4, 25] and stochastic optical reconstruction microscopy (STORM) [8], super resolution is accomplished by cumulative spatial localization of fluorescent molecules resulting from sequential imaging of sparse, stochastic subsets of fluorescence emitters, i.e., by interrogating the *on*-state ensemble molecule by molecule and recording these individual emitters on a grid detector, e.g., a CCD camera. Thousands of frames are imaged in a stream of individual pictures that include only a few, preferably well-separated molecules, and the coordinates of the *on*-state individual molecules are subsequently established by off-line localization analysis. In other words, in single-molecule localization nanoscopy, the ***temporal*** separation enables the ***spatial*** distinction of individual fluorescence emitters. The difference between the two aspects of the same principle is that in PALM, imaging is performed on bioengineered fluorescent proteins, whereas in STORM much smaller molecules, i.e., organic fluorescent dyes, are interrogated [26–33]. The basic principles behind the stochastic nanoscopy techniques are the same for STORM and PALM: they rely on the fact that a single fluorescent source (e.g., a fluorescent molecule) can be localized with nanometer precision by fitting an appropriate function (e.g., a two-dimensional Gaussian function) to determine the center of mass – the centroid – of the diffraction-limited, blurred wide-field image formed by the compound optical microscope. The super-resolved final image is a reconstructed image of the centroids of all validated fluorescent molecules in the individual images. The precision with which the centroid can be determined depends on the number of photons collected; in practice, this can be a few tens of nanometers or better.

STORM and PALM techniques currently provide the best spatial resolution in nanoscopy but have relatively less time resolution than parallelized scanning in the RESOLFT mode, which involves the use of many thousands of doughnuts [34]. STORM was one of the first applications of super-resolution techniques for imaging proteins in brain synapses with nanometric precision [35].

As with the other modalities of super-resolution microscopy, commercial firms have released various microscopes suitable for single-molecule stochastic localization nanoscopy: N-STORM from Nikon, GSD from Leica, Zeiss Elyra P1 from Zeiss, or Vutara 350 from Bruker. I should like to add that building a STED or a RESOLFT nanoscope requires considerable skill and is relatively expensive because of the type of lasers needed; constructing a SIM is not trivial, but building a PALM or STORM instrument on the basis of an existing commercial epifluorescence microscope is an accessible project for the biologist wishing to improve the resolution of the instrument beyond the diffraction limit.

In addition to the stochastic photoswitching single-molecule localization techniques like PALM, STORM, and GSDIM, the method termed “point accumulation imaging in the nanoscale topography” (PAINT) is based on the binding of a fluorescent-tagged ligand to the protein or other molecule to be identified [36–38]. Free ligands do not contribute to the image because of the combined effects of oblique illumination and their very fast diffusion in 3-D, resulting in their inefficient excitation. The PAINT technique was initially applied to the study of lipid bilayers and subsequently evolved to imaging proteins in living cells, giving rise to the “universal PAINT” (uPAINT) improved method [39]. This technique has found very interesting applications in the study of the dynamics of the synapse [40–42].

Here I would like to share various protocols adopted (and many of them adapted to particular applications; very few are developed by ourselves) in our laboratory for the imaging of synaptic proteins in primary cultures of neuronal cells. These protocols are by no means intended to cover in a comprehensive manner the current state of affairs in the field. Only one super-resolution modality is discussed; approaches based on quantum-dot technology will not be dealt with. The protocols are intended to provide useful hints and tips and refer the reader to the original sources in the use of one particular super-resolution technique, namely, single-molecule stochastic nanoscopy, as applied to the study of proteins in a specialized substructure of the neuron, the synapse. Understanding synapse function will require detailed knowledge of the spatiotemporal organization of its constituent molecules, and neurotransmitter receptors, transporters, enzymes, and scaffolding proteins alike – outnumbered by a considerable variety of lipid species – are major actors in the synaptic scenario.

A substantial section of the protocols is devoted to the preparation of dissociated neuronal cell cultures. As any model system, this preparation has its pros and cons. It is a compromise between the complexity of the 3-D architecture of the brain and the apparent simplicity of isolated molecules. The 2-D array of living neuronal cells adhered to a thin cover glass offers the possibility of addressing, with unprecedented time and space resolution, rather sophisticated questions on the changes occurring in the synapse at the molecular level. Physiological and pathological conditions can be met at the Petri dish and subjected to pharmacological intervention under controlled conditions. A step further in nanoscopy of synaptic structures in a living animal was taken by the STED imaging of dendritic spines in an intact live mouse brain [23]. In parallel, another leading-edge technology, optogenetics [43, 44] – which shares with nanoscopy the temporally and spatially resolved use of light as its major tool – is increasingly delivering information on the function of neuronal networks, including complex circuit behavior in the whole animal. It will not be long before the super-resolution imaging techniques are combined with optogenetic techniques, in real time, to correlate vis-à-vis structure and function at cellular, subcellular, and multicellular (neuronal network) levels in a living organism.

2 Materials

2.1 For Neuronal Primary Cultures

Pregnant rodent carrying embryonic stage 18 (E18) pups (preferably rat for hippocampal neuronal cultures) or E16 from mouse (more difficult to dissect, for mixed neuronal–glial cultures)

Distilled, deionized, sterile, filtered water

Methanol, spectroscopic grade

Ethanol, spectroscopic grade

Binocular stereoscopic microscope for the dissection of brains and hippocampi

Cell culture incubator at 35.5°C, preferably triple-gas, with automatic humidity control and automatic self-cleaning cycles

CO_2 cylinder

Water bath, 37°C

Compressed butane or natural gas burner

Chemical fume hood for preparing glass coverslip cleaning solution and undertaking cleaning

Corrosion-resistant ceramic staining rack for holding coverslips (Thomas Scientific, cat. no. 8542E40)

Hemocytometer for counting cells

Tabletop centrifuge

Freezing vials (Nunc CryoTubes 352350)

Sterile plasticware:

Large-volume filter units (Stericup-GP filter, Millipore SCGPU02RE)

Small-volume filter units for syringes (Steriflip-GP, Millipore SCGP00525)

5-, 10-, and 25-mL serological pipettes

60-mm tissue culture dishes

25-mL tissue culture flasks (Primaria, BD Biosciences 353824)

15- and 50-mL conical centrifuge tubes

Trypan blue, 0.4% solution

Sodium dodecyl sulfate, 0.1% solution

10 mg/mL DNase in CMF-HBSS (Roche Applied Science 10104159)

Trypsin (0.5%)/0.2% EDTA (Sigma-Aldrich 59418C; Invitrogen 1525300-054)

Alternatively, trypsin (2.5% wt/v) (Invitrogen 15090-046) stock kept at −20°C

Trypsin inhibitor

Polylysine (30,000–70,000 MW, Sigma-Aldrich P2636 or P7405), 1 mg/mL solution in borate buffer

Borate buffer (boric acid + tetraborate) 0.1 M, pH 8.5

Sorensen's phosphate buffer (133 mM)

0.133 M Na_2HPO_4

0.133 M KH_2PO_4

Mix 71.5 mL of Na_2HPO_4 and 28.5 mL of KH_2PO_4 to obtain pH 7.2

CMF-HBSS medium ($Ca2^+$, Mg^{2+}, and bicarbonate-free Hank's balanced salt solution [BSS] buffered with 10 mM HEPES, pH 7.3)

10× Hanks' BSS (Invitrogen 14185-052)

1 M HEPES buffer, pH 7.3 (Invitrogen 15630-080)

Hanks' Plus (HBSS+) medium with antibiotic

HBSS (without Ca^{2+} and Mg^{2+}).

HEPES (Sigma-Aldrich) 10 mM.

Glucose 33.3 mM.

Gentamicin (Sigma-Aldrich), 5 μg/mL.

Store up to 1 month at 4°C.

Neuronal dissociation medium

Hanks' balanced salt solution plus (HBSS, GIBCO BRL 24020-091)

0.3% (w/v) BSA (Sigma-Aldrich or Thermo Scientific)

0.25% trypsin (Gibco, Invitrogen Life Science, Carlsbad; Sigma T-9253)

Alternatively, trypsin/EDTA premixed solution (Biochrom L2143)

Trypsin inhibitor solution

25 mg albumin (bovine: Sigma A-4503)

25 mg trypsin inhibitor (Sigma T-9253)

10 mL 10% FCS medium (5% FCS medium), warmed to 37°C

Plating medium

MEM supplemented with glucose (0.6% wt/vol) and containing 10% (vol/vol) horse serum or 5% (vol/vol) fetal bovine serum:

D-glucose (Sigma-Aldrich G8769)

MEM with Earle's salts and L-glutamine (Invitrogen Life Science 11095-080)

5% (v/v) bovine fetal serum (Invitrogen 16000) or 10% horse serum (Thermo Scientific)

Neuronal maintenance N2 medium [45]

MEM containing the N2 supplement. It is prepared by combining nine parts MEM supplemented with:

Glucose (0.6% wt/vol) and one part 10× N2 supplement

10× N2 supplement contains the following ingredients prepared in MEM:

10 mM sodium pyruvate (Sigma P2256)

1 mM putrescine (Sigma P5780)

0.2 mM progesterone (Sigma P8783)

0.3 mM selenium dioxide (Sigma 200107)

1 mg/mL bovine transferrin (Sigma T1428)

50 mg/mL insulin (Sigma I5500)

Alternatively,
Neurobasal medium for hippocampal neuronal culture [45]

Neurobasal medium (GIBCO BRL 10888-022; Invitrogen 21103-049) supplemented with:

2% (v/v) B27 serum-free supplement (Invitrogen 17504-044; GIBCO BRL 17504-044)

GlutaMAX-I supplement (Invitrogen 35050-061)

100 U/mL penicillin

100 μg/mL streptomycin

1 mg/mL ovalbumin

Microsurgical tools (dry-heat sterilize in a metallic instrument sterilization container with lid)

10-cm diameter glass Petri dishes to dissect embryos

Sharp-ended scissors for caesarean section of adult rodent

100-mm straight Graefe tissue forceps with teeth for caesarean section

No. 2 Dumont tweezers 0.3 × 0.13 tip (Agar Sci. or Electron Microscopy Sciences)

No. 5 Dumont 45° tip forceps (F.S.T., Fine Science Tools)

Ophthalmological straight scissors (e.g., McPherson-Vannas scissors (Asico)) or 105-mm Castroviejo corneal scissors (Katalyst Surgical)

Rounded-end micro-spatula

Stainless steel instrument sterilization container with lid (MedicalExpo)

15-mL tubes

Pasteur pipettes, sterile (glass or disposable plastic ones)

Glass coverslips and material for cleaning them

Carefully select coverslips one by one, discarding those with scratches or digs (which generate diffraction and stray light), and submit them to the cleaning procedure (to avoid dirt and/or fluorescent background):

Glass coverslips (e.g., 18-mm round No. 1.5 coverslips, Warner Instruments CS18-R CS-18R15)

Toxic gas fume hood

N_2 cylinder or central air/air compressor system with filter

Magnetic/hot stir plate and magnetic bar

Coverslip-staining jars with lid

Ceramic coverslip-staining rack (e.g., Thomas Scientific 8542E40)

12-well cell culture plates

Metal plastic-coated tweezers or plastic tongs

250-mL tall beaker

250-mL flat bottom wide beaker ("crystallization-type")

Plastic squirting bottle with methanol

Millipore-filtered, deionized, and distilled water

Ammonium hydroxide

Hydrogen peroxide

2.2 Plasmids and Transfection Material

As its name indicates, PALM localization super-resolution microscopy is based on the use of photoactivatable fluorescent proteins (see review in [46]). If this technique is to be employed for the study of synaptic proteins, a variety of probes are already available. Several commercial firms have a wide selection of such plasmids covering the visible excitation spectrum (e.g., Addgene, https://www.addgene.org/fluorescent-proteins/ containing 3,300 plasmids from Michael Davidson's collection, Florida State University; BioVision, http://www.biovision.com/) and specific target organelles, cytoskeletal proteins, or subcellular structures (Golgi apparatus, mitochondria, centrioles, actin, F-actin, centrosomes, lysosomes, etc.).

Lipid carrier transfection techniques applied to primary neuronal cell cultures have relatively low efficiency. On average, only 10–20 neurons can be identified in a coverslip. For most purposes, this is sufficient to image the relevant areas in super-resolution mode.

Viral vectors, and in particular lentivirus and adenovirus, yield higher efficiencies. Techniques for constructing, packing, and delivering adeno-associated viral vectors and lentivirus are outside the scope of these protocols. The reader is referred to Ref. [47] for a review on the application of these techniques in primary neuronal cultures with 85–90% efficiency.

Self-labeling of proteins using Halo tags, SNAP, dihydrofolate reductase (DHFR) targeting with trimethoprim conjugates, or CLIP tags is a recent alternative in super-resolution studies (e.g., [48]). A variety of fluorescent dyes can be used in single-molecule super-resolution microscopy utilizing the SNAP-tag labeling system, as successfully reported for live-cell STORM nanoscopy of histone H2B proteins using rhodamine and green and tetramethylrhodamine dyes [49]. Dy549 and CF640 have been recommended as the best choices for single-molecule tracking studies. The possibility of dual-color single-molecule imaging of SNAP-tagged fusion proteins is also analyzed in a recent study [50].

The conventional fluorescent protein heterologous expression requires a few components:

1. Plasmid expressing the protein of interest engineered in a construct with the photoconvertible fluorescent protein (e.g., BFP, CFP, GFP, EGFP, YFP, mCherry, mTurquoise2, tdEos,

tdTomato, mRuby, mEmerald, DRONPA, LifeAct, mTurquoise2)

2. Opti-MEM I reduced serum medium (Invitrogen)
3. Lipofectamine 2000 reagent (Invitrogen 11668) for transfection with the plasmid of choice

2.3 Extrinsic Fluorescent Probes

Appropriate extrinsic fluorescent probes and fluorescent-labeled antibodies for tagging the protein(s) of interest when using STORM/GSDIM techniques are also available from various sources. Small organic dyes have much higher photon outputs than fluorescent proteins (e.g., 6,000 photons emitted by a Cy5 dye in contrast to 490 photons detected from a mEos protein [51]). An important advancement in this topic was the appreciation that conventional organic fluorophores could be used for super-resolution imaging in cell biology [27, 30].

1. ATTO-TEC GmbH, Siegen, Germany (http://www.atto-tec.com/), has a collection of fluorescent probes covering the 350–750-nm excitation range and fluorescent-labeled lipids (DPPE, DOPE, PPE, DMPE) conjugated to ATTO dyes, which exhibit high quantum yield and photostability for localization super-resolution microscopy, some of them with a large Stokes shift, as well as "quenching labels," designed as energy transfer acceptors in experiments using Förster's resonance energy transfer (FRET). The ATTO dyes for the red region of the excitation spectrum are particularly advantageous in localization nanoscopy (e.g., ATTO-647N exhibits fluorescence emission intensity more than twice the strength of Cy5).
2. The ATTO dyes are also commercialized by various other sources (Sigma-Aldrich, Alomone Labs, Abberior, Invitrogen Life Sciences, Miltenyi Biotec).
3. Synaptic Systems (https://www.sysy.com/) has a large collection of purified antibodies against a wide variety of synaptic proteins (Ca^{2+} binding proteins, neurotransmitter transporters, neurotransmitter receptors, SNARE proteins, synaptic adhesion proteins, glial proteins, snRNPs, postsynaptic proteins like gephyrin, Homer, PSDs).
4. Alomone Labs (http://www.alomone.com/) specializes in fluorescent-labeled antibodies against a wide selection of synaptic proteins, in particular ion channels and neurotransmitter receptors/transporters, GPCRs, pumps, and proteins involved in Ca^{2+} homeostasis.
5. Secondary antibodies labeled with Alexa Fluor or ATTO fluorophores, from the same sources quoted in **steps 1–4** above. Miltenyi Biotec provides a custom service to derivatize antibodies with ATTO STAR dyes from Abberior.

2.4 Solutions for Fixation of Cultured Neurons

1. Phosphate buffer saline (PBS, Sörensen buffer), pH 7.3–7.4.
2. Fixative solution: 4% paraformaldehyde (PFA) (e.g., USB Corp. 19943) in PBS, pH 7.4, containing 4% sucrose.
3. Blocking solution: 10% normal serum containing 0.1% Triton X-100/PBS (normal serum from the host species of the secondary antibodies is recommended for blocking).
4. Incubation solution: 3% serum/0.1% Triton X-100/PBS (Triton may be omitted; normal serum from the host species of the secondary antibodies is recommended for blocking).

2.5 Mounting Media

2.5.1 Fixed Specimens

1. Polyvinyl alcohol (PVA) (MW 25,000) (9002-89-5, www.polysciences.com), a low-toxicity resin soluble in water at 96°C.
2. The commercial mounting medium VECTASHIELD (H-1000, Vectorlab) has been recommended for specimens labeled with Alexa Fluor^{647} (Life Technologies), especially for 3-D imaging in STORM [52]. Olivier and coworkers also used a combination of VECTASHIELD and 95% glycerol-50 mM Tris buffer, pH 8.
3. The commercial mounting medium ProLong® Gold Antifade Mountant (Thermo Fisher Sci. P10144) is a curing medium with a refractive index of 1.47 which allows long-term storage of the sample. It is available with or without DAPI nuclear stain. It is not recommended for imaging fluorescent proteins, in which case ProLong® Diamond is suggested by the manufacturers. The use of the former mounting medium has been reported in STED intracellular imaging of pneumococcal autolysin LytA-infected cells [53], but to my knowledge it has not been used in imaging of the synapse.

2.5.2 Live Cells

There have been important developments on this front. Some were inherited from FRET know-how, some newly introduced to adapt to the demands of super-resolution microscopy, especially in relation to the control of photobleaching transitions, blinking frequency, and ultimately on–off transitions. In STORM-type nanoscopy, initial studies were performed with fluorescent probes capable of undergoing photoswitching transitions which arrested them in the off state. This was the case with cyanine dyes which can be induced to undergo transitions in the presence of a second fluorophore [8]. An important development in the field was the recognition that conventional fluorophores could also undergo photoswitching [27, 28, 30], leading to the development of a simplified and more universal single-molecule imaging procedure, direct STORM (dSTORM) [27]. This called for appropriate buffer systems [26]. There are excellent reviews on the use and pros and cons of different buffers and reducing–oxidizing additives [51]. In all cases: filter through a 0.22-μm filter to reduce fluorescent contaminations. Some media can be stored at –20°C, but those containing oxidative/reducing enzyme are better prepared fresh. The pH of the STORM buffers is critical and should be carefully controlled.

Standard buffer I

120 mM NaCl

3 mM KCl

2 mM $CaCl_2$

2 mM $MgCl_2$

10 mM glucose (Sigma-Aldrich)

10 mM HEPES (Sigma-Aldrich)

Adjust pH to 7.35

This imaging buffer can be stored up to 1 month at 4°C.

STORM buffer I, an *oxygen*-scavenging buffer on the basis of a PBS buffer, pH 7.4, containing:

5% glucose

10 U mL^{-1} glucose oxidase

200 U mL^{-1} catalase

100 MEA

Buffer recommended for Alexa Fluor[647]

Same buffer, containing:

5% glucose

4 U mL^{-1} glucose oxidase

80 U mL^{-1} catalase

100 mM glutathione

Oxygen-scavenging medium for improving dye stability in single-molecule experiments [54]

Protocatechuic acid (PCA) (Sigma-Aldrich).

Protocatechuate 3,4-dioxygenase (PCD) (Sigma-Aldrich).

Trolox (97%).

COT (98%).

NBA (99%) (Sigma-Aldrich).

Resuspend in DMSO and add to Tris-polymix buffer to a final concentration of 2 mM.

Simple PBS containing 10 mM β-mercaptoethylamine (MEA) adjusted to pH 7.4

This oxygen-scavenging buffer should be prepared fresh before use. Dissolve the MEA (cysteamine, Sigma-Aldrich 30070) in PBS and adjust to pII 7.4. MEA concentration can bc varicd between 10 and 100 mM.

3 Super-Resolution Microscopy Setup

1. Low-profile quick-change chamber (e.g., Warner Instruments model W4 64-0368 RC-41LP for 18-mm round coverslips); other models exist for 12-, 15-, and 25-mm round coverslips. ("Low-profile" design is intended to facilitate low entry angle for microelectrodes in electrophysiological recordings.)
2. Alternatively, plastic sample chambers having the size of a microscope slide with glass coverslip bottoms (Labtek, Nalge Nunc International Corp) can be used. They come in different sizes; we find the eight-chamber model (Labtek Prod. No. 155411) to be the most convenient. We prefer the quick chamber (alternative 1 above), because the flat chambers allow perfusion and rapid change of medium of the cells; they are apt for electrophysiological recording, are easier to keep at a given temperature, and can be conveniently cleaned and thus used indefinitely.
3. Vibration isolation system connected to a nitrogen gas tank or air compressor (we use the 780 Performance Series CleanTop Optical tabletop model 781-432-02R mounted on a Micro-G 4-Post Gimbal Piston isolator system with safety tie bars from TMC vibrational control).
4. Inverted fluorescence microscope (we use a Nikon TE-2000 stand with multiple optical bench/optical/laser additional components).
5. Objective lens(es). Minimally, one high-quality objective lens. TIRF plan-apochromatic oil immersion objective 60×/1.49 N.A. or 100×/1.49 N.A. will produce excellent results. Oil immersion objectives provide higher N.A., but water-immersion objectives, with lower N.A., reduce spherical aberrations resulting from refractive index mismatches of living cells in aqueous media.
6. Stage. Having tried various sophisticated (and particularly expensive) piezoelectric-driven stages as replacement for the original *X,Y* stage of the epifluorescence microscope, we settled for a purely mechanical, highly stable stage driven by very sturdy micrometer screws (Physik Instrumente GmbH, Karlsruhe, Germany, model MN-545.2MN). This is an important element for minimizing drift.

7. Epifluorescence light source (we use a LED system with four different excitation wavelengths, from the firm Tolket, Argentina) for initial wide-field microscopy. Of course a conventional Hg arc source suffices.
8. Activation laser. A 405-nm UV laser, or a powerful LED or Hg lamp source, for activation in case of PALM super-resolution microscopy (in our microscope setup, the laser excitation sources are via the TIRF port simultaneously with the LED or Hg arc excitation from wide-field illumination port).
9. A 532-nm DPSS laser (our main excitation source for red emission STORM super-resolution microscopy, 300 mW). Excitation source for various rhodamine-based fluorochromes, tetramethyl-rhodamine, Cy3, Alexa Fluor^{532}, Alexa Fluor^{546}, Alexa Fluor^{555}, Alexa Fluor^{568}, Texas Red, HcRed, Cherry and related fluorescent proteins, etc. There are various commercial sources for this type of DPSS laser: Coherent, RGB, and Quantum Laser.
10. A 561-nm diode-pumped solid-state (DPSS) laser (alternative excitation source for tdEos fluorescent protein, Ds Red fluorescent protein, Alexa Fluor^{568}, phycoerythrins, etc. Commercial sources of this laser: Cobolt Jive, RGB Laser, and GEM 561 from Quantum Laser).
11. A 660-nm DPSS laser. Excitation source for Cy5, Cy5.5, Alexa Fluor^{660}, DyLight^{649}, FluoSpheres Dark Red fluorescent microspheres, etc.
12. Filter cubes matching the wavelength characteristics of the above illumination sources.
13. Sensitive electron-multiplying charge-coupled-device (CCD) camera. We use the iXon + 897 EM-CCD camera (Andor Technology, Belfast, UK) and the QEM512SC Photometrics (Tucson, AZ) CCD camera. Both have the same chip with a pixel size of 16 μm and a high quantum efficiency (>90% peak QE).
14. In the last few years, the fast complementary metal-oxide semiconductor (sCMOS) camera sensors and ancillary technology (most importantly specific software development) have found a niche for super-resolution microscopy applications [55]. High-throughput applications requiring fast acquisition rates benefit from the use of these cameras.
15. For co-localization studies, a dual-viewer instrument (e.g., DV2 splitter, Photometrics) positioned on the emission side before the CCD camera enables one to simultaneously project the emission of two different channels onto the same CCD camera (see, e.g., [56]), thus enabling one to acquire two spatially identical but spectrally different images simultaneously. We

have a similar instrument built at the workshop of the Max Planck Institute for Biophysical Chemistry in Göttingen.

16. Computer. Super-resolution image acquisition and storage in the single-molecule localization modalities (PALM, STORM/GSDIM) is computer CPU and storage-space demanding. Acquiring 10,000–50,000 images at the highest CCD camera frame rates (few milliseconds/frame) is preferably done by streaming the image data directly to memory. It is thus best to have a fast CPU, at least 16 GB of storage memory, a fast and robust hard disk (2–4 TB), and a high-resolution monitor to focus the specimen and for image inspection and subsequent analysis.

4 Software

1. Camera control and image acquisition: In the case of the Andor cameras, the software packages Solis and the IQ CCD camera control software from Andor Technology are simple, convenient tools developed by the manufacturer. We control this camera with the image acquisition and hardware control software SlideBook (Intelligent Imaging Innovations, Boulder, CO) (www.intelligent-imaging.com).
2. In the case of the Photometrics QuantEM:512SC back-illuminated EMCCD camera, we also use SlideBook to drive the acquisition.
3. The open-source image processing package ImageJ (https://imagej.nih.gov/ij/), developed by Dr. Wayne S. Rasband at the National Institutes of Health, is a very convenient software package with several possibilities for analyzing super-resolution microscopy images.
4. One such possibility, developed specifically for PALM stochastic localization microscopy, is QuickPALM developed by Dr. Ricardo Henriques, currently at the MRC in University College London (https://code.google.com/p/quickpalm/).
5. MATLAB software (Image Processing Toolbox, MathWorks) is a very comprehensive and versatile software package which can be adapted to essentially all aspects of image analysis (localization, tracking, statistics, etc.)
6. The École Polytechnique Fedérale de Lausanne (EPFL) has compiled and maintains a catalog of available software for stochastic localization super-resolution microscopy and deconvolution. At the time of writing this protocol, there were 54 different software packages listed under http://bigwww.epfl.ch/smlm/software/.

7. The software package Localizer https://bitbucket.org/pdedecker/localizer, developed and regularly updated by Prof. Dr. Peter Dedecker and colleagues at the Department of Chemistry of the University of Leuven, Belgium, is a modular and very comprehensive software package for both static localization and single-particle tracking (SPT) running under Igor Pro (WaveMetrics, Inc.).
8. Google Inc. and Drs. Stefan Bollmann and Felix Rüdinger (Univ. Aachen and Biozentrum at the Departments of Biotechnology and Biophysics of the University of Würzburg, Germany) have developed rapidSTORM, a project containing a software package for the analysis of stochastic localization microscopy images and with capabilities for hardware control (http://www.super-resolution.biozentrum.uni-wuerzburg.de/research_topics/rapidstorm/).
9. Ovésny and coworkers from the Charles University in Prague, Czech Republic, have developed ThunderSTORM, a software package for PALM and STORM nanoscopies with Bayesian statistical analysis [57]. http://code.google.com/p/thunder-storm/.
10. For SPT analysis of molecular motions, a MATLAB-based single-particle tracking package analysis is freely available from http://physics.georgetown.edu/matlab/.

5 Methods

5.1 Cell Culture

5.1.1 Before Use of the Cell Culture Hood

1. If UV lamp is routinely used overnight, turn it off. Turn on hood lights and air filter.
2. Wear gloves. Clean them with 70% ethanol. Do the same with working surfaces, tube and pipette holders, and other materials to be introduced under the hood.

5.1.2 During Use of the Cell Culture Hood

1. Do not interpose hands or arms interrupting clean air flow on sterile material.
2. Avoid touching working surfaces with your hands or arms. If you inadvertently do so, clean surface with ethanol.
3. Follow safety rules and throw waste in appropriate container as you follow the procedures.
4. Return Petri dishes/flasks to the CO_2 incubator. Check gas and water levels and tightness of doors.

5.1.3 After Using the Cell Culture Hood

1. If you are using glass pipettes, immerse the used ones in the storage/wash cylinder, cotton-filled side up.
2. Turn off Bunsen or alcohol burner if you used one.

3. Wipe working surface of the hood/tube racks, etc., with 0.1% SDS followed by 70% ethanol.
4. Check the occupancy of waste material and dispose of container when 80% of its volume is full.
5. Turn off gas valve if you used a gas Bunsen burner.
6. Close hood.
7. Turn off air filter.
8. Turn on UV lamp.

6 Cleaning and Poly-L-Lysine Coating of the Coverslips for Neuronal Cell Primary Culture

Coverslips are a most important but often neglected component in light microscopy. One should bear in mind that they sit immediately adjacent to the main optical element of the light microscope – the objective lens – and as such they are not only the physical support for the adhered cells but also part of the imaging system itself. Two main considerations for optimally contributing to the image quality are (1) their physical and chemical composition (high-performance thin borosilicate glass) and (2) the uniformity of their thickness (e.g., 0.17 mm for No. 1.5 glass coverslips). If the coverslip departs too much from the optimal thickness of 170 nm, axial shift of the best focus may result. This affects spatial resolution (particularly in the *z*-axis) and reduces the fluorescence intensity. The rather painstaking cleaning procedure – not without risk – is worth the while for obtaining aberration-free, good-quality super-resolution images.

6.1 Simplified Procedure for Cleaning Glass Coverslips for Routine Experiments

1. Lay glass coverslips in a flat bottom crystallizer flask and cover them with a small volume of 1 N HCl. Leave overnight in acid.
2. On the next day, rinse abundantly with distilled water followed by 70% EtOH.
3. Coverslips may be stored in 95% EtOH and sterilized by passing them briefly over the flame of a Bunsen burner before coating them with poly-L-lysine.

6.2 Cleaning Coverslips for Critical Super-Resolution Microscopy and Poly-L-Lysine Coating

1. Turn on the air extraction of the chemical fume hood. Wear gloves, goggles, and lab coat. Avoid exposure of skin and most importantly eyes to the highly corrosive glass cleaning solution used in **steps 4** or **5**.
2. Put the coverslips in a ceramic staining rack, immerse into the glass coverslip-staining jars, and fill the jars with acetone. Incubate for 1 h, drain the acetone, and wash ten times with deionized water.

3. Prepare the etch solution by mixing 125 mL Millipore-filtered, distilled, and deionized water, 25 mL ammonium hydroxide, and 25 mL hydrogen peroxide in a 250-mL *tall* cylindrical beaker sitting on top of the stirring hot plate. Heat the mixture to 80°C. Place the staining rack with coverslips into the etch solution with metal plastic-coated tweezers or plastic tongs. Leave the coverslips overnight with gentle stirring to avoid spilling.
4. Lift the rack carefully with the metal plastic-coated tweezers or plastic tongs and transfer it into a flat bottom ("crystallizer-type") beaker filled with Millipore-filtered, deionized, and distilled water. Rinse the rack profusely, replacing the water several times.
5. An alternative to the etch solution is the "piranha" solution, requiring shorter incubation times: Put 60 mL of hydrogen peroxide solution (30% in water) and slowly add 100 mL of sulfuric acid (final ratio of acid to hydrogen peroxide solution is 5:3). The mixture will heat up.
6. When using the "piranha" solution, it is convenient to carry out the operation in glass coverslip-staining jars with lids. Immerse the ceramic coverslip rack with the coverslips into the jars, close the lids, and place the jar on the hot stirring plate at 80°C for about 1 h. Wash coverslips as in **step 4**.
7. Neutralize residual acid remaining on the coverslips after "piranha" treatment by filling the coverslip-staining jars with 0.1 M KOH, incubate 10 min, and wash ten times with Millipore-filtered, deionized, and distilled water.
8. Carefully hold and remove the cleaned coverslips from the rack with plastic forceps and rinse both sides by squirting methanol from a squeeze plastic bottle.
9. Blow excess methanol by gently blowing N_2 or filtered air from a plastic pipette tip connected via rubber tubing to either a N_2 cylinder or a central air/air compressor system with filter, respectively.
10. A pipette tip can be connected with tubing to the air supply and used as an aid in blowing away the methanol.
11. Connect the burner to the butane (or natural gas) and light. Pass the coverslip quickly through the topmost portion of the flame, and repeat this process a couple of times for a few seconds passage under the flame to avoid cracking.
12. An even shorter cleaning procedure is to use a plasma cleaner set at 30 W for 15 min. Pressure inside the cleaning chamber is set at 100–200 mTorr. Both atmospheric and compressed oxygen can be used.

13. Store the cleaned coverslips preferably in vertical position in a staining rack kept in a clean vessel with lid until use.
14. Move the necessary amount of racks to suit your experiments to the cell culture hood. As an additional precaution, if coverslips have been stored for a prolonged period, dip them in spectroscopic-grade methanol, flame briefly to dry, let cool for a few seconds, and then place them in Petri dishes or 6-well culture plates.
15. Immediately before use, prepare the poly-L-lysine (e.g., Sigma P7405) solution at a final concentration of 10 mg/mL in 0.1 M borate buffer, pH 8.5, and sterilize by filtration. Add a few drops of this solution spreading with a sterile tip or swirling the plate to cover the surface of the coverslip. Incubate for 1 h at 37°C in the cell culture incubator or leave them covered in the hood with the air blower off at room temperature for 12–24 h.
16. Wash coverslips thrice with sterile water, for a total of 3 h. Leave the coverslips in plating medium in the incubator until ready to plate cells.
17. Remove the water and add neuronal plating medium. Place the coverslips in the incubator to reach temperature.

7 The Hippocampal Embryonic Neuronal Cell Culture

The hippocampal neuronal cell cultures constitute a useful in vitro model system to study physiological and structurally relevant issues in neurobiology, from neuronal development, the birth of the axon – the "symmetry-breaking event" [58] in the establishment of neuronal polarity, synaptogenesis, spinogenesis, localization, and trafficking of synaptic proteins (the purpose of these protocols) and organelles – to changes in dendritic spine morphology associated with long-term potentiation, purportedly linked to learning, memory, and several other physiologically relevant processes. The procedure outlined in these protocols is circumscribed to the culture of neuronal cells obtained from embryonic day 18 (E18) rat hippocampi, the stage at which pyramidal neurons have already developed, following a procedure involving caesarean section of the parental rodent, extraction and sacrifice of the pups, enzymic-mechanical dissociation of the nervous tissue, and establishment of the neuronal cell only or mixed neuronal–glial cell cultures in a Petri dish. The original procedure developed by Gary Banker and colleagues [59] and its subsequent improvements (see, e.g., [45]) have been field-tested over the course of the last few decades and by now are well established, having generated abundant literature on the subject, thus facilitating comparisons between laboratories. The procedure is essentially the same for cerebral cortex neuronal cell cultures, except that the cortex is much easier and faster to dissect. In this sense, rat hippocampi are easier to dissect than mouse hippocampi, but this difficulty is counterbalanced by the fact that mice open the door to studying knockout animals. There is abundant literature raising relevant questions on various neurobiological topics addressed to the neuronal cell model using embryonic stage 18 (E18) rodent pups. There are also procedures for the cell culture of adult brain tissue and for cells cultured on other simple substrates (plastic, meshes, with or without special coatings) or even patterned (in the form of microdots, meshes, channels) substrates. Here I describe a simple procedure to establish high-density hippocampal primary neuronal cell cultures on poly-L-lysine-coated glass coverslips. The reader is encouraged to read the classical original work [59] and/or its well-illustrated updated protocol [45]. Comprehensive protocols on this and related subjects have appeared more recently, e.g., Cold Spring Harbor Protocols (http://cshprotocols.cshlp.org/) (see also [60–62]), especially for implementing the hybrid Banker's sandwich cultures of low-density neuronal cells growing on a feeder layer of astroglial cells in an adjacent coverslip, providing nutrients and growth factors that diffuse through the medium to reach the neuronal layer. The sandwich cultures provide the best results for super-resolution imaging because the low-density neuronal cells and especially their neurites, lie flat on a single plane with very low contamination from glial cells: however, these cultures are more involved, requiring the preparation of glial feed cultures 2 weeks in advance. Also, the preparation of the coverslips is lengthier and more elaborate.

8 Dissection of Brain Tissue for Primary Embryonic Neuronal Cell Culture

1. Prepare all necessary materials, including stereomicroscope, sterile surgical instrumentation, media, 70% ethanol in plastic squirting bottle, ice, etc., on clean labtop area for the first part of the procedure and under the laminar flow hood for the second half.
2. Place a 60-mm Petri dish on the stereomicroscope stage and fill up to 3/4 of its volume with ice-cold medium for dissecting embryos' heads (CMF-HBSS or HBSS+ medium with antibiotic).
3. Anesthetize an E18-period pregnant rodent with CO_2 and euthanize following the protocol authorized in your research establishment (e.g., head dislocation).
4. Perform caesarean section by opening the abdominal wall with sharp-end scissors, with an incision from anus to diaphragm

region. Remove embryos and place in a 100-mm glass dish filled with ice-cold CMF-HBSS. The remainder of the operation should be conducted in a sterile environment, ideally in a laminar flow hood. For this purpose, transfer the Petri dish containing embryos to the hood.

5. The second half of the procedure is performed under sterile conditions. Wash embryos in sterile cold dissection medium and transfer to clean glass Petri dish. Decapitate embryos by cutting the head behind the ears, and proceed to remove their brains. To do this, hold the head of an embryo with forceps at the level of the brainstem, and cut the skin and skull through the midline with fine scissors under the stereomicroscope. Lift the skin and skull to expose the brain. Transfer the brain into a 60-mm dish containing ice-cold dissection medium or CMF-HBSS using a micro-spatula. Using eye-surgical (iridectomy) fine spring scissors, proceed to separate hemispheres. The hippocampus can be recognized on the medial internal surface by its characteristic hippocampal fissure, with blood vessels marking the separation with adjacent entorhinal cortex.
6. Dissect meninges and adherent choroid plexus to expose the hippocampus using fine shanks forceps. Flip the brain upside down such that the hypothalamus is facing up. Dissect the hippocampus from adjoining entorhinal cortex, cutting first parallel to the hippocampal fissure and secondly using transverse cuts at its rostral and caudal ends using the iridectomy fine spring scissors while holding the rest of the tissue with fine Dumont forceps [59]. Always maintain the dissected hippocampi under cold medium.

9 Enzymic Digestion and Mechanical Dissociation of Hippocampal Embryonic Tissue

9.1 According to MacGillavry and Blanpied [63]

1. Chop embryonic day 18 (E18) hippocampal tissue into pieces smaller than the opening of a 200-μL pipette tip, and pipette the small pieces of hippocampal tissue into a 15-mL sterile plastic tube containing 4 mL ice-cold HBSS dissociation medium.
2. Rinse tissue with 10 mL ice-cold dissection medium.
3. Centrifuge tissue for 30 s at 120 × *g*, at RT, and remove supernatant with a Pasteur pipette.
4. Resuspend hippocampal tissue in 4 mL of trypsin/EDTA solution (prewarmed to 37°C) and digest hippocampal tissue 5–8 min at 37°C, slowly agitating from time to time. During incubation, dissolve 3 mg DNase in 3 mL trypsin/EDTA solution.

5. Centrifuge cells for 30 s at 120 × *g*, at RT, and remove supernatant with a Pasteur pipette.
6. Incubate with filter-sterilized, prewarmed trypsin inhibitor solution for 3–5 min at 37°C.
7. Centrifuge for 1 min at 120 × *g* at room temperature, and aspirate supernatant with a Pasteur pipette. If cells are still stringy with DNA and the supernatant cannot be removed, add a more concentrated, filter-sterilized solution of DNase (e.g., 2 mg in 500 μL for 4 mL tissue), and incubate until supernatant can be removed (1–2 min).
8. Gently resuspend tissue on ice in 2 mL dissection medium, avoiding air bubbles. First, use a fire-polished glass Pasteur pipette, and then use a fire-polished Pasteur pipette with a smaller tip diameter. Finally, use a third pipette with an even smaller diameter.

9.2 According to Kaech and Banker [45]

1. When all of the hippocampi have been removed, place them in a 15-mL conical centrifuge tube and bring the volume to 4.5 mL with CMF-HBSS. Add 0.5 mL of 2.5% trypsin and incubate for 15 min in a water bath at 37°C.
2. Gently remove trypsin solution, leaving the hippocampi at the bottom of the tube; add 5 mL of CMF-HBSS and let stand for 5 min at room temperature. Repeat this twice to allow residual trypsin to diffuse from the tissue. Bring the final volume to 2–3 mL.
3. Dissociate the hippocampi by repeatedly pipetting them up and down in a Pasteur pipette. Begin with five to ten passes through a regular pipette and then continue with five to ten passes through a Pasteur pipette that has been flame-polished so that its tip diameter is narrowed by half. Expel the suspension forcefully against the wall of the tube to minimize foaming. By the end, there should be no chunks of tissue left. The diameter of the flame-polished pipette is important. A too narrow tip will damage the cells; a too wide tip will not fully dissociate the tissue. Pipette only as many times as is necessary to obtain a homogeneous solution with no tissue pieces.
4. Determine the cell density by adding a drop of the cell suspension to a hemocytometer. Also determine the total yield, which should be 400,000–500,000 cells per hippocampus. The viability, according to exclusion of trypan blue, should be 85–90%. Adjust the cell density to about 1–1.5 million cells/mL.
5. Using a micropipette, add the desired number of cells to each of the dishes containing the poly-L-lysine-treated coverslips in neuronal plating medium.

6. After 3–4 h, examine the dishes to ensure that most of the cells have attached and then establish the mixed neuron–glial hybrid cultures as described under procedure 12 below.

10 Preparation of Astroglial Feeder Cells [45]

1. Kill three or four postnatal rat pups following the procedure described above.
2. Remove the brains and place them in a dish containing CMF-HBSS. The tissue needs to remain submerged at all times.
3. Remove the cerebral hemispheres and carefully strip away the meninges with fine forceps. Make sure to remove meningeal tissue penetrating deep in cerebral fissures, which contains fibroblasts that can overgrow astroglial cultures.
4. Transfer the hemispheres to a drop of CMF-HBSS and chop the tissue as finely as possible with scissors.
5. Transfer the tissue pieces to a 50-mL conical centrifuge tube in a final volume of 12 mL CMF-HBSS and add 1.5 mL each of 2.5% trypsin and 1% (wt/vol) DNase. Incubate in a 37°C water bath for 5 min, swirling the tube occasionally.
6. Triturate by passing the solution up and down 10–15 times in a 10-mL pipette. Return to the water bath for another 10 min, swirling occasionally. Triturate another 10–15 times with a 5-mL pipette until most chunks disappear.
7. Pass the cell suspension through a cell strainer to remove chunks of undissociated tissue and collect in a 50-mL conical tube containing 15 mL glial medium.
8. Centrifuge the cells for 5–10 min at 120 × *g* to remove enzymes and lysed cells. Discard the supernatant and resuspend the cell pellet in 15–20 mL glial medium. Count cells using a hemocytometer – the yield should be in the order of 10^7 cells per brain.
9. Plate 7.5×10^6 cells per 75-mL flask and add glial medium to a final volume of 15–20 mL.
10. After 1 day, feed cultures with glial medium. First, swirl the flask to remove loosely attached cells, then aspirate off the medium, and replace with fresh medium.
11. Feed the culture every 2–3 days. Before removing the medium, slap the flask 5–10 times against your hand to dislodge loosely attached cells.
12. Contaminating cells such as O2A progenitors and microglia (macrophages) sometimes grow on the surface of the astrocyte monolayer. Removing loosely attached cells at this step helps to

minimize their number. Microglia (recognizable as round, phase-bright, loosely attached cells) release cytokines that are highly neurotoxic. Preparations with high numbers of microglia should be discarded.

13. Harvest the astroglia when the cells are near confluence (usually within 7–10 days). Rinse each flask briefly with 3–5 mL of trypsin/EDTA, then add 2–3 mL of trypsin/EDTA, and incubate at 37°C until the first cells just begin to lift off (usually in less than 2 min). Add 5 mL glial medium to stop the trypsin enzymic digestion, release cells by repetitive pipetting, transfer the cell suspension to a conical centrifuge tube, and pellet the cells at 120 × *g* for 7 min. Resuspend in glial medium and passage into 60-mm dishes (10–30 dishes per flask).
14. Extra cells can be resuspended in cell-freezing medium at ~2×10^6/mL, frozen in cryotubes and stored in liquid nitrogen, and then thawed and plated as needed.
15. One to two days before seeding the neuronal cells, remove the medium from the glial feeder cultures and add 6 mL neuronal maintenance medium, either N2 or Neurobasal/B27, for preconditioning.

11 Counting the Number of Neuronal Cells

1. Remove a 10 μL sample of cells and dilute into 90 μL of 0.4% trypan blue stain. Mix.
2. Take a 10 μL aliquot of the cell suspension and pipette into a hemocytometer, determining the number of cells per unit volume.

12 Plating the Cells

12.1 Neuronal Cell-Only Cultures

1. Plate 50,000–70,000 cells/mL of the initial plating medium in each well (or ~150,000 cells/well for low-density cultures suitable for imaging). Return multiwell to the incubator and let the cells plate for 2 h at 37°C in *plating medium*.
2. After 2 h, aspirate the medium and replace with *medium for culturing hippocampal neurons* (Neurobasal medium containing B27 serum-free supplement) and incubate cells at 37°C.
3. Aspirate half the volume of medium twice a week, replacing the volume with fresh medium. Some researchers do not carry out this step, maintaining the original medium.

12.2 Neuronal–Glial Cell Hybrid Sandwich Cultures

1. As stated in procedure 10 above, 1–2 days before establishing the hybrid culture, the medium from the glial feeder culture should be removed and replaced by neuronal maintenance medium, either N2 or Neurobasal/B27 for preconditioning.
2. Follow the same steps as in Sect. 12.1 above, except that glial cell feeder coverslips with spacers are transferred to the plastic multiwell (or Petri dish) containing the neuronal culture and placed upside down on their spacer feet. Make sure the upper glial coverslips are submersed in the N2 or Neurobasal/B27 medium.
3. Three days after plating, add cytosine arabinoside (1-b-D-arabinofuranosylcytosine) to a final concentration of 5 mM to inhibit glial proliferation.

13 Transfection of Dissociated Hippocampal Cultured Neurons (Modified from Ref. [63])

The predominant type of cell in the hippocampus is the pyramidal neuron. Interneurons are fewer in number and can be distinguished on a morphological basis from the pyramidal cells. Of further aid in the distinction between different neuronal subtypes are a few specific markers for some of the interneurons.

1. Transfect neuronal cell cultures 1–3 days before imaging using Lipofectamine 2000. This reagent has a low transfection efficiency (<1%), but the apparent drawback may prove convenient for very dense cultures.
2. Dilute 0.5–2.0 μg plasmid DNA in 25 μL of serum-free medium or Opti-MEM I reduced serum medium.
3. Dilute 1.0 μL Lipofectamine 2000 in 25 μL serum-free medium or Opti-MEM I reduced serum medium. Allow the tubes to stand for 5 min at room temperature.
4. Combine the contents of the two tubes containing the suspensions (Lipofectamine and DNA), mix them vigorously by pipetting with a micropipette, and leave them for 20–30 min at room temperature to allow the DNA–Lipofectamine complexes to form.
5. Retrieve the multiwell containing the coverslips with the cells from the incubator, and add the DNA–Lipofectamine suspension dropwise (about 50 μL per well). Mix by gently swirling the plate by hand or sitting it on a rotary shaker. Return the multiwell plate to the incubator and let it stand for 90–120 min.
6. Retrieve the multiwell from the incubator, remove coverslips, and place them in a new multiwell plate. If a mixed coculture with glial cells is used [62], place the coverslips with the cultured neuronal cells facing the glial monolayer below

them. Return to the incubator and retrieve for microscopy 24–48 h later.

7. The use of viral vectors and of adenovirus and lentivirus in particular is very advantageous for introducing foreign constructs in neuronal cells in culture. The reader is referred to recent protocols for the preparation of the virus vectors [47] and transfection of neuronal cells [64].
8. Similarly the "self-labeling techniques" (Halo tag, SNAP tags, etc.) mentioned under Materials are gaining momentum and could find wider application in the labeling of synaptic proteins.

14 Fixation and Immunostaining

Fixatives are carcinogenic. Wear gloves and other appropriate protective equipment and avoid contact with skin or eyes.

Strong and/or prolonged fixation is detrimental for fluorescence microscopy. Inclusion of glutaraldehyde in a cocktail with paraformaldehyde, as is customary in electron microscopy, is particularly critical; glutaraldehyde concentrations >1% practically abolish emission from fluorescent proteins.

Fix cells in 4% (wt/vol) paraformaldehyde (electron microscopy grade is recommended, e.g., from Electron Microscopy Sciences) fixative solution following the protocol below.

14.1 Method A

1. Wash coverslips twice with 0.1 M (Sörensen's) phosphate buffer saline (PBS, Invitrogen), pH 7.3–7.4.
2. Fix cells with 3 mL/well paraformaldehyde solution for 15–20 min at RT or 4°C (on ice).
3. Wash three times with PBS for 10–20 min.
4. Incubate for 30 min with blocking solution (4% goat normal serum, NGS, Jackson ImmunoResearch Laboratories, West Grove).
5. Put a piece of Parafilm on a wet Whatman paper inside a Petri dish and apply 200 μL of primary antibody solution in incubation solution (appropriate dilution must be determined experimentally). Lay coverslips upside down on antibody-containing solution and incubate for 1–2 h at RT.
6. Transfer coverslips to multiwell plate and wash three times with PBS for 10 min.
7. Repeat **steps 4–7** with secondary antibody (optimal dilution must be determined experimentally).

14.2 Method B. Fixation-Permeabilization for Immunostaining Intracellular Proteins

1. Fix cells as in **steps 1–3** above.
2. Incubate for 30 min with blocking solution (4% goat normal serum) containing 0.1% Triton X-100 (Sigma) in PBS for 15 min.
3. Wash with PBS for 15 min.

14.3 Method C

1. Fix cells with 3 mL/well paraformaldehyde solution at RT.
2. Immediately transfer to freezer and leave at −20°C for 5 min.
3. Without washing, transfer coverslip to spectroscopic-grade methanol at −20°C and leave for 15 min in the freezer at this temperature.
4. Wash thrice with PBS.
5. Permeabilize with 0.2% Triton X-100 solution in PBS for 10 min.
6. Block in 10% goat serum in PBS for 1 h.
7. Dilute primary antibody in blocking solution. Incubate sample with primary antibody at final concentrations which vary according to antibody (e.g., 1:200 for anti-glutamate receptor antibody, 1:1.000 for myc, etc.). The optimal dilution should be found by the combination of manufacturer's recommendation + one's own experience.
8. Wash thrice with PBS.
9. Incubate with fluorescent-labeled secondary antibody for 45 min–1 h at room temperature. Dilutions also vary (e.g., 1:500 for Alexa Fluor^{488}, 1:1,000 for Alexa Fluor^{647}).
10. Wash thrice with PBS buffer and mount for microscopy as described in Sect. 16.
11. In the case of fluorescent proteins, if the neuronal specimens are to be imaged after fixation, bear in mind that neurons transfected with Venus-FP or Tomato-FP withstand fixation better than GFP or EGFP.
12. It is advisable to run a "control" experiment using a marker of the glutamatergic nerve terminals (e.g., vesicular glutamate transporter 1, VGLUT1) and a dendritic marker (e.g., microtubule-associated protein 2 [MAP2]) to ensure the quality of the specimen and the entire fixation/immunostaining procedure.
13. Incubate coverslips with a polyclonal, guinea-pig anti-VGLUT1 antibody (e.g., AB5905, Chemicon-Millipore, Billerica, MA) (1:2,000 dilution) and a polyclonal, rabbit anti-MAP2 antibody (AB5622, Chemicon-Millipore) (1:2,000 dilution) in PBS with 4% NGS (blocking solution) for 3 h.
14. Wash with PBS for 15 min.

15. Incubate coverslips with goat anti-guinea-pig IgG (H + L) antibody conjugated with, e.g., Alexa Fluor^{594} (A-11076, Invitrogen) (1:1,000 dilution in blocking solution) and goat anti-rabbit IgG (H + L) antibody conjugated with Alexa Fluor^{488} (1:1,000 dilution) for 60 min.
16. Rinse thrice with PBS for a total of 30 min.

14.4 Staining Specific Synaptic Proteins and Counterstaining Whole Neurons, Axonal, and Dendritic Arborizations

1. A variety of presynaptic protein markers is available: anti-vGlut1 (Antibodies Inc., NeuroMab) labels glutamatergic terminals and antibodies against Vamp, synapsin 1, synaptotagmin, synaptophysin (Abcam), synapsin (Synaptic Systems), synaptopodin, Bassoon, Munc18, etc.
2. Similarly, various antibodies against postsynaptic proteins produce very strong signals, like anti-PSD-95 (Synaptic Systems, NeuroMab), anti-extracellular epitopes of various subunits of the NMDA receptor (Alomone Labs), GluN1 and GluNR2 obligatory pore-forming subunits (Chemicon mAb397), nicotinic acetylcholine receptor subunits (Alomone Labs), etc.
3. Various synaptic proteins can be visualized using fluorescent peptides (e.g., for actin, F-actin-binding peptides, F-tractin from Clontech, phalloidin conjugated to various fluorophores).
4. In addition to specific neuronal protein markers, it is convenient to counterstain with antibodies against MAP-2 (a neuronal dendritic marker, Novus Biological) and the tau protein to identify each type of neurite.
5. To differentiate between inhibitory and excitatory synapses, one can employ antibodies against vGAT (presynaptic)/gephyrin (postsynaptic) for inhibitory synapses, occurring predominantly on dendritic shafts, and vGlut1 (pre)/PSD-95 (postsynaptic) for excitatory synapses, mostly present on dendritic spine heads.
6. For quantifying dendritic spines, the developmental protein drebrin can be used as a general marker (1:500 anti-drebrin from Med. Biol. Labs followed by a 1:500 dilution of a fluorescent anti-mouse IgG antibody).

15 Excitation Modalities in Super-Resolution Microscopy

In order to efficiently detect single molecules in the focal imaging plane, one must suppress fluorescence emission from those emitters that are out of plane. The illumination modality plays a key role in accomplishing this task. The position at which the laser beam enters the back aperture of the objective lens determines the angle at which the light will excite the sample. In conventional wide-field epifluorescence microscopy, the laser beam is symmetrically centered and parallel to the z-axis of the objective lens (and the microscope); the resulting illuminated field is an inverted cone-shaped volume of the cell above and below the focus. Displacement of the laser beam off the central axis of the objective lens will result in oblique illumination, a condition which reduces the volume of the cone of light exciting fluorophores and scattering in the cell above the focus. If the incident excitation laser beam is positioned at an angle larger than the critical angle $\alpha = 61^\circ$, where $\sin(\alpha) = \eta_w/\eta_g$, the ratio of the refractive indices of water (the medium in which the cells are embedded and that of the cells proper) and immersion oil/cover glass (the two media that the light must transverse before hitting the cell), respectively, we meet the condition of another wide-field technique, total internal reflection fluorescence (TIRF) microscopy (see, e.g., [65]). In this mode of illumination, an evanescent electromagnetic field is used to selectively illuminate and excite fluorophores in a restricted volume of the specimen, immediately adjacent to the glass–water interface. The incident light is reflected back into the high refractive index medium at the interface, whereas a non-propagating evanescent wave is used to excite the fluorophores.

The good and the bad thing about TIRF illumination is that the evanescent wave, decaying exponentially from the interface, penetrates to depths between 80 and 200 nm (depending on the incident angle) into the low-refractive-index aqueous medium containing the specimen. Thus only fluorescent molecules in a narrow disk of about 100-nm height are excited. When using coverslip-adhered cells and epi-illumination in an inverted fluorescence microscope, this is undoubtedly the best choice for interrogating fluorescent molecules in the ventral, glass-attached cell membrane, rendering an improvement in contrast of about fourfold over conventional wide-field illumination, and considerably reducing the background fluorescence. In the case of primary cultures of neuronal cells, TIRF microscopy is best for imaging molecules in the ventral surface of the soma, the growth cones at early culture stages (D2–D5), and fully developed dendritic spines at culture stages D15–D25. TIRF illumination in the form of a standing wave can further produce a marked improvement in spatial resolution [66, 67]. The most dramatic improvement in axial resolution has been recently reported by George Patterson and colleagues at the NIH: TIRF illumination at varying penetration depths, produced by tuning the angle of the TIRF beam in combination with photobleaching. The latter preferentially affected the fluorescence from the shallow layers, allowing observation of deeper layers. Images constructed by sequential imaging and photobleaching the fluorescent molecules rendered, upon reconstruction, an axial resolution of ~20 nm [68]. Although primarily intended for illumination of cell-surface components, some illumination modalities such as the so-called "highly inclined and laminated optical" (HILO) sheet microscopy [69] or "selective-plane illumination" microscopy (SPIM) [70] have also been specifically developed for single-molecule imaging deeper inside cells, achieving improved signal-to-noise ratio. HILO illumination is also at the root of the so-called uPAINT super-resolution technique, which makes use of oblique illumination from a slightly below the critical angle laser beam to reject background fluorescence of unbound fluorescent ligands. These techniques are ideal for studying plasma membrane-resident proteins, cell-surface neurotransmitter receptors being ideal targets among synaptic proteins. The uPAINT technique is particularly attractive for dynamic measurements, such as following the trajectories of ligands bound to surface proteins, degree of confinement, and organization of the target proteins in clusters [40, 41].

16 Preparation of the Samples for Imaging

Super-resolution microscopy shares with conventional microscopy many of the requirements for identifying structures illuminated with visible light and interrogated with fluorescence. The first common requisite with wide-field microscopy is to make objects visible. This can be accomplished by using extrinsic fluorescent probes that tag proteins or subcellular structures or by constructing chimeras of, e.g., the protein of interest with a foreign fluorescent protein, introducing the construct into the cell via the resulting plasmid, and letting the cell express the fluorescent product. The technique for labeling proteins, lipids, and other cellular chemical components have been worked out in great detail, and the use of fluorescent proteins in fluorescence microscopy, a younger field, has also reached a stage of maturity. In addition to these and other aspects common with conventional light microscopy, super-resolution techniques have demanded new developments in sample preparation, coverslip handling, protein tagging, embedding media, oxygen-scavenging requirements, mounting of the specimens, etc. Sample preparation is an important step for successful nanoscopy.

The commonly used cyanine and Alexa Fluor dyes are known to undergo dark state transitions over a broad range of time scales. Fluorophore dark states have been attributed to *cis–trans* isomerization within the conjugated polyene cores, the buildup of triplet states, charge-relay mechanisms, photo-ionization, photooxidation, and the absorbance of short-wavelength photons from the first excited singlet state [71]. Several procedures have been developed to curb this phenomenon, including the use of oxygen-scavenging imaging buffers containing oxygen-scavenging enzymes for wet live specimens or dry coatings for fixed specimens. Other requirements in some super-resolution experiments arise from the need to match the high numerical aperture of the immersion lenses with the aqueous mounting media, calling for nontoxic embedding media capable of suppressing the refractive index mismatch. One such mounting medium is based on the use of 2,2′-thiodiethanol, which, by being miscible with water at any ratio, allows fine adjustment of the average refractive index of the sample from that of water (1.33) to that of immersion oil (1.52). The thiol compound thus enables super-resolution imaging deep inside fixed specimens with objective lenses of the highest available aperture angles [72].

16.1 Fixed Specimens

1. Coat the coverslip with the fixed neuronal cells with a few microliter (this will depend on the diameter of the coverslip) of a 1% aqueous solution of polyvinyl alcohol in Millipore-filtered distilled water (25,000 MW PVA, Sigma-Aldrich 184632; Polysciences 02975-00).
2. Spin the coverslip in a spin coater (e.g., SCS G3 spin coater, Kisco; spin coater model WS-400 from Laurell Technologies).
3. A simple procedure developed in our laboratory replaces a commercial spin coater with a tabletop centrifuge. Remove the rotor from the tabletop centrifuge. Place a strip of Parafilm on the shaft of the tabletop centrifuge and adhere firmly. Place the round coverslip with the wet side up on the Parafilm, adhere gently but firmly, close the lid of the centrifuge, and spin for 30 s–1 min. The PVA film will form a uniform dry thin film. Let dry for about five additional minutes before imaging. The thin PVA film suffices to prevent oxygen quenching to the biological sample. Dried samples in the dark have been successfully imaged after several months of storage in the dark.
4. The commercial mounting medium VECTASHEILD has been recommended for STORM super-resolution imaging with Alexa Fluor dyes, with optimal results with Alexa Fluor^{647} [52]. The authors state that this medium yields images comparable or superior to those obtained with more complex buffers, especially for 3-D super-resolution imaging. VECTASHIELD has a refraction index of ca. 1.45 measured at 590 nm. A mixture of 25% VECTASHIELD–75% Tris-glycerol buffer was also suggested by these authors.

16.2 Live Specimens

1. Wash the coverslip containing the neuronal cells with extracellular imaging buffer and place in coverslip chamber containing imaging buffer to cover the cells (a few microliter, depending

on coverslip diameter and model of chamber but sufficient to keep the specimen wet even after a long imaging session).

2. If available, maintain the temperature of the specimen relatively constant. One of our setups combines the super-resolution microscope with a Peltier-element thermostated holder for temperature-controlled patch-clamp recordings (Luigs & Neumann, Ratingen, Germany).

17 Single-Molecule Stochastic Localization Microscopy

Since the publication of the first step-by-step protocols on stochastic localization microscopy only a few years ago [73], several methodological advances have occurred. Among them are the explosion in the number of available fluorescent proteins for PALM microscopy and the recognition that most fluorophores undergo off–on dark–light transitions, thus enormously broadening the possible probes that can be used in nanoscopy. In this section, a formula-less practical protocol is provided for the implementation of stochastic localization super-resolution microscopy of synaptic components in the PALM or STORM/GSDIM modalities. More detailed protocols can be found in the literature (e.g., [63, 74]).

17.1 Image Acquisition

1. Set the image acquisition conditions to be used in the actual experiment using a standard "sea of fluorescence" specimen and set the incidence angle of the TIRF beam for oblique illumination of the sample.
2. Knowledge of the number of photons detected by the camera after background subtraction is necessary to calibrate both the latter and the precision with which a single molecule is localized. Calibrate camera photoelectron conversion – the number of photons per image pixel intensity units – a parameter that is required to calculate the number of photons emitted per molecule and thus estimate the localization precision of the recorded single molecules. Calibration can be obtained as follows: (1) record images under uniform Koehler illumination at several different intensities, covering the dynamic linear range of the camera, at a given CCD camera gain setting; (2) calculate the mean pixel value and variance of the uniformly illuminated region at each intensity; (3) plot variance as a function of mean pixel value; and (4) the slope of the variance as a function of mean pixel value is a good approximation of the ratio of recorded pixel value per detected photon [75].
3. Switch to the actual specimen and turn on transmitted light, coarse focusing the specimen. At this stage, one may take a snapshot image in a contrast-enhancing mode (phase contrast,

DIC interference contrast, Hoffmann contrast microscopy) to record the fine overall morphology of the cells.

4. Switch to fluorescence epi-illumination, turning on a standard light source (Hg arc, LED). Use the minimal excitation intensity compatible with localizing a region of interest (ROI) through the eyepiece or preferably use the CCD camera in your imaging software to carry out the procedure digitally using the focusing mode at a reasonably high camera gain.
5. In the case of PALM localization microscopy, carry out these preliminary steps imaging the unconverted state of the fluorescent protein using the appropriate filter combination. As MacGillavry and Blanpied [63] warn, one should avoid long exposures, since photoactivatable fluorescent proteins are already activated and partially depleted at wavelengths used for wide-field epifluorescence visualization. For visualizing unconverted green Eos, a standard EGFP filter set is recommended. For photoactivation and imaging of converted red Eos, use a filter cube containing a dual-band excitation filter (405 and 561 nm), a dual-band dichroic to reflect these laser lines, and a red-pass emission filter [76].
6. Set out all acquisition parameters (camera gain, frame rate) to maximize efficient performance and check microscope hardware conditions (filter combinations, projection lens resulting in adequate camera pixel size, etc.) and switch to camera acquisition mode. Select a region of interest (ROI), minimizing the void areas outside the cell(s), thus maximizing acquisition rates and, correspondingly, minimizing off-line analysis time. The actual pixel size of our Andor iXon-897 or Photometrics CCD cameras is 16 μm. We use the 1.5× projection lens of the Nikon TE-2000 to obtain a 160-nm pixel size in the image.
7. Switch off LED or Hg lamp and turn on laser(s). After the initial burst of fluorescence, individual blinking events will appear stochastically, and their number should be modified by varying the intensities of the excitation and activation lasers. We do this manually with precision potentiometers to reach a number of blinking events which are adequate for off-line analysis. Since the acquisition rate is about 10 ms/frame, and the refreshing rate of the visualization mode is much slower than this, it is a matter of experience acquired by trial and error to set the relative intensities of the two lasers (essential for PALM imaging) or the single laser (in STORM/GSDIM mode) to acquire the series of several thousand individual frames that will make the "movie." In the SlideBook software package, we set the "streaming mode" to record the individual images and store them directly in memory. A standard "movie" consists of a stack of 5,000–30,000 individual frames

depending on the size of the chosen ROI and the brightness and stability of the specimen.

8. In SlideBook, the compound image is recorded as a *.sld file, which can be read directly by ImageJ Fiji or converted and exported as a TIF or MATLAB file for subsequent image analysis.

17.2 Image Analysis

At the time of writing this protocol, there are a few dozen software packages for analyzing super-resolution single-molecule localization microscopy images, as listed in Sect. 4 above. It is difficult to recommend a given one because the applications vary, and the know-how of the microscopists differs. It is advisable to read the literature on nanoscopy applied to specific problems and select approaches similar to those that the experimentalist is tackling, trying a few. Some software packages are more user-friendly and demand less knowledge of computer programming than others. Some software packages have been conceived as "key in hand" and require little effort from the experimentalist. There is always a compromise between the size of the interrogated area, speed, computer "muscle" devoted to the analysis, accuracy required to identify and localize single molecules, and whether it is a still, fixed specimen or living cell.

17.2.1 Fixed Specimens

If one inspects the stream of images collected in the acquisition step in a typical stochastic localization experiment, each frame should contain a sparse number of blurred spots stochastically appearing and disappearing in successive frames. This results from the turning on and off of the individual molecules undergoing transitions between light-emitting and dark states. The movie is, however, a series of wide-field photomicrographs with diffraction-limited spots from which one must extract x and y positions at the molecular scale. In other words, the main task of the stochastic super-resolution microscopy analysis is to extract with sub-diffraction, nanometer precision the centroid of the photon-emitting particle (the molecule) from the blurred, diffraction-limited intensity profile of the spots recorded by the CCD. This is accomplished by fitting the PSF of the intensity distribution of each spot with an appropriate function (Gaussian, 2-D Gaussian, elliptical Gaussian, etc.). This will result in a 2-D map of all localizations in the captured stream of data. Most software analysis packages produce a table of x and y coordinates of all centroids and provide ancillary modules to eliminate uncertain localizations, correct for drift, etc.

1. Select a ROI size in your streamed stack of image frames in the file format suitable for the next image analysis program.
2. Run the initial analysis of the raw data files using a software package of your choice, e.g., QuickPALM [77] in ImageJ

(NIH, USA, http://rsb.info.nih.gov/ij/). QuickPALM produces immediate graphic outputs which enable one to assess the quality of the experimental data and obtain preliminary parameters. Localizer (BitBucket Localizer), written by Dedecker and coworkers [78, 79]: http://www.igorexchange.com/project/Localizer, implemented in Igor Pro (http://www.wavemetrics.com), also offers the possibility of rapidly evaluating the quality of the experimental data. The first output is a list of all sub-pixel localizations generated through a least-square fitting of the PFSs, resulting from application of a generalized likelihood ratio test algorithm specifically designed to identify PSF-shaped, that is, Gaussian-like, spots. There is a choice between symmetrical (round-shaped) and asymmetrical-shaped (elliptical) 2-D Gaussian functions describing the PSFs. Calculate the standard deviation of the PSFs.

3. If, e.g., ThunderSTORM (GitHub thunderstorm) (https://code.google.com/p/thunder-storm/) [57] was employed in the ImageJ environment, choose an appropriate filter (e.g., Wavelet, B-spline) for image filtering and perform approximate localization of molecules using the modules local maximum and PSF:integrated Gaussian for sub-pixel localization of molecules. ThunderSTORM has provision for performing lateral drift correction using cross correlation.
4. Estimate the drift. Hardware measures preventing drift (thermostated room, high-quality vibrational isolation system, stable stage) or correcting it online also via hardware (automatic focal drift correction) in combination with short acquisition times are the best alternatives. Inclusion of fiduciary markers (e.g., 50 nm gold, nanoshell particles or fluorescent beads) in the specimen is also helpful. Most super-resolution software packages include a drift correction algorithm, which may or may not require fiduciary markers. In Localizer [78] the consecutive drift estimations are interpolated using a polynomial fit and used to determine the approximate position of the sample relative to the starting point at each time point. The resulting offset is subtracted from the fluorophore positions to obtain the drift-corrected positions. Other methods of drift correction for single-molecule localization microscopy include sub-pixel cross correlation of 2-D histograms binned for single-molecule localizations every *n* number of frames [10, 80] and also see a comprehensive treatment of drift correction for various super-resolution techniques in [81].
5. Estimate the precision of the single-molecule localizations. The accuracy of the localizations, the result of model fitting to the emission pattern, is a measure of the confidence in the estimated position of the single-molecule localizations. It scales inversely with the square root of the number of photons

collected per molecule, N, and increases with the amount of background noise, given by the standard deviation, σ, of the spatial response function of the optical system to an infinitely small object (i.e., the point-spread function, PSF). This can be approximated by $\sigma/N^{1/2}$ in the absence of negligible background [82].

6. Exclude poorly localized molecules by selecting only localizations that have a high photon count (e.g., >100) and acceptable localization precision (e.g., <30 nm).
7. Produce a rendering of the spatial distribution of the collection of localized molecules. This can be done in multiple ways, ranging from the simple x and y point distribution of all centroids superimposed on a black and white depiction of the wide-field image to color-coded distributions depicting density, time of appearance, or any other property comparing representative samples under different experimental conditions.
8. Perform statistical analysis of the distribution of single molecules. Even an elementary property such as the number of particles per unit area may provide information on the functional arrangement or other relevant properties of proteins in the synapse. Knowledge of the photophysical properties of the fluorophores is important in determining the actual number of proteins in quantitative PALM super-resolution experiments [83].
9. Various other analytical tools can be employed to extract relevant information from images of synaptic proteins in fixed neuronal cultures. We are interested in the supramolecular organization of neurotransmitter receptors in the cell-surface membrane, a property which can be interrogated with various statistical analytical tools like Ripley's K function [84, 85] and other tests of two-dimensional spatial point pattern distribution [86, 87]. We have used Ripley's K function to analyze the nanocluster organization of adult muscle-type nicotinic acetylcholine receptors [20]. The use of Ripley-like functions has improved considerably and is successfully applied to the analysis of protein nanoclusters in the immunological synapse imaged with PALM and STORM techniques [88]. The pair correlation or cross correlation function $G(r)$ estimates the probability of finding another particle at a distance r and compares the data to the expected values for a random distribution of particles. $Gs(r) > 1$ correlations can be attributed to clustering of labeled molecules [89–91] and hence can be employed to determine the length scale of the clustering.

17.2.2 Live Specimens

The basic questions in a live-cell experiment concern the time-dependent evolution of some physiologically relevant dynamic phenomenon. In single-molecule super-resolution experiments, this immediately refers to the displacement of the molecule(s) in a functionally adequate time window, to establish molecular trajectories by single-particle tracking (SPT) analysis, determine the speed of the motions (the mean square displacement [MSD]), and derive the apparent diffusion coefficient D of the molecules.

1. In addition to particle detection **steps 1–3** as in fixed specimens, the single-molecule analysis in a dynamic time-lapse sequence is more complex, since in addition to the detection of the centroid of connected components of segmented objects, the analysis should include application of modules which build the trajectories of the connected components, perform additional steps to reject potentially overlapping false maxima in motion, and validate the trajectories using physically meaningful models that predict the particle positions in a frame based on the positions in the past.
2. The MATLAB-based SPT analysis from http://physics.georgetown.edu/matlab/, the equivalent module in Localizer, or the software package ThunderSTORM can be applied to track individual molecules that appear in consecutive frames and assemble them into a molecular trajectories. In brief, the essential requirement in this step is to set an a priori maximum allowable frame-to-frame displacement, i.e., the tracking radius of the molecule along its trajectory, on the basis of the expected diffusion coefficient D of the given molecule.
3. Calculate the MSD of the population(s) of molecules.
4. Calculate the mean diffusion coefficient for each track. In two dimensions, the relationship between the MSD and the effective diffusion coefficient (D_{eff}) is given by MSD $= 4D_{\text{eff}}\,\Delta t$, where Δt is the interval between observations. For freely diffusing molecules, the MSD increases linearly with time, whereas for confined diffusion (as found when molecules are confined within a cytoskeletal corral of filamentous actin, for instance), this relationship plateaus after an initial linear rise. In such cases, one can calculate the instantaneous D_{eff} from the slope of the first few time points of the linear fit of the MSD versus time curve at $2\Delta t$, $3\Delta t$, and $4\Delta t$, referred to as short-range diffusion coefficient or $D_{2\text{–}4}$ [92]. $D_{2\text{–}4}$ is convenient because it can be determined independently of the motional modes. On average, $D_{2\text{–}4}$ is larger than the values of D determined using Eqs. 5–12 of Kusumi and coworkers [92–94] by a factor of 1.2 in the case of constrained or restricted diffusion.
5. Produce graphical displays of single-molecule trajectories. For very high molecular densities, the complete set of trajectories

may produce overcrowded and uninformative graphical representations, and it may be advisable to select identical intervals under different experimental conditions for comparison. Color-coded individual molecular trajectories on the basis of frame intervals can help to identify changes in the behavior of molecular mobility and can also help to detect artifactual lateral drift.

6. It is useful to construct a "probability map" in the form of a 2-D Gaussian with standard deviation σ_{xy} with pixel values proportional to the probability that a molecule can be found at that location [95].

18 Summing Up

These protocols are intended to provide practical guidelines to perform super-resolution imaging of single molecules in synapses using the most inexpensive of currently available nanoscopy techniques, namely, the stochastic single-molecule localization approach. The methodology is relatively easy to implement using an existing fluorescence microscope as the basic body on which to mount the additional building blocks, i.e., the laser lines, CCD camera, filters, and other optical components. As stated in Sect. 3, the objective lens – and its quality – is the most vital individual piece of equipment. The mechanical stability of the optical system is also critical, and investing in a good optical isolation table in combination with a purely mechanical stage pays off good dividends. The other actor in the play is, of course, the specimen. I have reserved these final lines to confess that (at least in our hands) primary neuronal cells from the hippocampus are not easy to maintain in culture, but they provide a unique middle ground between slices and biochemically isolated molecules for in vitro studies of the latter in situ. The field is developing at a fast pace, and the techniques are likely to need frequent update.

References

1. Hell SW, Wichmann J (1994) Breaking the diffraction resolution limit by stimulated emission: stimulated-emission-depletion fluorescence microscopy. Opt Lett 19(11):780–782
2. Klar TA, Hell SW (1999) Subdiffraction resolution in far-field fluorescence microscopy. Opt Lett 24(14):954–956
3. Hell SW (2007) Far-field optical nanoscopy. Science 316(5828):1153–1158. doi:10.1126/science.1137395
4. Betzig E, Patterson GH, Sougrat R, Lindwasser OW, Olenych S, Bonifacino JS, Davidson MW, Lippincott-Schwartz J, Hess HF (2006) Imaging intracellular fluorescent proteins at nanometer resolution. Science 313(5793):1642–1645. doi:10.1126/science.1127344
5. Betzig E (2015) Single molecules, cells, and super-resolution optics (Nobel Lecture). Angew Chem Int Ed Engl. doi:10.1002/anie.201501003
6. Moerner WE (2007) Single-molecule chemistry and biology special feature: new directions in single-molecule imaging and analysis.

Proc Natl Acad Sci U S A 104(31): 12596–12602

7. Moerner WE (2015) Single-molecule spectroscopy, imaging, and photocontrol: foundations for super-resolution microscopy (Nobel Lecture). Angew Chem Int Ed Engl. doi:10.1002/anie.201501949
8. Rust MJ, Bates M, Zhuang X (2006) Subdiffraction-limit imaging by stochastic optical reconstruction microscopy (STORM). Nat Methods 3(10):793–795. doi:10.1038/nmeth929
9. Bates M, Huang B, Dempsey GT, Zhuang X (2007) Multicolor super-resolution imaging with photo-switchable fluorescent probes. Science 317(5845):1749–1753
10. Huang B, Wang W, Bates M, Zhuang X (2008) Three-dimensional super-resolution imaging by stochastic optical reconstruction microscopy. Science 319(5864):810–813
11. Bates M, Huang B, Zhuang X (2008) Super-resolution microscopy by nanoscale localization of photo-switchable fluorescent probes. Curr Opin Chem Biol 12(5):505–514
12. Gustafsson MG (2000) Surpassing the lateral resolution limit by a factor of two using structured illumination microscopy. J Microsc 198(Pt 2):82–87
13. Gustafsson MG (2005) Nonlinear structured-illumination microscopy: wide-field fluorescence imaging with theoretically unlimited resolution. Proc Natl Acad Sci U S A 102 (37):13081–13086
14. Hofmann M, Eggeling C, Jakobs S, Hell SW (2005) Breaking the diffraction barrier in fluorescence microscopy at low light intensities by using reversibly photoswitchable proteins. Proc Natl Acad Sci U S A 102(49):17565–17569. doi:10.1073/pnas.0506010102
15. Kwon J, Hwang J, Park J, Han GR, Han KY, Kim SK (2015) RESOLFT nanoscopy with photoswitchable organic fluorophores. Sci Rep 5:17804. doi:10.1038/srep17804. http://www.nature.com/articles/srep17804#supplementary-information
16. Ta H, Keller J, Haltmeier M, Saka SK, Schmied J, Opazo F, Tinnefeld P, Munk A, Hell SW (2015) Mapping molecules in scanning far-field fluorescence nanoscopy. Nat Commun. doi:10.1038/ncomms8977
17. Eggeling C, Willig KI, Barrantes FJ (2013) STED microscopy of living cells – new frontiers in membrane and neurobiology. J Neurochem 126(2):203–212. doi:10.1111/jnc.12243
18. Willig KI, Barrantes FJ (2014) Recent applications of superresolution microscopy in neurobiology. Curr Opin Chem Biol 20:16–21. doi:10.1016/j.cbpa.2014.03.021
19. Chereau R, Tonnesen J, Nagerl UV (2015) STED microscopy for nanoscale imaging in living brain slices. Methods. doi:10.1016/j.ymeth.2015.06.006
20. Kellner RR, Baier CJ, Willig KI, Hell SW, Barrantes FJ (2007) Nanoscale organization of nicotinic acetylcholine receptors revealed by stimulated emission depletion microscopy. Neuroscience 144(1):135–143. doi:10.1016/j.neuroscience.2006.08.071
21. Nagerl UV, Willig KI, Hein B, Hell SW, Bonhoeffer T (2008) Live-cell imaging of dendritic spines by STED microscopy. Proc Natl Acad Sci U S A 105(48):18982–18987
22. Urban NT, Willig KI, Hell SW, Nagerl UV (2011) STED nanoscopy of actin dynamics in synapses deep inside living brain slices. Biophys J 101(5):1277–1284. doi:10.1016/j.bpj.2011.07.027
23. Berning S, Willig KI, Steffens H, Dibaj P, Hell SW (2012) Nanoscopy in a living mouse brain. Science 335(6068):551. doi:10.1126/science.1215369
24. Yang X, Xie H, Alonas E, Liu Y, Chen X, Santangelo PJ, Ren Q, Xi P, Jin D (2016) Mirror-enhanced super-resolution microscopy. Light Sci Appl 5, e16134. doi:10.1038/lsa.2016.134
25. Hess ST, Girirajan TP, Mason MD (2006) Ultra-high resolution imaging by fluorescence photoactivation localization microscopy. Biophys J 91(11):4258–4272. doi:10.1529/biophysj.106.091116
26. Vogelsang J, Kasper R, Steinhauer C, Person B, Heilemann M, Sauer M, Tinnefeld P (2008) A reducing and oxidizing system minimizes photobleaching and blinking of fluorescent dyes. Angew Chem Int Ed Engl 47 (29):5465–5469
27. Heilemann M, van de Linde S, Schuttpelz M, Kasper R, Seefeldt B, Mukherjee A, Tinnefeld P, Sauer M (2008) Subdiffraction-resolution fluorescence imaging with conventional fluorescent probes. Angew Chem Int Ed Engl 47 (33):6172–6176
28. van de Linde S, Sauer M, Heilemann M (2008) Subdiffraction-resolution fluorescence imaging of proteins in the mitochondrial inner membrane with photoswitchable fluorophores. J Struct Biol 164(3):250–254
29. Heilemann M (2009) Light at the end of the tunnel. Angew Chem Int Ed Engl 48 (22):3908–3910
30. Heilemann M, van de Linde S, Mukherjee A, Sauer M (2009) Super-resolution imaging with

small organic fluorophores. Angew Chem Int Ed Engl 48(37):6903–6908

31. Sibarita JB, Heilemann M (2012) Targeting, resolving and quantifying cellular structures by single-molecule localization microscopy. EMBO Rep 13(12):1043–1045. doi:10.1038/embor.2012.177
32. van de Linde S, Heilemann M, Sauer M (2012) Live-cell super-resolution imaging with synthetic fluorophores. Annu Rev Phys Chem 63:519–540. doi:10.1146/annurev-physchem-032811-112012
33. Fernandez-Suarez M, Ting AY (2008) Fluorescent probes for super-resolution imaging in living cells. Nat Rev Mol Cell Biol 9 (12):929–943
34. Chmyrov A, Keller J, Grotjohann T, Ratz M, d'Este E, Jakobs S, Eggeling C, Hell SW (2013) Nanoscopy with more than 100,000 'doughnuts'. Nat Methods 10(8):737–740. doi:10.1038/nmeth.2556
35. Dani A, Huang B, Bergan J, Dulac C, Zhuang X (2010) Superresolution imaging of chemical synapses in the brain. Neuron 68(5):843–856. doi:10.1016/j.neuron.2010.11.021
36. Sharonov A, Hochstrasser RM (2006) Wide-field subdiffraction imaging by accumulated binding of diffusing probes. Proc Natl Acad Sci U S A 103(50):18911–18916
37. Sharonov A, Hochstrasser RM (2007) Single-molecule imaging of the association of the cell-penetrating peptide Pep-1 to model membranes. Biochemistry 46(27):7963–7972
38. Sharonov A, Bandichhor R, Burgess K, Petrescu AD, Schroeder F, Kier AB, Hochstrasser RM (2008) Lipid diffusion from single molecules of a labeled protein undergoing dynamic association with giant unilamellar vesicles and supported bilayers. Langmuir 24 (3):844–850
39. Giannone G, Hosy E, Levet F, Constals A, Schulze K, Sobolevsky AI, Rosconi MP, Gouaux E, Tampe R, Choquet D, Cognet L (2010) Dynamic superresolution imaging of endogenous proteins on living cells at ultra-high density. Biophys J 99(4):1303–1310
40. Giannone G, Hosy E, Sibarita JB, Choquet D, Cognet L (2013) High-content super-resolution imaging of live cell by uPAINT. Methods Mol Biol 950:95–110. doi:10.1007/978-1-62703-137-0_7
41. Nair D, Hosy E, Petersen JD, Constals A, Giannone G, Choquet D, Sibarita JB (2013) Super-resolution imaging reveals that AMPA receptors inside synapses are dynamically organized in nanodomains regulated by PSD95. J Neurosci 33(32):13204–13224. doi:10.1523/jneurosci.2381-12.2013
42. Papadopulos A, Harper CB, Mavlyutov TA, Niranjan D, Glass NR, Cooper-White JJ, Sibarita JB, Choquet D, Davletov B, Meunier FA, Chazeau A, Mehidi A, Nair D, Gautier JJ, Leduc C, Chamma I, Kage F, Kechkar A, Thoumine O, Rottner K, Choquet D, Gautreau A, Sibarita JB, Giannone G (2014) Nanoscale segregation of actin nucleation and elongation factors determines dendritic spine protrusion. EMBO J 33(23):2745–2764. doi:10.1523/jneurosci.3757-14.201510.15252/embj.201488837
43. Deisseroth K (2015) Optogenetics: 10 years of microbial opsins in neuroscience. Nat Neurosci 18(9):1213–1225. doi:10.1038/nn.4091, http://www.nature.com/neuro/journal/v18/n9/abs/nn.4091.html#supplementary-information
44. Cho YK, Li D (2016) Optogenetics: basic concepts and their development. Methods Mol Biol 1408:1–17. doi:10.1007/978-1-4939-3512-3_1
45. Kaech S, Banker G (2006) Culturing hippocampal neurons. Nat Protoc 1(5):2406–2415. doi:10.1038/nprot.2006.356
46. Lippincott-Schwartz J, Patterson GH (2003) Development and use of fluorescent protein markers in living cells. Science 300 (5616):87–91
47. Ding B, Kilpatrick DL (2013) Lentiviral vector production, titration, and transduction of primary neurons. Methods Mol Biol 1018:119–131. doi:10.1007/978-1-62703-444-9_12
48. Schroder J, Benink H, Dyba M, Los GV (2009) In vivo labeling method using a genetic construct for nanoscale resolution microscopy. Biophys J 96(1):L1–L3
49. Klein T, Loschberger A, Proppert S, Wolter S, van de Linde S, Sauer M (2011) Live-cell dSTORM with SNAP-tag fusion proteins. Nat Methods 8(1):7–9. doi:10.1038/nmeth0111-7b
50. Bosch PJ, Correa IR Jr, Sonntag MH, Ibach J, Brunsveld L, Kanger JS, Subramaniam V (2014) Evaluation of fluorophores to label SNAP-tag fused proteins for multicolor single-molecule tracking microscopy in live cells. Biophys J 107(4):803–814. doi:10.1016/j.bpj.2014.06.040
51. Henriques R, Griffiths C, Hesper Rego E, Mhlanga MM (2011) PALM and STORM: unlocking live-cell super-resolution. Biopolymers 95(5):322–331. doi:10.1002/bip.21586

52. Olivier N, Keller D, Rajan VS, Gonczy P, Manley S (2013) Simple buffers for 3D STORM microscopy. Biomed Opt Express 4 (6):885–899. doi:10.1364/boe.4.000885
53. Mellroth P, Daniels R, Eberhardt A, Ronnlund D, Blom H, Widengren J, Normark S, Henriques-Normark B (2012) LytA, major autolysin of *Streptococcus pneumoniae*, requires access to nascent peptidoglycan. J Biol Chem 287(14):11018–11029. doi:10.1074/jbc.M111.318584
54. Aitken CE, Marshall RA, Puglisi JD (2008) An oxygen scavenging system for improvement of dye stability in single-molecule fluorescence experiments. Biophys J 94(5):1826–1835. doi:10.1529/biophysj.107.117689
55. Almada P, Culley S, Henriques R (2015) PALM and STORM: into large fields and high-throughput microscopy with sCMOS detectors. Methods 88:109–121. doi:10.1016/j.ymeth.2015.06.004
56. Grover JR, Veatch SL, Ono A (2015) Basic motifs target PSGL-1, CD43, and CD44 to plasma membrane sites where HIV-1 assembles. J Virol 89(1):454–467. doi:10.1128/jvi.02178-14
57. Ovesný M, Krizek P, Borkovec J, Svindrych Z, Hagen GM (2014) ThunderSTORM: a comprehensive ImageJ plug-in for PALM and STORM data analysis and super-resolution imaging. Bioinformatics 30(16):2389–2390. doi:10.1093/bioinformatics/btu202
58. Kaech S, Banker G, Stork P (2007) Putting on the RITz. Sci STKE 2007(416):pe71. doi:10.1126/stke.4162007pe71
59. Banker GA, Cowan WM (1977) Rat hippocampal neurons in dispersed cell culture. Brain Res 126(3):397–425
60. Kaech S, Huang CF, Banker G (2012) General considerations for live imaging of developing hippocampal neurons in culture. Cold Spring Harb Protoc 2012(3):312–318. doi:10.1101/pdb.ip068221
61. Kaech S, Huang CF, Banker G (2012) Long-term time-lapse imaging of developing hippocampal neurons in culture. Cold Spring Harb Protoc 2012(3):335–339. doi:10.1101/pdb.prot068239
62. Kaech S, Huang CF, Banker G (2012) Short-term high-resolution imaging of developing hippocampal neurons in culture. Cold Spring Harb Protoc 2012(3):340–343. doi:10.1101/pdb.prot068247
63. MacGillavry HD, Blanpied TA (2013) Single-molecule tracking photoactivated localization microscopy to map nano-scale structure and dynamics in living spines. Curr Protoc Neurosci 2(220):2.20.1–2.20.19. doi:10.1002/0471142301.ns0220s65
64. Karra D, Dahm R (2010) Transfection techniques for neuronal cells. J Neurosci 30 (18):6171–6177. doi:10.1523/jneurosci.0183-10.2010
65. Axelrod D (2001) Total internal reflection fluorescence microscopy in cell biology. Traffic 2(11):764–774
66. Chung E, Kim YH, Tang WT, Sheppard CJ, So PT (2009) Wide-field extended-resolution fluorescence microscopy with standing surface-plasmon-resonance waves. Opt Lett 34 (15):2366–2368
67. Chung E, Kim D, So PT (2006) Extended resolution wide-field optical imaging: objective-launched standing-wave total internal reflection fluorescence microscopy. Opt Lett 31(7):945–947
68. Fu Y, Winter PW, Rojas R, Wang V, McAuliffe M, Patterson GH (2016) Axial superresolution via multiangle TIRF microscopy with sequential imaging and photobleaching. Proc Natl Acad Sci U S A 113(16):4368–4373. doi:10.1073/pnas.1516715113
69. Tokunaga M, Imamoto N, Sakata-Sogawa K (2008) Highly inclined thin illumination enables clear single-molecule imaging in cells. Nat Methods 5(2):159–161. doi:10.1038/nmeth1171
70. Cella Zanacchi F, Lavagnino Z, Perrone Donnorso M, Del Bue A, Furia L, Faretta M, Diaspro A (2011) Live-cell 3D super-resolution imaging in thick biological samples. Nat Methods 8(12):1047–1049. doi:10.1038/nmeth.1744
71. Dave R, Terry DS, Munro JB, Blanchard SC (2009) Mitigating unwanted photophysical processes for improved single-molecule fluorescence imaging. Biophys J 96(6):2371–2381. doi:10.1016/j.bpj.2008.11.061
72. Staudt T, Lang MC, Medda R, Engelhardt J, Hell SW (2007) 2,2′-thiodiethanol: a new water soluble mounting medium for high resolution optical microscopy. Microsc Res Tech 70 (1):1–9. doi:10.1002/jemt.20396
73. Gould TJ, Verkhusha VV, Hess ST (2009) Imaging biological structures with fluorescence photoactivation localization microscopy. Nat Protoc 4(3):291–308. doi:10.1038/nprot.2008.246
74. Gould TJ, Hess ST, Bewersdorf J (2012) Optical nanoscopy: from acquisition to analysis. Annu Rev Biomed Eng 14:231–254. doi:10.1146/annurev-bioeng-071811-150025
75. Thompson RE, Larson DR, Webb WW (2002) Precise nanometer localization analysis for

individual fluorescent probes. Biophys J 82 (5):2775–2783. doi:10.1016/s0006-3495(02) 75618-x

76. MacGillavry HD, Song Y, Raghavachari S, Blanpied TA (2013) Nanoscale scaffolding domains within the postsynaptic density concentrate synaptic AMPA receptors. Neuron 78(4): 615–622. doi:10.1016/j.neuron.2013.03.009
77. Henriques R, Lelek M, Fornasiero EF, Valtorta F, Zimmer C, Mhlanga MM (2010) QuickPALM: 3D real-time photoactivation nanoscopy image processing in ImageJ. Nat Methods 7(5):339–340
78. Dedecker P, Duwé S, Neely RK, Zhang J (2012) Localizer: fast, accurate, open-source, and modular software package for superresolution microscopy. J Biomed Opt 17(12):126008. doi:10.1117/1.jbo.17.12.126008
79. Ruckebusch C, Bernex R, Allegrini F, Sliwa M, Hofkens J, Dedecker P (2015) Mapping pixel dissimilarity in wide-field super-resolution fluorescence microscopy. Anal Chem 87 (9):4675–4682. doi:10.1021/ac504295p
80. Geisler C, Hotz T, Schonle A, Hell SW, Munk A, Egner A (2012) Drift estimation for single marker switching based imaging schemes. Opt Express 20(7):7274–7289. doi:10.1364/oe.20.007274
81. Hu YS, Cang H, Lillemeier BF (2016) Super-resolution imaging reveals nanometer- and micrometer-scale spatial distributions of T-cell receptors in lymph nodes. Proc Natl Acad Sci U S A. doi:10.1073/pnas.1512331113
82. Yildiz A, Selvin PR (2005) Fluorescence imaging with one nanometer accuracy: application to molecular motors. Acc Chem Res 38 (7):574–582
83. Annibale P, Vanni S, Scarselli M, Rothlisberger U, Radenovic A (2011) Quantitative photo activated localization microscopy: unraveling the effects of photoblinking. PLoS One 6(7), e22678. doi:10.1371/journal.pone.0022678
84. Ripley BD (1977) Modelling spatial patterns. J R Stat Soc Series B Stat Methodol 39 (2):172–212. doi:10.2307/2984796
85. Ripley BD (1979) Tests of "randomness" for spatial point patterns. J R Stat Soc Series B Stat Methodol 41(3):368–374
86. Appleyard ST, Witkowski JA, Ripley BD, Shotton DM, Dubowitz V (1985) A novel procedure for pattern analysis of features present on freeze-fractured plasma membranes. J Cell Sci 74:105–117
87. Kiskowski MA, Hancock JF, Kenworthy AK (2009) On the use of Ripley's K-function and its derivatives to analyze domain size. Biophys J 97(4):1095–1103. doi:10.1016/j.bpj.2009.05.039
88. Owen DM, Rentero C, Rossy J, Magenau A, Williamson D, Rodriguez M, Gaus K (2010) PALM imaging and cluster analysis of protein heterogeneity at the cell surface. J Biophotonics 3(7):446–454
89. Sengupta P, Jovanovic-Talisman T, Skoko D, Renz M, Veatch SL, Lippincott-Schwartz J (2011) Probing protein heterogeneity in the plasma membrane using PALM and pair correlation analysis. Nat Methods 8(11): 969–975
90. Sengupta P, Jovanovic-Talisman T, Lippincott-Schwartz J (2013) Quantifying spatial organization in point-localization superresolution images using pair correlation analysis. Nat Protoc 8(2):345–354
91. Veatch SL, Machta BB, Shelby SA, Chiang EN, Holowka DA, Baird BA (2012) Correlation functions quantify super-resolution images and estimate apparent clustering due to over-counting. PLoS One 7(2), e31457
92. Kusumi A, Sako Y, Yamamoto M (1993) Confined lateral diffusion of membrane receptors as studied by single particle tracking (nanovid microscopy). Effects of calcium-induced differentiation in cultured epithelial cells. Biophys J 65(5):2021–2040
93. Suzuki K, Ritchie K, Kajikawa E, Fujiwara T, Kusumi A (2005) Rapid hop diffusion of a G-protein-coupled receptor in the plasma membrane as revealed by single-molecule techniques. Biophys J 88(5):3659–3680. doi:10.1529/biophysj.104.048538
94. Almarza G, Sánchez F, Barrantes FJ (2014) Transient cholesterol effects on nicotinic acetylcholine receptor surface mobility. PLoS One 9(6), e100346. doi:10.1371/journal.pone.0100346
95. Patterson G, Davidson M, Manley S, Lippincott-Schwartz J (2010) Superresolution imaging using single-molecule localization. Annu Rev Phys Chem 61:345–367

Chemical Methods for Monitoring Protein Fatty Acylation

Xiaoqiu Yuan and Howard C. Hang

Abstract

Protein fatty acylation encompasses different forms of lipidation and plays critical roles in regulating protein membrane binding and trafficking, stability, and activity. Due to the importance of protein fatty acylation in many different biological processes, there has been significant interest in sensitive detection and enrichment methods. To facilitate the analysis of protein fatty acylation in biology, fatty acid analogs bearing alkyne or azide tags have been developed that enable fluorescent imaging and proteomic profiling of modified proteins using bioorthogonal chemistry methods. In this chapter, we will briefly introduce various kinds of protein fatty acylation, their regulation and function, as well as associations with human diseases. The focus of the chapter will be on metabolic labeling using chemical reporters of protein fatty acylation and other complementary approaches to analyze protein S-fatty acylation, including ABE, acyl-RAC, and acyl-PEG exchange methods.

Keywords Bioorthogonal chemistry, Chemical reporters, Fatty acylation, N-Myristoylation, S-Palmitoylation

1 Introduction

The covalent attachment of fatty acids to proteins provides an important mechanism to alter the hydrophobicity of proteins and regulate their membrane affinity and activity in all kingdoms of life (Fig. 1) [1–3]. Protein fatty acylation can be divided into different subclasses depending on the type of lipid modification and linkage to protein side chains. In bacteria, fatty acids are primarily attached to the N-terminus of canonical lipoproteins through N-acyl S-diacylglyceryl group [4] but can also be directly attached to lysine (Lys) or cysteine (Cys) side chains [4, 5]. Plants and eukaryotes have different forms of fatty-acylated proteins that can be divided into N-myristoylation, S-palmitoylation, O-fatty acylation, and Lys fatty acylation (Fig. 1). Protein S-fatty acylation is reversible and can act like a switch to regulate protein activity, association with specific membrane domains, and other proteins as well as controlled interplay with other posttranslational modifications (Fig. 2). Due to the importance of protein fatty acylation in

Arun K. Shukla (ed.), *Chemical and Synthetic Approaches in Membrane Biology*, Springer Protocols Handbooks, (2017) 199–217, DOI 10.1007/8623_2016_3, Published online: 28 September 2016

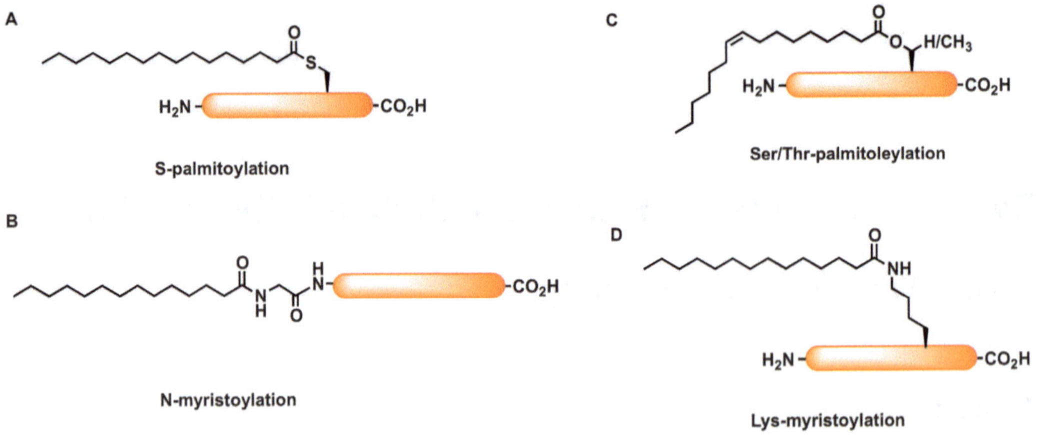

Fig. 1 Different types of fatty-acylated proteins. (**a**) S-Palmitoylation. (**b**) N-Myristoylation. (**c**) Palmitoylation on Ser or Thr. (**d**) Lys myristoylation

regulating protein function in many different biological processes, there has been significant interest in sensitive detection and enrichment methods. Unlike phosphorylation and other posttranslational modifications, no robust antibodies specific to lipid-modified peptides or proteins are readily available. Bioinformatic predictions of fatty acylation are also limited since some lipid modifications, S-fatty acylation in particular, do not have clear consensus motifs (Table 1). Historically, the analysis of protein fatty acylation relied on radioactive lipid labeling [15, 16]. However, this classical method is cumbersome, potentially hazardous, and has relatively low sensitivity that often requires long exposure times, from days to weeks. More sensitive detection methods for protein fatty acylation have been recently developed and described in this chapter.

1.1 Lipid Chemical Reporters for Bioorthogonal Detection

Bioorthogonal chemistry developed in the recent years has permitted improved detection and large-scale analysis of many different protein and nucleic acid modifications [1, 17]. The two-step strategy involves metabolic labeling with chemical reporters and then detection using bioorthogonal chemistry with secondary reagents such as fluorophores or affinity tags (Fig. 3) [18–20]. The advantages of separating incorporation and detection are that (1) sterically demanding fluorophores or affinity tags would not interfere with enzymatic incorporation and (2) bioorthogonal reaction with appropriately functionalized fluorophores or biotin probes provides more sensitive and immediate detection compared to radioactive methods. The azide or alkyne chemical reporters are typically designed to mimic natural metabolites (i.e., lipids) to facilitate recognition by enzymes in vitro and in cells (Table 2). For example, alk-12 efficiently labels N-myristoylated and S-fatty-acylated proteins [20], whereas longer-chain fatty acid reporters such as alk-16 preferentially label S-fatty-acylated proteins [20, 29, 30] and can be

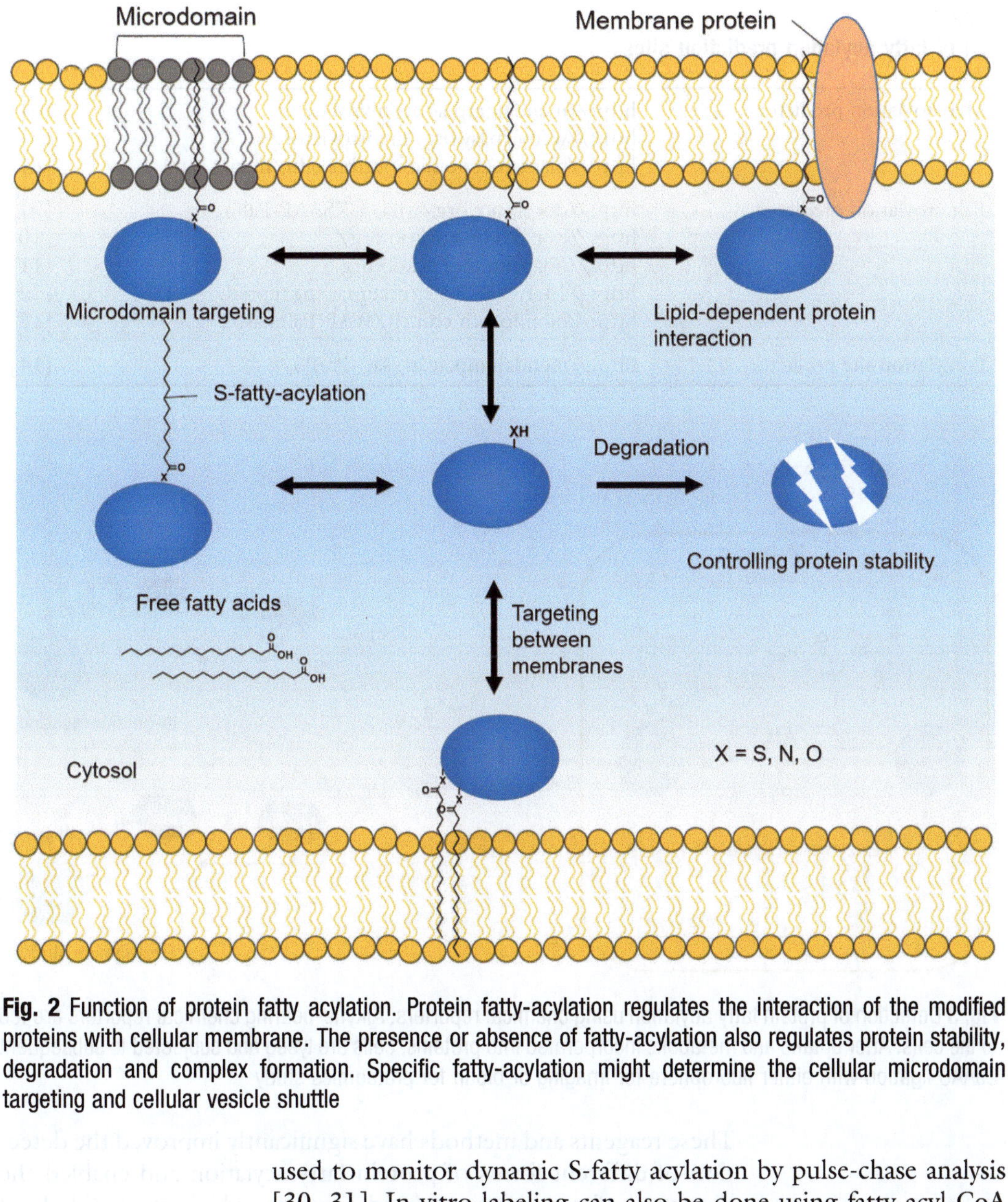

Fig. 2 Function of protein fatty acylation. Protein fatty-acylation regulates the interaction of the modified proteins with cellular membrane. The presence or absence of fatty-acylation also regulates protein stability, degradation and complex formation. Specific fatty-acylation might determine the cellular microdomain targeting and cellular vesicle shuttle

used to monitor dynamic S-fatty acylation by pulse-chase analysis [30, 31]. In vitro labeling can also be done using fatty acyl-CoA reporter analogs. Although both azide and alkyne reporters have been used in the past, alkyne chemical reporters and azide reagents provide less nonspecific labeling and better signal to noise with Cu (I)-catalyzed azide-alkyne cycloaddition (CuAAC) (Fig. 4) [32]. In addition to N-myristoylation and S-palmitoylation, alkyne fatty acid reporters can be used to monitor Lys fatty acylation [33]. Beyond fluorescence detection, chemical reporter-labeled proteins can be functionalized with affinity tags for the identification and quantitative proteomic analysis of fatty-acylated proteins (Fig. 3).

Table 1
List of fatty acylation prediction sites

Myristoylation predictor	http://mendel.imp.ac.at/myristate/	[6]
	http://web.expasy.org/myristoylator/	[7]
	http://plantsp.genomics.purdue.edu/myrist.html	[8]
Palmitoylation predictor	http://doc.aporc.org/wiki/CKSAAP-Palm	[9]
	http://csspalm.biocuckoo.org/	[10]
	http://nbapalm.biocuckoo.org/	[11]
	http://14.139.227.92/mkumar/palmpred/	[12]
	http://bioinfo.ncu.edu.cn/WAP-Palm.aspx	[13]
Prenylation site predictor	http://mendel.imp.ac.at/sat/PrePS/	[14]

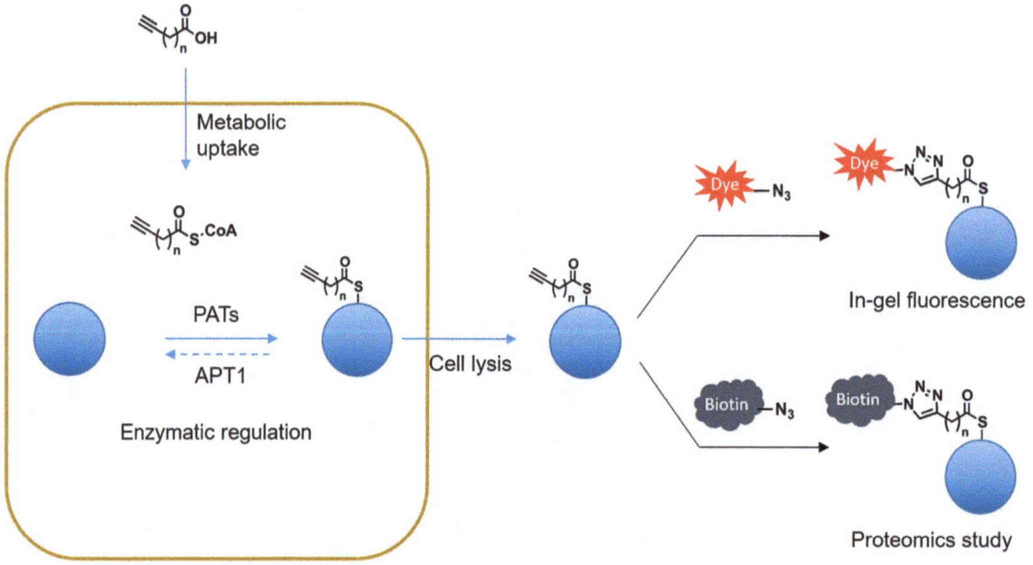

Fig. 3 Detection of protein fatty acylation using chemical reporters. Alkyne-bearing chemical reporters are fed to the cells. After uptake and metabolic incorporated into proteins, cells are lysed and subjected to subsequent CuAAC ligation with either fluorophore for imaging or biotin for proteomics study

These reagents and methods have significantly improved the detection of different forms of protein fatty acylation and enabled the discovery of novel fatty-acylated proteins, which are highlighted with a few examples below.

1.2 Selective Chemical Labeling Methods for Protein S-Fatty Acylation

Complementary chemical labeling methods have been developed for analyzing protein S-fatty acylation (Fig. 1a). Unlike N-/O-acylation, thioester linkage is sensitive to nucleophilic reagents that can be exploited for detection and enrichment. Drisdel and Green originally developed acyl-biotin exchange (ABE) chemistry to take advantage of the free thiol that is liberated by hydroxylamine (NH_2OH) to be labeled with a Cys-reactive reagent [34]. In the

Table 2
List of chemical reporters for fatty acylation

Acylation	Reporter chemical structure	References
S-Palmitoylation	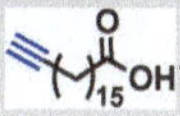, alk-16 or ODYA	[21, 22]
	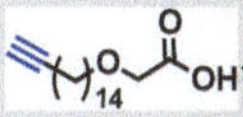, HDYOA	[23]
	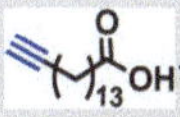, alk-14	[20, 22]
	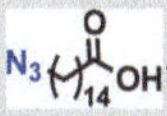, az-15	[20, 24]
	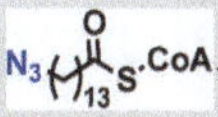, az-14-CoA	[25]
N-Myristoylation	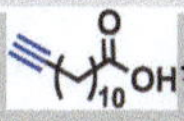, alk-11	[22]
	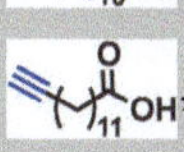, alk-12	[20, 22, 26]
	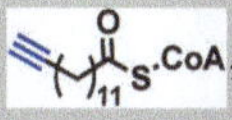, alk-12-CoA	[26]
	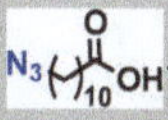, az-11	[26]
	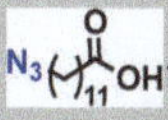, az-12	[20, 24, 27]
	, az-11-CoA	[26, 28]

first step, cell extracts are treated with N-ethylmaleimide (NEM) to block free Cys residues. One-half of the sample is then treated with NH_2OH to remove fatty acids from Cys residues, whereas the other half is the control sample. Reactive biotin reagents like HPDP-Biotin (N-[6-(Biotinamido)hexyl]-3′-(2′-pyridyldithio)propionamide) can then be used to label newly liberated Cys residues. Biotinylated proteins are then captured on a streptavidin column and eluted. Eluted proteins can be either visualized by sodium dodecyl sulfate (SDS)-PAGE or analyzed by proteomics (Fig. 4a). An S-palmitoylated protein should be present in the NH_2OH-treated sample but absent from the control sample.

ABE is readily adapted to immunoblotting techniques and is also adaptable to mass spectrometric-based identification of individual S-acylated proteins [35–37]. However, the detection of biotinylated proteins requires repeated protein precipitations, SDS neutralization, and streptavidin pulldown. An alternative to ABE that uses the detection of S-acylated proteins via acyl-resin-assisted capture (acyl-RAC) instead of biotinylation was proposed [38–40]. Specifically, acyl-RAC method was initially described as a methodology to identify S-nitrosylation sites in

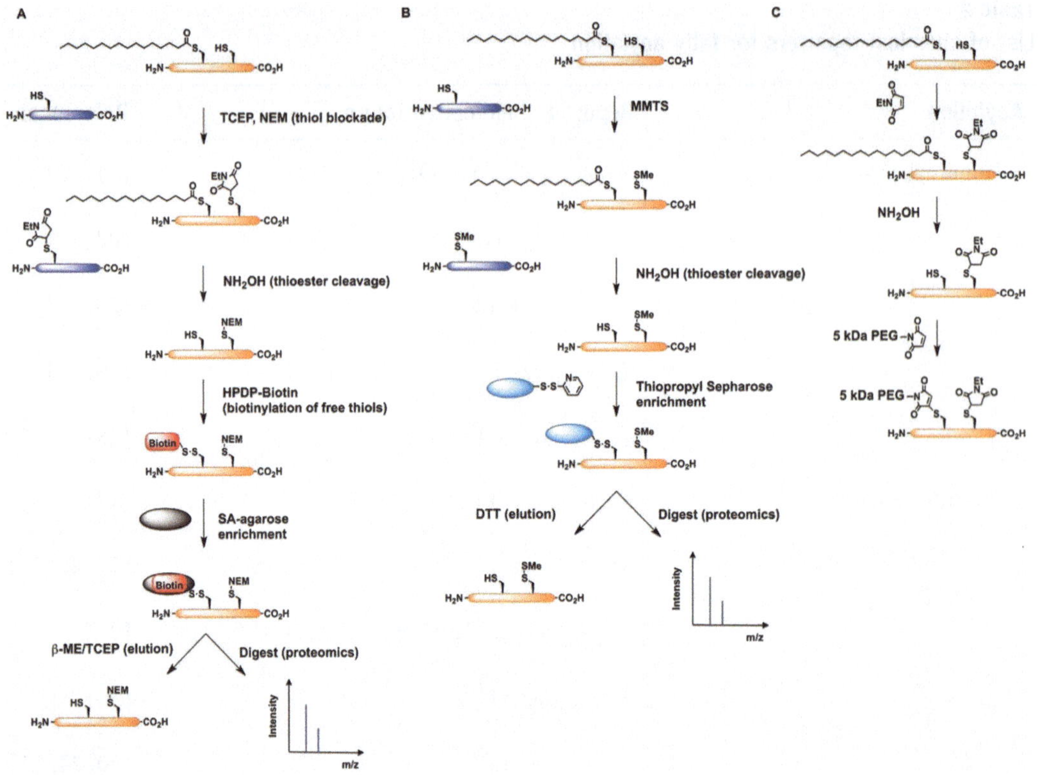

Fig. 4 Methods for analysis of fatty-acylated proteins. (**a**) Acyl-biotin exchange (ABE). (**b**) Acyl-RAC. (**c**) Acyl-PEG exchange (APE)

proteins and later developed to be applied on S-fatty acylation. Coupled with mass spectrometry proteomics, acyl-RAC can be also used to enrich and identify S-acylproteome (Fig. 4b). Compared with ABE, which involves multiple steps of precipitation and resolubilization of samples, acyl-RAC method uses Thiopropyl Sepharose to eliminate the biotin enrichment steps so the purification scheme is relatively simplified [41]. This method can also be generalized to study other cysteine-based reversible modifications, including disulfide formation, S-nitrosylation (SNO), and S-glutathionylation (SSG), by various reducing reagents such as ascorbate, glutaredoxin, and DTT or NH_2OH, respectively [42]. Acyl-RAC has been used to detect S-fatty acylation of overexpressed and endogenous H-Ras in mammalian cell system in an NH_2OH-dependent manner. Double mutants, C181 and 184S, which are not S-acylated, were not detected, which further validate the specificity of acyl-RAC method [39]. Coupled with mass spectrometry and isobaric labeling method, acyl-RAC was able to identify new S-acylated substrates, such as the β-subunit of the protein translocating system (Sec61b), ribosomal protein S11 (Rps11), and microsomal glutathione S-transferase 3 (MGST3) [39].

To evaluate endogenous levels of S-fatty acylation, our laboratory developed a mass-shift labeling method that exploits the NH_2OH sensitivity of thioesters and selective reactivity of Cys residues for site-specific alkylation with maleimide-functionalized polyethylene glycol reagents, termed acyl-PEG exchange (APE) (Fig. 4c) [43], which is similar to acyl-PEG switch method [44]. APE induces a mass shift on S-acylated proteins that can be readily monitored by Western blot analysis of target proteins and circumvents the need for metabolic labeling or affinity enrichment of proteins. More importantly, acyl-PEG exchange method reveals the ratio of unmodified versus S-fatty-acylated proteins or multiple sites of S-fatty acylation, which is critical for understanding how quantitative differences in S-fatty acylation level control protein function and associated cellular phenotypes. APE provides a sensitive and readily accessible method of evaluating endogenous S-fatty acylation levels and should facilitate the quantitative analysis of this dynamic lipid modification in diverse cell types and tissues [43].

Compared to the chemical reporter method, S-acylation-specific labeling methods do not require metabolic labeling so they can be used to analyze S-acylated proteins in cell lysates, tissues, and biofluids [41]. ABE has been used to analyze global protein palmitoylation in the yeast, which identified 35 new S-palmitoyl proteins including many soluble N-ethylmaleimide-sensitive factor attachment protein receptor (SNARE) proteins and amino acid permeases as well as other participants in cellular signaling and membrane trafficking pathway [35]. Nonetheless, it should be noted that other non-fatty-acylated proteins can be modified with thioesters (i.e., E2 and E3 ubiquitin-conjugating enzymes) and also be labeled. These methods and chemical reporter labeling are thus complementary and are best used together to fully validate protein S-fatty acylation.

1.3 S-Fatty Acylation

Protein S-fatty acylation is the posttranslational modification of long-chain fatty acids onto Cys residues through a thioester linkage [45] (Fig. 1a). This form of fatty acylation is frequently referred to as S-palmitoylation, since 16-carbon palmitic acid is the predominant fatty acid. However, other long-chain and unsaturated fatty acids have also been observed [46]. Historically, radioactive labeling using [^{3}H]palmitate had been the major method to study palmitoylation. G-protein-coupled receptors (GPCRs) [47], Gα subunits [48], Src family kinases [49], Ras proteins [50], and soluble N-ethylmaleimide-sensitive factor attachment protein receptor (SNARE) [51] are among the list of proteins that were identified using this approach. Protein S-fatty acylation is a reversible process, a key property that distinguishes it from other lipid modifications. Pulse-chase experiments on N-Ras with [^{3}H]palmitate highlighted the reversibility of S-fatty acylation (half-life ~20 min) [52], which is 10–20 times faster than turnover of the protein [53, 54]. However, marked differences can exist in the turnover rates of different S-fatty acylation sites in the same protein [55]. Indeed, rates of palmitate turnover at individual Cys residues can be dramatically different. For example, although S-palmitoylation at the proximal site of the β-adrenergic receptor is remarkably stable, S-palmitoylation at the distal site is rapidly turned over [55]. Major functions of protein S-fatty acylation include targeting proteins to discrete intracellular membrane compartments, control of protein stability, and modulation of protein-protein interactions [45]. Furthermore, the reversibility of S-fatty acylation offers spatial and temporal control of protein function, which regulates stable association and release of peripheral membrane proteins from ordered lipid bilayers (Fig. 2).

While protein S-palmitoylation was first documented more than 30 years ago [56], the enzymes responsible for this modification took over 20 years to identify [57, 58]. S-Fatty acylation is enzymatically mediated by a family of zinc finger DHHC (zDHHC) motif-containing protein acyltransferases (DHHC-PATs) that catalyze S-palmitoylation. There are seven DHHC-PATs in budding yeast [59], five in fission yeast [60], and more than 20 DHHC-PATs in mammalian species, all of which are multi-pass transmembrane proteins. Early work on DHHC-PAT enzymatic activity toward postsynaptic density protein of 95 kDa (PSD-95) showed that DHHC-PATs 2, 3, 7, and 15 were the most active [61]. The analysis *Schizosaccharomyces pombe* DHHC-PATs showed that regulated expression of the Erf2/4 DHHC-PAT complex enhanced meiotic entry of fission yeast, which was primarily mediated through site-specific S-palmitoylation of small GTPase Rho3 [60]. More recent genetic studies have implicated DHHC-PATs in different aspects of mammalian physiology and diseases including neurological disorders and cancer [62]. Notably, recent enrichment and mass spectrometry-based proteomics approaches have significantly expanded the scope and potential functions of dynamic protein S-fatty acylation [41, 63].

1.4 N-Myristoylation

Protein N-myristoylation describes the covalent modification of N-terminal glycine (Gly) residues with a 14-carbon fatty acid (myristic acid) [64] (Fig. 1b). N-Myristoylation is mediated by N-myristoyltransferase (NMT), 50–60 kDa monomeric enzyme that catalyzes the transfer of myristate from myristoyl-CoA to suitable peptides and proteins. The structure of the yeast *Saccharomyces cerevisiae* NMT1p was solved as a ternary complex and revealed how myristoyl-CoA and peptide substrates bind to the enzyme [65]. N-Myristoylation is necessarily preceded by proteolysis to reveal an N-terminal glycine, the only completely conserved motif across all known NMT substrates. N-Myristoylation typically targets to proteins to membranes that are further regulated by additional binding motifs, such as polybasic regions, represented by the myristoylated Ala-rich C-kinase substrate (MARCKS) protein [66]. Alternatively, X-ray crystal structures of c-Abl 1b isoform show that the N-terminal myristoyl modification can be buried in hydrophobic pocket and induce an autoinhibited conformation [67]. These studies underscore the critical role that N-myristoylation plays not only in proper localization of proteins but also regulating their activity. N-Myristoylation can also occur posttranslationally during apoptosis, when cell death-induced proteolysis reveals N-terminal glycine residues for NMT modification [68].

Recent chemical proteomics approaches have identified many N-myristoylated proteins during normal cellular growth and apoptosis [27, 69], which significantly expands the scope of proteins with diverse functions that undergo cotranslational or posttranslational N-myristoylation, such as c-Abl [67], c-Src [70], Bid [71], PAK2 [72], and gelsolin [73], to name a few. Although usually non-eukaryotic species lack NMT that is required for this modification, there are instances of pathogen-host interaction around N-myristoylation. Indeed, since first discovered in the early 1980s, many viral proteins have been found to be myristoylated by NMTs of their eukaryotic hosts [6]. N-Myristoylation of viral proteins has roles in assembly, structure, budding, viral entry, and intracellular host interactions. Furthermore, bacteria effectors injected into host cells through type III secretion system (TTSS) can also be myristoylated [74]. A recent study found that *Shigella flexneri* virulence factor invasion plasmid antigen J (IpaJ) cleaves N-myristoylated glycines of eukaryotic proteins as a pathogenic mechanism in host cells [75, 76]. Specifically, mass spectrometry showed that IpaJ cleaved the peptide bond between N-myristoylated Gly2 and Asn3 of human ARF1, thereby providing a new mechanism for host secretory inhibition by a bacterial pathogen [75, 76]. These approaches should help uncover potential roles of N-myristoylation in host-pathogen interactions, cancer, and neurodegenerative disease.

1.5 O-Fatty Acylation

Protein O-acylation is a less common form of fatty acylation at serine or threonine residues. A notable example of O-acylation is O-palmitoleoylation on Wnt proteins (Fig. 1c). Wnts are secreted signaling molecules that are covalently modified with palmitoleic acid (16:1) at a serine residue [77, 78]. Other O-acylated proteins include ghrelin [79, 80], PLTX-II [81], and histone H4 protein [82]. Serine-3 of ghrelin is acylated with an eight-carbon fatty acid, octanoate, which is required for its endocrine actions. Notably, Wnt has been studied extensively because first of all, Wnt protein family is one of the largest secreted protein families that are responsible for extracellular signaling in virtually all animal development stages and secondly posttranslational fatty acylation is essential for its activity. Interestingly, mass spectrometry studies and recent bioorthogonal fatty acid labeling on Wnt3a have shown that Ser209 is O-palmitoleoylated rather than Cys77 [77, 78]. The enzyme that is responsible for the modification is called porcupine, a member of membrane-bound O-acyl transferase (MBOAT). Chemical reporter methods also demonstrated that porcupine is S-palmitoylated, which negatively regulates Wnt signaling [83].

1.6 ε-N-Lys Acylation

Fatty acylation on Lys side chain via formation of an amide bond is another potentially regulated lipid modification (Fig. 1d). Tumor necrosis factor alpha (TNF-α) precursor and interleukin-1-α were shown to undergo myristoylation on specific inner lysine residues [84, 85]. ε-N-Lys acylation has been long thought to be non-hydrolyzable due to the irreversible nature of amide bond linkage, but the recent discovery of SIRT6, a sirtuin family NAD^+-dependent protein deacetylase that can also remove long-chain fatty acids on Lys side chains, reignites interests on dynamics regulation of fatty N-acylation. SIRT6 promotes the secretion of TNF-α by removing the fatty acyl modification on K19 and K20 of TNF-α [33]. Subsequently, in vitro studies of all sirtuins to hydrolyze 13 different acyl groups have revealed the specificity of sirtuin enzymes for deacylating Lys modified with a broad range of fatty acids [86, 87]. Interestingly, free fatty acids (FFAs) could stimulate deacetylation activity of SIRT6 toward long-chain acylated peptides. Crystal structure of SIRT6 reveals that, different from the other four members of mammalian sirtuin family, there is a large hydrophobic pocket that can accommodate long-chain fatty acyl groups rather than shorter acetyl groups. The new findings of protein deacylation enzyme activity provide a road map to uncover the biological functions of protein ε-N-Lys fatty acylation modifications with previously unknown functions.

2 Materials

HeLa cells, HEK293T cells, or other cultured cells of interest were obtained from ATCC.

DMEM (Thermo Fisher). FBS and 2% charcoal-stripped fetal bovine serum (GE Healthcare Life Sciences, cat. 12676029)

50 mM alkyne fatty acid chemical reporter stock in DMSO alk-16 (Sigma-Aldrich, cat. O8382, and also see Charron et al. [20])

Brij97 (Sigma-Aldrich, cat. P6136). Triton X-100 (Sigma-Aldrich, cat. X100)

RIPA buffer (Thermo Fisher, cat. 89900)

N-Ethylmaleimide, NEM (Sigma-Aldrich, cat. 04259)

Hydroxylamine, HA (Sigma-Aldrich, cat. 159417)

HPDP-Biotin (Thermo Fisher, cat. 21341)

Streptavidin Agarose (SA-agarose, Thermo Fisher, cat. 20349)

Methyl methanethiosulfonate (MMTS, Thermo Fisher, cat. 23011)

Thiopropyl Sepharose (GE Healthcare Life Sciences, cat. 17-0420-01)

DTT (Thermo Fisher, cat. R0861)

Protease inhibitor cocktail (Roche, cat. 11697498001)

PMSF (Sigma-Aldrich, cat. P7626)

EDTA (Sigma-Aldrich, cat. EDS)

Benzonase (Sigma-Aldrich, cat. E1014)

BCA assay (Thermo Fisher, cat. 23227)

Neutralized tris(2-carboxyethyl)phosphine (TCEP, Thermo Fisher, cat. 77720)

Methoxy polyethylene glycol-maleimide (mPEG-Mal, 5 or 10 kDa, Sigma-Aldrich, 63187/712469)

Anti-calnexin (1:2000 ab22595, Abcam)

Anti-Pan Ras (1:500, Ras10, Millipore)

Anti-mouse IFITM3 (1:1000, ab15592, Abcam)

Anti-FLAG (1:1000, F1804, Sigma-Aldrich)

Anti-HA (1:1000, ab9134, Abcam)

HRP-conjugated anti-HA (3F10, Roche)

HRP-conjugated goat anti-rabbit (DC03L, Calbiochem)

HRP-conjugated goat anti-mouse (ab97023, Abcam)

3 Methods

3.1 Metabolic Labeling of Chemical Reporters of Protein Fatty Acylation in Living Cells

HeLa and HEK293T cells were cultured in DMEM (Thermo Fisher) and supplemented with 10% fetal bovine serum (FBS). Cells are maintained in a humidified 37°C incubator with 5% CO_2. Trypan blue exclusion was used to determine cell viability. Cells are treated with alk-16 (Sigma-Aldrich. cat. O8382, and also see Charron et al. [20]) at concentrations from 20 to 100 μm (final concentration) or DMSO control in 2% charcoal-stripped fetal bovine serum (GE Healthcare Life Sciences, cat. 12676029). Charcoal-stripped fetal bovine serum removes lipids in serum, minimizing uptake competition of natural lipids with chemical reporters. Cells are then incubated with chemical reporters for 1–3 h at 37°C depending on how fast protein S-acylation turn-overs. So this should be titrated for each protein of interest. Harvest the cells by scraping them into suspension and centrifuging 3 min at 1,000× *g*, 4°C.

3.1.1 Fluorescent Profiling of Fatty-Acylated Proteomes in Whole Cell Lysates

Chemical reporter-labeled cells (e.g., ~1 × 10^6 HeLa cells) are lysed by adding 50 μl of 4% SDS buffer with EDTA-free protease inhibitors and 1 μl (250 U) of Benzonase to the cell pellet. After quantifying the protein concentration by a standard BCA assay, aliquots of equal amounts of protein (~50 μg) of each sample are dispensed into 1.5 ml microcentrifuge tubes, and bring the volumes to 44.5 μl with 4% SDS buffer plus EDTA-free protease inhibitors. 5.5 μl of the click chemistry master mix (Table 3) is added to each protein sample for a final volume of 50 μl, and vortexed to mix. Click chemistry master mix should be prepared fresh each time, and $CuSO_4{\cdot}5H_2O$ solids should be used to prepare mix each time to avoid loss of reactivity during solution storage (oxidation of Cu(I)). After mixing, samples are incubated for 1 h at room temperature, and a chloroform/methanol (CM) precipitation is performed to remove residual dye and excessive compounds. After redissolving protein pellet into 4% SDS buffer, 4× loading buffer (Thermo Fisher, cat. 84788) and β-ME (or DTT) are added for SDS-PAGE analysis.

3.1.2 Fluorescent Visualization of Fatty Acylation of Protein of Interest

To analyze fatty acylation of the protein of interest, proteins labeled with chemical reporters are immunoprecipitated prior to click chemistry ligation with azido-rhodamine and SDS-PAGE. Similar to the previous method about global metabolic labeling, metabolically labeled cells are lysed in 1% Brij97 (Sigma-Aldrich, cat. P6136) buffer with EDTA-free protease inhibitors. Centrifuge 5 min at 1,000× *g*, 4°C, to remove cellular debris. Note that protease inhibitors containing EDTA are not compatible with click chemistry. Quantify protein concentration using BCA assay, and dispense 500 μg of protein for each sample into 1.5 ml microcentrifuge tubes. After bringing the volumes to 100 μl with 1% Brij97 buffer with EDTA-free protease inhibitors, protein samples are incubated with antibody-conjugated agarose beads, followed by washing three times with lysis buffer. Resuspend beads with 22.25 μl of lysis buffer, and add 2.75 μl of the click chemistry reagent master

Table 3
Click chemistry master mix recipe

Name	Concentration	Volume
Azido-/alkynyl-rhodamine	5 mM	1 μl
TCEP[a]	50 mM	1 μl
TBTA[b] in 1:4 (v/v) DMSO/butanol	2 mM	2.5 μl
$CuSO_4$	50 mM	1 μl

[a]Tris(2-carboxyethyl)phosphine hydrochloride (TCEP)
[b]Tris[(1-benzyl-1H-1,2,3-triazol-4-yl)methyl] amine (TBTA)

mix (Table 3) to each sample. After incubation for 1 h, beads are washed three times with RIPA buffer (Thermo Fisher, cat. 89900). 1 × loading buffer with 2-ME or tris(2-carboxyethyl)phosphine (TCEP) is added to the beads, and boil on 95°C heat block for 5 min.

3.1.3 Chloroform/Methanol Protein Precipitation

For chloroform/methanol protein precipitation step, 200 μl methanol, 75 μl chloroform, and 150 μl water are added in each 50 μl sample tube followed by vortexing and centrifuging for 15 min at 20,000× *g* at 4°C. Reagents should be ice-cold in this step. Upper aqueous phase is then discarded, leaving the lower organic phase and the white layer of protein between the two layers. The lower organic layer contains the pink-colored rhodamine. 1 ml of ice-cold methanol is added to each sample, and the mixture is centrifuged for 10 min at 20,000× *g*, 4°C. Protein pellet will sink to the bottom of the tube. Carefully remove all of the liquid by pipetting, leaving the white protein pellet at the bottom. Pay attention not to disturb the pellet. Protein pellet is then washed three times by adding 1 ml of ice-cold methanol and inverting the tube followed by centrifuging for 10 min at 20,000× *g*, 4°C. Methanol is then removed carefully by pipetting, leaving behind the protein pellet. The tube is left open for 20 min at room temperature to evaporate methanol. 50 μl of 4% SDS buffer with EDTA-free protease inhibitors is added to redissolve the protein pellets.

3.2 Acyl-Biotin Exchange Method for S-Fatty Acylation Detection and Enrichment

Mammalian cells are growing in suitable media and dishes. Cells are harvested after they reach certain confluency, and then Triton X-100 (Sigma-Aldrich, cat. X100) is added to 1.7% for detergent solubilization. Incubate with end-over-end rotation at 4°C for 1 h. Remove particulates and unbroken cells by low-speed centrifugation (250× *g*, 4°C, 5 min). Chloroform/methanol (CM) precipitation is performed using the method above. To each protein pellet, add 300 ml 4% SDS buffer with 10 mM NEM (Sigma-Aldrich, cat. 04259). The mixture is incubated for 10 min at 37°C with occasional agitation of the tube to dissolve pellet. The protein denaturation accompanying this step facilitates access of NEM to Cys that are buried within the folded protein interior. 900 ml of lysis buffer with 1 mM NEM, 1 × protease inhibitor, 1 mM PMSF, and 0.2% Triton X-100 is added to each tube. The resulting solution is transferred to 1.5 ml screw cap centrifuge tubes and incubated on rocker overnight at 4°C with gentle rocking. NEM is removed from samples by three sequential CM precipitations. After the third and final CM precipitation, protein pellet is dissolved in 250 ml 4% SDS buffer at 37°C for 10 min. Samples are diluted fivefold with the addition of 960 ml of the hydroxylamine (Sigma-Aldrich, cat. 159417) containing buffer followed by incubation at room temperature for 1 h with end-over-end rotation. CM precipitation is performed, and each resulting protein pellet is dissolved in 240 μl 4% SDS buffer with addition of 960 μl HPDP-Biotin (Thermo Fisher, cat. 21341) for room temperature incubation for 1 h. Then three CM precipitations are performed to remove unreacted HPDP-Biotin before loading to Streptavidin Agarose (SA-agarose, Thermo Fisher, cat. 20349) affinity incubate at room temperature for 30 min (dilute SDS to 0.1% before adding samples to Streptavidin Agarose). Supernatant is centrifuged and transferred to new tubes containing Streptavidin Agarose for end-over-end rotation incubation for 90 min. Elute bound proteins through reduction of the protein-biotin disulfide linkage or on-beads digestion for proteomics study.

3.3 Acyl-RAC Method for S-Fatty Acylation Detection and Enrichment

Free cysteines can be blocked by either NEM similar to ABE method or methyl methanethiosulfonate (MMTS, Thermo Fisher, cat. 23011). Take MMTS blocking reagent as an example; crude protein lysates obtained from cell lysis were purified to get rid of cell debris and quantified with BCA assay using BSA as the standard. Equal amounts of protein are diluted to a concentration of 2 mg/ml in blocking buffer (100 mM HEPES, 1.0 mM EDTA, 2.5% SDS, 0.1% MMTS, pH 7.5) and incubated at 40°C for 10 min with frequent vortexing. Proteins were precipitated using chloroform/methanol or cold acetone precipitation. Following centrifugation of the solution at 5,000× *g* for 10 min, the pellet is extensively washed with 70% acetone, resuspended in 300 μl of binding buffer (100 mM HEPES, 1.0 mM EDTA, 1% SDS, pH 7.5), and added to 40 μl of prewashed Thiopropyl Sepharose (GE Healthcare Life Sciences, cat. 17-0420-01). To this, mixture is added 40 μl of either 2 M NH_2OH (freshly prepared in water from HCl salt and brought to pH 7.5 with concentrated NaOH) or 2 M NaCl. Binding reactions are carried out on a rocker at room temperature for 2–4 h. Resins are washed several times before elution with buffer containing 50 mM DTT (Thermo Fisher, cat. R0861) for immunoblotting and proteomics study.

3.4 Acyl-PEG Exchange Method for S-Fatty Acylation Detection

Cell samples are lysed with 4% sodium dodecyl sulfate (SDS, Fisher) in triethanolamine (TEA) buffer (pH 7.3, 50 mM triethanolamine, 150 mM NaCl) containing 1× protease inhibitor cocktail (Roche, cat. 11697498001), 5 mM PMSF (Sigma-Aldrich, cat. P7626), 5 mM EDTA (Sigma-Aldrich, cat. EDS), and 1,500 units/ml Benzonase (Sigma-Aldrich, cat. E1014). The protein concentration of the cell lysate is then measured using a BCA assay (Thermo Fisher, cat. 23227) and adjusted to 2 mg/ml with lysis buffer. Typically, 200 μg of total protein in 92.5 μl of lysis buffer is treated with 5 μl of 200 mM neutralized tris(2-carboxyethyl)phosphine (TCEP, Thermo Fisher, cat. 77720) for a final concentration of 10 mM TCEP for 30 min with nutation. N-Ethylmaleimide (NEM, Sigma-Aldrich, cat. 04259), 2.5 μl from freshly made 1 M stock in ethanol, is added to a final concentration of 25 mM and incubated for 2 h at room temperature. Reductive alkylation of the proteins is then terminated by chloroform/methanol precipitation. The reactions are then mixed by inversion and centrifugation at 20,000× *g* for 5 min at 4°C. To pellet the precipitated proteins, the aqueous layer is removed, 1 ml of prechilled methanol is added, and the Eppendorf tube is inverted several times and centrifuged at 20,000× *g* for 3 min at 4°C. The supernatant is then decanted, and the protein pellet is washed once more with 800 μl of prechilled methanol, centrifuged again, and dried using a speed vacuum. To ensure complete removal of NEM from the protein pellets, the samples are resuspended with 100 μl of TEA buffer containing 4% SDS, warmed to 37°C for 10 min, briefly (~5 s) sonicated, and subjected to two additional rounds of chloroform/methanol precipitations as described above.

For hydroxylamine (NH_2OH) cleavage and methoxy polyethylene glycol-maleimide (mPEG-Mal) alkylation, the protein pellet is resuspended in 30 μl TEA buffer containing 4% SDS and 4 mM EDTA and treated with 90 μl of 1 M neutralized NH_2OH dissolved in TEA buffer pH 7.3, containing 0.2% Triton X-100 to obtain a final concentration of 0.75 M NH_2OH. Note that in this step, protease inhibitor cocktail or PMSF should be omitted, as these reagents can interfere with the NH_2OH reactivity. Control samples not treated with NH_2OH were diluted in 90 μl TEA buffer with 0.2% Triton X-100. Samples were incubated at room temperature for 1 h with nutation. The samples were then subjected to chloroform/methanol precipitation as described above and resuspended in 30 μl TEA buffer containing 4% SDS and 4 mM EDTA, warmed to 37°C for 10 min, briefly (~5 s) sonicated, and treated with 90 μl TEA buffer with 0.2% Triton X-100 and 1.33 mM methoxy polyethylene glycol-maleimide (mPEG-Mal, 5 or 10 kDa, Sigma-Aldrich, 63187/712469) for a final concentration of 1 mM mPEG-Mal. Samples were incubated for 2 h at room temperature with nutation before a final chloroform-methanol-H_2O precipitation. Dried protein pellets are resuspended in 50 μl 1 × Laemmli buffer (Bio-Rad) and then heated for 5 min at 95°C. Typically, 15 μl of the sample is loaded in 4–20% Criterion-TGX Stain-Free polyacrylamide gels (Bio-Rad), separated by SDS-PAGE, and analyzed by Western blot. For Western blots, primary antibodies used were anti-calnexin (1:2000 ab22595, Abcam), anti-Pan Ras (1:500, Ras10, Millipore), anti-mouse IFITM3 (1:1000, ab15592, Abcam) anti-FLAG (1:1000, F1804, Sigma-Aldrich), anti-HA (1:1000, ab9134, Abcam), and HRP-conjugated anti-HA (3F10, Roche). Secondary antibodies used are HRP-conjugated goat anti-rabbit (DC03L, Calbiochem) and HRP-conjugated goat anti-mouse (ab97023, Abcam). Protein detection is performed with ECL detection reagent (GE Healthcare Life Sciences) on a Bio-Rad ChemiDoc MP Imaging System.

References

1. Hang HC, Linder ME (2011) Exploring protein lipidation with chemical biology. Chem Rev 111:6341–6358. doi:10.1021/cr2001977
2. Resh MD (2006) Trafficking and signaling by fatty-acylated and prenylated proteins. Nat Chem Biol 2:584–590. doi:10.1038/nchembio834
3. Blaskovic S, Blanc M, van der Goot FG (2013) What does S-palmitoylation do to membrane proteins? FEBS J 280:2766–2774. doi:10.1111/febs.12263
4. Kovacs-Simon A, Titball RW, Michell SL (2011) Lipoproteins of bacterial pathogens. Infect Immun 79:548–561. doi:10.1128/IAI.00682-10
5. Al-Quadan T, Price CT, London N et al (2011) Anchoring of bacterial effectors to host membranes through host-mediated lipidation by prenylation: a common paradigm. Trends Microbiol. doi:10.1038/nrclinonc.2015.209
6. Maurer-Stroh S, Eisenhaber F (2004) Myristoylation of viral and bacterial proteins. Trends Microbiol 12:178–185. doi:10.1016/j.tim.2004.02.006
7. Bologna G, Yvon C, Duvaud S, Veuthey A-L (2004) N-Terminal myristoylation predictions by ensembles of neural networks. Proteomics 4:1626–1632. doi:10.1002/pmic.200300783
8. Podell S, Gribskov M (2004) Predicting N-terminal myristoylation sites in plant proteins. BMC Genomics 5:37. doi:10.1186/1471-2164-5-37
9. Wang X-B, Wu L-Y, Wang Y-C, Deng N-Y (2009) Prediction of palmitoylation sites using the composition of k-spaced amino acid pairs. Protein Eng Des Sel PEDS 22:707–712. doi:10.1093/protein/gzp055
10. Ren J, Wen L, Gao X et al (2008) CSS-Palm 2.0: an updated software for palmitoylation sites prediction. Protein Eng Des Sel 21:639–644. doi:10.1093/protein/gzn039
11. Xue Y, Chen H, Jin C et al (2006) NBA-Palm: prediction of palmitoylation site implemented in Naïve Bayes algorithm. BMC Bioinformatics 7:458. doi:10.1186/1471-2105-7-458
12. Kumari B, Kumar R, Kumar M (2014) PalmPred: an SVM based palmitoylation prediction method using sequence profile information. PLoS One 9, e89246. doi:10.1371/journal.pone.0089246
13. Shi S-P, Sun X-Y, Qiu J-D et al (2013) The prediction of palmitoylation site locations using a multiple feature extraction method. J Mol Graph Model 40:125–130. doi:10.1016/j.jmgm.2012.12.006
14. Maurer-Stroh S, Eisenhaber F (2005) Refinement and prediction of protein prenylation motifs. Genome Biol 6:R55. doi:10.1186/gb-2005-6-6-r55
15. Berthiaume L, Peseckis SM, Resh MD (1995) Synthesis and use of iodo-fatty acid analogs. Methods Enzymol 250:454–466
16. Schlesinger MJ, Magee AI, Schmidt MF (1980) Fatty acid acylation of proteins in cultured cells. J Biol Chem 255:10021–10024
17. Hang HC, Wilson JP, Charron G (2011) Bioorthogonal chemical reporters for analyzing protein lipidation and lipid trafficking. Acc Chem Res 44:699–708. doi:Research Support, N.I.H., Extramural Research Support, Non-U.S. Gov't
18. Grammel M, Hang HC (2013) Chemical reporters for biological discovery. Nat Chem Biol. doi:10.1038/nchembio.1296
19. Prescher JA, Bertozzi CR (2005) Chemistry in living systems. Nat Chem Biol 1:13–21. doi:10.1038/nchembio0605-13
20. Charron G, Zhang MM, Yount JS et al (2009) Robust fluorescent detection of protein fatty-acylation with chemical reporters. J Am Chem Soc 131:4967–4975. doi:10.1021/ja810122f
21. Charron G, Wilson J, Hang HC (2009) Chemical tools for understanding protein lipidation in eukaryotes. Curr Opin Chem Biol 13:382–391. doi:Review
22. Hannoush RN, Arenas-Ramirez N (2009) Imaging the lipidome: ω-alkynyl fatty acids for detection and cellular visualization of lipid-modified proteins. ACS Chem Biol 4:581–587. doi:10.1021/cb900085z
23. Yount JS, Charron G, Hang HC (2011) Bioorthogonal proteomics of 15-hexadecynyloxyacetic acid chemical reporter reveals preferential targeting of fatty acid modified proteins and biosynthetic enzymes. Bioorg Med Chem. doi:10.1016/j.bmc.2011.03.062
24. Hang HC, Geutjes E-J, Grotenbreg G et al (2007) Chemical probes for the rapid detection of fatty-acylated proteins in mammalian cells. J Am Chem Soc 129:2744–2745. doi:10.1021/ja0685001
25. Kostiuk MA, Corvi MM, Keller BO et al (2008) Identification of palmitoylated mitochondrial proteins using a bio-orthogonal azido-palmitate analogue. FASEB J Off Publ Fed Am Soc Exp Biol 22:721–732. doi:10.1096/fj.07-9199com

26. Heal WP, Wickramasinghe SR, Leatherbarrow RJ, Tate EW (2008) N-Myristoyl transferase-mediated protein labelling in vivo. Org Biomol Chem 6:2308–2315. doi:10.1039/b803258k
27. Martin DDO, Vilas GL, Prescher JA et al (2008) Rapid detection, discovery, and identification of post-translationally myristoylated proteins during apoptosis using a bio-orthogonal azidomyristate analog. FASEB J 22:797–806. doi:10.1096/fj.07-9198com
28. Heal WP, Wickramasinghe SR, Bowyer PW et al (2008) Site-specific N-terminal labelling of proteins in vitro and in vivo using N-myristoyl transferase and bioorthogonal ligation chemistry. Chem Commun 480–482. doi:10.1039/B716115H
29. Yount JS, Moltedo B, Yang Y-Y et al (2010) Palmitoylome profiling reveals S-palmitoylation-dependent antiviral activity of IFITM3. Nat Chem Biol 6:610–614. doi:10.1038/nchembio.405
30. Zhang MM, Tsou LK, Charron G et al (2010) Tandem fluorescence imaging of dynamic S-acylation and protein turnover. Proc Natl Acad Sci U S A 107:8627–8632. doi:10.1073/pnas.0912306107
31. Martin BR, Wang C, Adibekian A et al (2012) Global profiling of dynamic protein palmitoylation. Nat Methods 9:84–89. doi:10.1038/nmeth.1769
32. Thinon E, Hang HC (2015) Chemical reporters for exploring protein acylation. Biochem Soc Trans 43:253–261. doi:10.1042/BST20150004
33. Jiang H, Khan S, Wang Y et al (2013) SIRT6 regulates TNF-α secretion through hydrolysis of long-chain fatty acyl lysine. Nature 496:110–113. doi:10.1038/nature12038
34. Drisdel RC, Green WN (2004) Labeling and quantifying sites of protein palmitoylation. Biotechniques 36:276–285
35. Roth AF, Wan J, Bailey AO et al (2006) Global analysis of protein palmitoylation in yeast. Cell 125:1003–1013. doi:10.1016/j.cell.2006.03.042
36. Kang R, Wan J, Arstikaitis P et al (2008) Neural palmitoyl-proteomics reveals dynamic synaptic palmitoylation. Nature 456:904–909. doi:10.1038/nature07605
37. Yang W, Vizio DD, Kirchner M et al (2010) Proteome scale characterization of human S-acylated proteins in lipid raft-enriched and non-raft membranes. Mol Cell Proteomics 9:54–70. doi:10.1074/mcp.M800448-MCP200
38. Forrester MT, Thompson JW, Foster MW et al (2009) Proteomic analysis of S-nitrosylation and denitrosylation by resin-assisted capture. Nat Biotechnol 27:557–559. doi:10.1038/nbt.1545
39. Forrester MT, Hess DT, Thompson JW et al (2011) Site-specific analysis of protein S-acylation by resin-assisted capture. J Lipid Res 52:393–398. doi:10.1194/jlr.D011106
40. Ren W, Jhala US, Du K (2013) Proteomic analysis of protein palmitoylation in adipocytes. Adipocyte 2:17–28. doi:10.4161/adip.22117
41. Zhou B, An M, Freeman MR, Yang W (2014) Technologies and challenges in proteomic analysis of protein S-acylation. J Proteomics Bioinform 7:256–263. doi:10.4172/jpb.1000327
42. Guo J, Gaffrey MJ, Su D et al (2014) Resin-assisted enrichment of thiols as a general strategy for proteomic profiling of cysteine-based reversible modifications. Nat Protoc 9:64–75. doi:10.1038/nprot.2013.161
43. Percher A, Ramakrishnan S, Thinon E et al (2016) Mass-tag labeling reveals site-specific and endogenous levels of protein S-fatty acylation. Proc Natl Acad Sci U S A 113:4302–4307. doi:10.1073/pnas.1602244113
44. Howie J, Reilly L, Fraser NJ et al (2014) Substrate recognition by the cell surface palmitoyl transferase DHHC5. Proc Natl Acad Sci 111:17534–17539. doi:10.1073/pnas.1413627111
45. Chamberlain LH, Shipston MJ (2015) The physiology of protein S-acylation. Physiol Rev 95:341–376. doi:10.1152/physrev.00032.2014
46. Liang X, Nazarian A, Erdjument-Bromage H et al (2001) Heterogeneous fatty acylation of Src family kinases with polyunsaturated fatty acids regulates raft localization and signal transduction. J Biol Chem 276:30987–30994. doi:10.1074/jbc.M104018200
47. O'Brien PJ, Zatz M (1984) Acylation of bovine rhodopsin by [3H]palmitic acid. J Biol Chem 259:5054–5057
48. Linder ME, Middleton P, Hepler JR et al (1993) Lipid modifications of G proteins: alpha subunits are palmitoylated. Proc Natl Acad Sci U S A 90:3675–3679
49. Paige LA, Nadler MJ, Harrison ML et al (1993) Reversible palmitoylation of the protein-tyrosine kinase p56lck. J Biol Chem 268:8669–8674
50. Buss JE, Sefton BM (1986) Direct identification of palmitic acid as the lipid attached to p21ras. Mol Cell Biol 6:116–122
51. Hess DT, Slater TM, Wilson MC, Skene JH (1992) The 25 kDa synaptosomal-associated

protein SNAP-25 is the major methionine-rich polypeptide in rapid axonal transport and a major substrate for palmitoylation in adult CNS. J Neurosci Off J Soc Neurosci 12:4634–4641

52. Magee AI, Gutierrez L, McKay IA et al (1987) Dynamic fatty acylation of p21N-ras. EMBO J 6:3353–3357
53. Rocks O, Peyker A, Kahms M et al (2005) An acylation cycle regulates localization and activity of palmitoylated ras isoforms. Science 307:1746–1752. doi:10.1126/science.1105654
54. Rocks O, Gerauer M, Vartak N et al (2010) The palmitoylation machinery is a spatially organizing system for peripheral membrane proteins. Cell 141:458–471. doi:10.1016/j.cell.2010.04.007
55. Zuckerman DM, Hicks SW, Charron G et al (2011) Differential regulation of two palmitoylation sites in the cytoplasmic tail of the beta1-adrenergic receptor. J Biol Chem 286:19014–19023. doi:10.1074/jbc.M110.189977
56. Schmidt MFG, Schlesinger MJ (1979) Fatty acid binding to vesicular stomatitis virus glycoprotein: a new type of post-translational modification of the viral glycoprotein. Cell 17:813–819. doi:10.1016/0092-8674(79)90321-0
57. Bartels DJ, Mitchell DA, Dong X, Deschenes RJ (1999) Erf2, a novel gene product that affects the localization and palmitoylation of Ras2 in *Saccharomyces cerevisiae*. Mol Cell Biol 19:6775–6787
58. Deschenes RJ, Broach JR (1987) Fatty acylation is important but not essential for *Saccharomyces cerevisiae* RAS function. Mol Cell Biol 7:2344–2351. doi:10.1128/MCB.7.7.2344
59. Mitchell DA, Vasudevan A, Linder ME, Deschenes RJ (2006) Thematic review series: Lipid Posttranslational Modifications. Protein palmitoylation by a family of DHHC protein S-acyltransferases. J Lipid Res 47:1118–1127. doi:10.1194/jlr.R600007-JLR200
60. Zhang MM, Wu P-YJ, Kelly FD et al (2013) Quantitative control of protein S-palmitoylation regulates meiotic entry in fission yeast. PLoS Biol 11, e1001597. doi:10.1371/journal.pbio.1001597
61. Fukata M, Fukata Y, Adesnik H et al (2004) Identification of PSD-95 palmitoylating enzymes. Neuron 44:987–996. doi:10.1016/j.neuron.2004.12.005
62. Korycka J, Łach A, Heger E et al (2012) Human DHHC proteins: a spotlight on the hidden player of palmitoylation. Eur J Cell Biol 91:107–117. doi:10.1016/j.ejcb.2011.09.013
63. Peng T, Thinon E, Hang HC (2016) Proteomic analysis of fatty-acylated proteins. Curr Opin Chem Biol 30:77–86. doi:10.1016/j.cbpa.2015.11.008
64. Farazi TA, Waksman G, Gordon JI (2001) The biology and enzymology of protein N-myristoylation. J Biol Chem 276:39501–39504. doi:10.1074/jbc.R100042200
65. Bhatnagar RS, Fütterer K, Farazi TA et al (1998) Structure of N-myristoyltransferase with bound myristoylCoA and peptide substrate analogs. Nat Struct Mol Biol 5:1091–1097. doi:10.1038/4202
66. Matsubara M, Titani K, Taniguchi H, Hayashi N (2003) Direct involvement of protein myristoylation in myristoylated alanine-rich C kinase substrate (MARCKS)-calmodulin interaction. J Biol Chem 278:48898–48902. doi:10.1074/jbc.M305488200
67. Nagar B, Hantschel O, Young MA et al (2003) Structural basis for the autoinhibition of c-Abl tyrosine kinase. Cell 112:859–871. doi:10.1016/S0092-8674(03)00194-6
68. Martin DDO, Beauchamp E, Berthiaume LG (2011) Post-translational myristoylation: fat matters in cellular life and death. Biochimie 93:18–31. doi:10.1016/j.biochi.2010.10.018
69. Thinon E, Serwa RA, Broncel M et al (2014) Global profiling of co- and post-translationally N-myristoylated proteomes in human cells. Nat Commun 5:4919. doi:10.1038/ncomms5919
70. Patwardhan P, Resh MD (2010) Myristoylation and membrane binding regulate c-Src stability and kinase activity. Mol Cell Biol 30:4094–4107. doi:10.1128/MCB.00246-10
71. Zha J, Weiler S, Oh KJ et al (2000) Posttranslational N-myristoylation of BID as a molecular switch for targeting mitochondria and apoptosis. Science 290:1761–1765
72. Vilas GL, Corvi MM, Plummer GJ et al (2006) Posttranslational myristoylation of caspase-activated p21-activated protein kinase 2 (PAK2) potentiates late apoptotic events. Proc Natl Acad Sci 103:6542–6547. doi:10.1073/pnas.0600824103
73. Sakurai N, Utsumi T (2006) Posttranslational N-myristoylation is required for the anti-apoptotic activity of human tGelsolin, the C-terminal caspase cleavage product of human gelsolin. J Biol Chem 281:14288–14295. doi:10.1074/jbc.M510338200
74. Nimchuk Z, Marois E, Kjemtrup S et al (2000) Eukaryotic fatty acylation drives plasma membrane targeting and enhances function of several Type III effector proteins from *Pseudomonas syringae*. Cell 101:353–363. doi:10.1016/S0092-8674(00)80846-6

75. Burnaevskiy N, Peng T, Reddick LE et al (2015) Myristoylome profiling reveals a concerted mechanism of ARF GTPase deacylation by the bacterial protease IpaJ. Mol Cell 58:110–122. doi:10.1016/j.molcel.2015.01.040
76. Burnaevskiy N, Fox TG, Plymire DA et al (2013) Proteolytic elimination of N-myristoyl modifications by the Shigella virulence factor IpaJ. Nature 496:106–109. doi:10.1038/nature12004
77. Takada R, Satomi Y, Kurata T et al (2006) Monounsaturated fatty acid modification of Wnt protein: its role in Wnt secretion. Dev Cell 11:791–801. doi:10.1016/j.devcel.2006.10.003
78. Gao X, Hannoush RN (2014) Single-cell imaging of Wnt palmitoylation by the acyltransferase porcupine. Nat Chem Biol 10:61–68. doi:10.1038/nchembio.1392
79. Yang J, Brown MS, Liang G et al (2008) Identification of the acyltransferase that octanoylates ghrelin, an appetite-stimulating peptide hormone. Cell 132:387–396. doi:10.1016/j.cell.2008.01.017
80. Gutierrez JA, Solenberg PJ, Perkins DR et al (2008) Ghrelin octanoylation mediated by an orphan lipid transferase. Proc Natl Acad Sci 105:6320–6325. doi:10.1073/pnas.0800708105
81. Branton WD, Rudnick MS, Zhou Y et al (1993) Fatty acylated toxin structure. Nature 365:496–497. doi:10.1038/365496a0
82. Zou C, Ellis BM, Smith RM et al (2011) Acyl-CoA:Lysophosphatidylcholine Acyltransferase I (Lpcat1) catalyzes histone protein O-palmitoylation to regulate mRNA synthesis. J Biol Chem 286:28019–28025. doi:10.1074/jbc.M111.253385
83. Berthiaume LG (2014) Wnt acylation: seeing is believing. Nat Chem Biol 10:5–7. doi:10.1038/nchembio.1414
84. Stevenson FT, Bursten SL, Locksley RM, Lovett DH (1992) Myristyl acylation of the tumor necrosis factor alpha precursor on specific lysine residues. J Exp Med 176:1053–1062. doi:10.1084/jem.176.4.1053
85. Stevenson FT, Bursten SL, Fanton C et al (1993) The 31-kDa precursor of interleukin 1 alpha is myristoylated on specific lysines within the 16-kDa N-terminal propiece. Proc Natl Acad Sci 90:7245–7249
86. Feldman JL, Baeza J, Denu JM (2013) Activation of the protein deacetylase SIRT6 by long-chain fatty acids and widespread deacylation by mammalian sirtuins. J Biol Chem 288:31350–31356. doi:10.1074/jbc.C113.511261
87. Teng Y-B, Jing H, Aramsangtienchai P et al (2015) Efficient demyristoylase activity of SIRT2 revealed by kinetic and structural studies. Sci Rep 5:8529. doi:10.1038/srep08529

Quantum Dot Toolbox in Membrane Neurotransmitter Transporter Research

Lucas B. Thal, Danielle M. Bailey, Oleg Kovtun, and Sandra J. Rosenthal

Abstract

Quantum dot-based fluorescence techniques enable multi-scale molecular profiling ranging from real-time single molecule dynamics to expression trends in million-cell populations. In comparison to currently available probes, quantum dots are particularly well suited for such studies by virtue of their unique photophysical properties. We discuss in this chapter methodological components of what makes up the "Quantum Dot Toolbox" in neurotransmitter transporter studies along with specific work our group has published. First, we describe ensemble analysis of subcellular transporter localization and provide visualization of transporter residence in distinct cellular surface features. Second, we provide discussion on high content analysis of changes in transporter surface levels and give insight into the advantages of using quantum dot probes in flow cytometry. Third, we review the fundamental principles of subdiffraction-limit fluorescence microscopy and single molecule analysis of transporter surface dynamics. Included in this chapter are three protocols with experimental considerations specific to each technical section.

Keywords: Flow cytometry, Neurotransmitter transporters, Quantum dot-based fluorescence, Quantum dots, Subcellular transporter localization

1 Introduction

Quantum dots (QDs) are semiconductor nanocrystals typically ranging from 4 to 10 nm in diameter. Owing to their inorganic nature, QDs exhibit unique photophysical properties controlled by shape, size, and composition (e.g. broad absorption spectra, high extinction coefficients, high quantum yield, narrow Gaussian emission spectra, photochemically stable). Nearly two decades ago, QDs were introduced as a new class of fluorescent probes for biological imaging [1, 2]. In contrast to organic dyes and fluorescent proteins, QDs undergo minimal photodegradation under continuous excitation and are easily resolved at video imaging rates. Consequently, QD probes have been particularly useful in the field of molecular neuroscience, where tracking of QD-labeled transmembrane neuronal receptors has shed light on the molecular

Arun K. Shukla (ed.), *Chemical and Synthetic Approaches in Membrane Biology*, Springer Protocols Handbooks, (2017) 219–230, DOI 10.1007/8623_2016_12, © Springer Science+Business Media LLC 2017, Published online: 23 March 2017

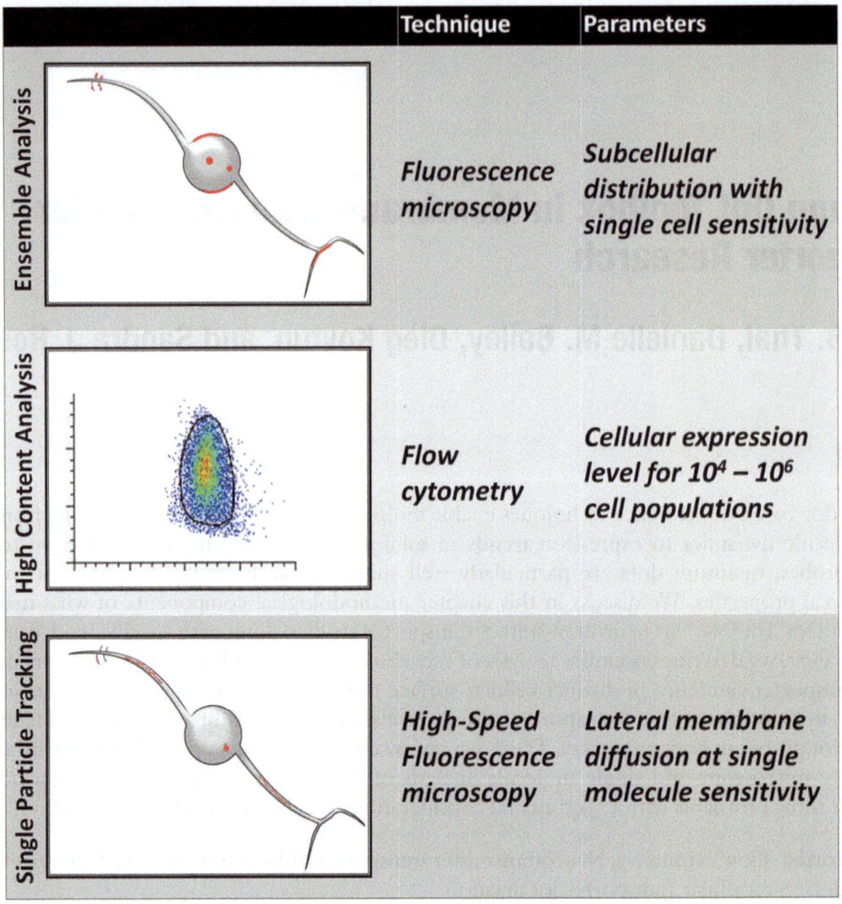

Fig. 1 Quantum dot toolbox in NTT studies. We provide fluorescence detection methodology, achieved sensitivity, and obtained molecular parameters for each tool box component

mechanisms underlying receptor trafficking regulation [3–6]. Our group focuses on the family of sodium-coupled transmembrane neurotransmitter transporters (NTTs), including serotonin transporter (SERT) and dopamine transporter (DAT). To address the lack of suitable antibodies against extracellular NTT epitopes, we pioneered the use of ligand-conjugated QDs for labeling NTTs and has expanded these constructs to high-resolution microscopy, flow cytometry-based high content assays, and single QD tracking (Fig. 1).

This chapter provides a set of detailed instructions and experimental considerations to allow readers to successfully implement our ligand-conjugated QD labeling approach in the study of NTTs and other neuronal transmembrane proteins. Specifically, we discuss the use of laser-scanning confocal microscopy, flow cytometry, and single particle tracking (SPT) to analyze localization, expression, and surface dynamics of NTTs, respectively.

2 Materials

1. No. 1.5 35-mm MatTek glass-bottom dishes (14 mm coverslip, MatTek Catalog # P35G-1.5-1.4-C)
2. Fluorobrite™ DMEM (ThermoFisher Catalog # A1896701)
3. 12 × 75 mm round bottom test tubes or 96-well round bottom microtiter plates
4. Phosphate-buffered saline (PBS; ThermoFisher Catalog #10010023)
5. Biotinylated NTT-specific ligand (1 mM stock in PBS)
6. Streptavidin-conjugated QDs (Sav-QDs; emission maximum at 655 nm; 1 μM stock; ThermoFisher Catalog #Q10123MP)
7. QD binding buffer (borate buffer supplemented with 215 mM sucrose, 0.05% sodium azide, 1–5% bovine serum albumin (BSA) and/or 1–2% fetal bovine serum (FBS))
8. Staining buffer (PBS supplemented with 1–2% FBS)
9. Adherent cells transiently expressing NTT of interest
10. Analysis software (Microsoft Excel, ImageJ, MATLAB, FlowJo)
11. 24-well flat bottom culture plates
12. Optional: poly-D-lysine (working concentration of 0.01 mg/mL) to pre-coat 24-well culture plates when working with weakly adherent cell lines
13. Cell Stripper™ non-enzymatic cell dissociation solution (Corning Catalog #25-056-CI)
14. Optional: 0.25% Trypsin-EDTA (ThermoFisher Catalog #25200056)
15. 2.0-mL microcentrifuge tubes
16. 5-Laser BD LSRII flow cytometer configured with 355, 405, 488, 535, and 633 nm excitation lines as well as 8 PMT detectors
17. Carl Zeiss LSM 5 Live configured with a 63× 1.4NA oil-immersion objective, solid-state laser diode (488 nm, 100 mW), and a linear array CCD

3 Subcellular Localization of QD-Bound Neurotransmitter Transporters

Ensemble fluorescence microscopy has emerged as an essential tool in biological research. Specifically, QD probes provide the user a spatial and visual guide for biomolecule accumulation and localization at subcellular levels. Our labeling approach based on

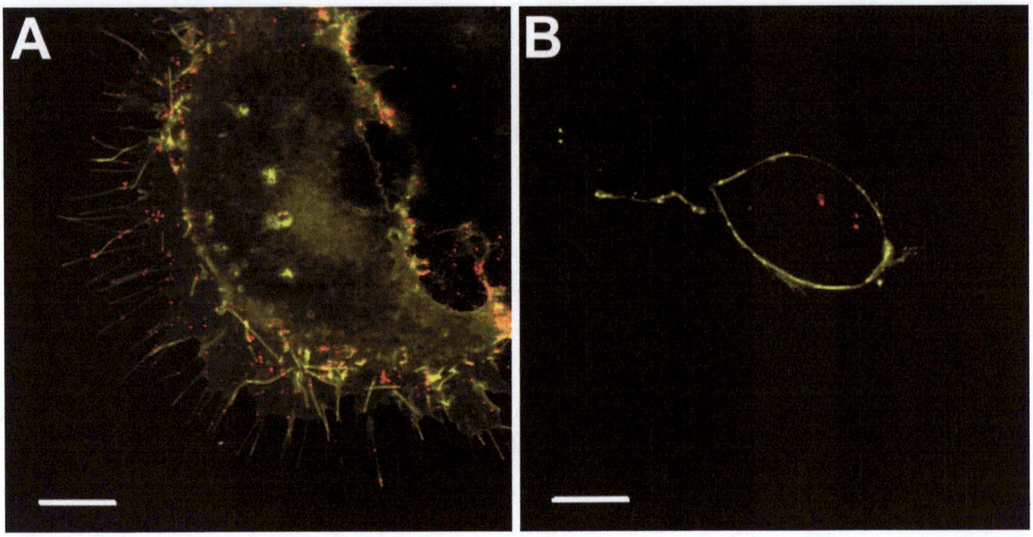

Fig. 2 Visualization of NTT subcellular localization using laser-scanning confocal microscopy. (**a**) QDs permit the user to readily identify whether NTTs are targeted to morphologically distinct membrane compartments (cell body, filopodia, lamellipodia). (**b**) QDs allow the user to observe and monitor the time course of NTT intracellular accumulation

antagonist-conjugated QDs has allowed us to visualize NTT delivery and localization within morphologically distinct subcellular compartments. Our group performed ensemble analysis utilizing a unique serotonin transporter (SERT) antagonist, IDT318, to demonstrate polarized expression of QD-serotonin transporter (QD-SERT) in *Xenopus laevis* oocytes [7]. We used the same ligand-QD system to show QD-SERT residing in ganglioside GM1-rich plasma membrane microdomains of immortalized rat serotoninergic RN46A neuron [8]. QD-based ensemble fluorescence microscopy has further enabled us to detect stimulated endocytosis of NTTs by monitoring the time course of QD internalization. By labeling dopamine transporters with streptavidin-conjugated QDs in conjugation with a biotinylated cocaine analog, IDT444, we detected internalization of QD-dopamine transporter (QD-DAT) complexes upon protein kinase C (PKC) activation [9].

We recommend using point scanning confocal microscope systems to generate high-resolution maps of NTT subcellular localization (Fig. 2). We provide in this section a general labeling approach for subcellular detection of NTTs in adherent cells (e.g., HeLa, HEK293, N2A, CAD, SH-SY5Y) with antagonist-conjugated QDs.

3.1 Protocol

1. Culture NTT-expressing cells in a No. 1.5 glass-bottom 35-mm MatTek dish. Optimal cell density for single-cell imaging is in the 0.5–5 × 10^4 cells/cm^2 range.
2. Prepare cells by aspirating complete growth medium and add 2 mL of 0.01–10 μM biotinylated NTT-specific ligand diluted in Fluorobrite™ DMEM. Alternatively, add a small aliquot of the 1 mM ligand stock solution directly to the complete growth medium.
3. Incubate the coverslip for 10–30 min in the cell culture incubator at 5% CO_2 and 37°C.
4. Aspirate biotinylated NTT-specific ligand solution and wash the coverslip 3 times with pre-warmed Fluorobrite™ DMEM.
5. Promptly add a solution of 0.5–2.5 nM Sav-QDs in pre-warmed Fluorobrite™ DMEM supplemented with 1–5% BSA and/or 1–2% FBS to the MatTek dish.
6. Incubate the MatTek dish for 5–10 min in the cell culture incubator at 5% CO_2 and 37°C.
7. Aspirate the QD solution and wash the MatTek dish 3–5 times with pre-warmed Fluorobrite™ DMEM.
8. Replenish the MatTek dish with pre-warmed Fluorobrite™ DMEM to prepare for imaging.
9. Immediately mount the MatTek dish on a heated microscope stage and acquire individual images using DIC and appropriate fluorescence bandpass filters.

3.2 Experimental Considerations

1. All QD labeling steps should be done with minimal exposure to light.
2. Use a complete set of controls including unstained cells, NTT-expressing cells labeled with Sav-QD only (-biotinylated NTT-specific ligand), non-expressing cells labeled with biotinylated ligand and QDs, and NTT-expressing cells pre-incubated with a high-affinity inhibitor to block the transporter–ligand interaction.
3. If the user is observing non-specific QD binding, decrease cell debris on coverslip by keeping cells in viable media, temperature, and atmosphere, increase percentage of BSA when incubating QDs, treat with other passivation reagents (casein, newborn calf serum, dehydrated fat-free milk) in addition to or replacing FBS, and/or increase number of wash steps.
4. If the user is experiencing low signal-to-noise ratio, adjust pinhole size, gain, and amplitude, increase excitation intensity, and/or adjust Sav-QD concentration in step 5 in the labeling protocol.

4 Flow Cytometric Detection of Quantum Dot-Labeled Neurotransmitter Transporters in Single Cells

Flow cytometry has emerged as one of the premier analytical tools that enables rapid multi-parameter molecular phenotyping and functional profiling with a single-cell sensitivity. With recent advances in hardware, reagent availability, and software, it is now possible to routinely implement polychromatic (up to 17 spectrally distinct fluorophores on a 5-Laser BD LSRII instrument) flow cytometry analyses. QDs are particularly well suited for polychromatic applications of flow cytometry aimed at detection of surface biomolecules due to broad absorption spectra and size tunable narrow emission spectra. We have recently utilized antagonist-conjugated QDs to monitor changes in DAT surface levels and DAT-mediated antagonist binding in response to external stimuli in HEK293 cells stably expressing transmembrane DAT molecules [10]. This report was an experimental demonstration that flow cytometric detection of QD-labeled cells represents an attractive alternative to the current "gold standard" techniques in the field of NTT research – radioligand uptake assay and Western blot – effectively eliminating the need for radiolabeled isotopes and time-consuming, expensive immunoblotting. In this section, we provide detailed guidelines for detecting surface NTT molecules transiently expressed in an adherent cell platform (e.g., HeLa, HEK293, N2A, CAD, SH-SY5Y) with ligand-conjugated QDs and implementing flow cytometric analysis of the resulting cell suspension.

4.1 Protocol

1. Allow NTT-expressing cells growing in a 24-well plate culture plate to reach 80–90% confluency.
2. Prepare cells by aspirating complete growth medium and add 500 μL of a 0.01–10 μM biotinylated NTT-specific ligand diluted in PBS.
3. Incubate the 24-well plate for 10–30 min in the cell culture incubator at 5% CO_2 and 37°C.
4. Carefully aspirate the ligand solution, dissociate the cells from the plate bottom using Cell Stripper™ (use 0.25% Trypsin-EDTA for lifting strongly adherent cells), and transfer the resulting cell suspension to 2.0 mL microcentrifuge tubes.
5. Pellet the cells by centrifugation at 1,000–3,000 rpm for 5 min.
6. Discard supernatant by gentle aspiration and resuspend the labeled cells in 500 μL of cold QD binding buffer containing 0.5–2.5 nM streptavidin-conjugated QDs. Incubate on ice for 5–10 min.

7. Pellet the cell by centrifugation at 1,000–3,000 rpm for 5 min. Discard supernatant by aspiration and wash the cells by resuspending in cold staining buffer followed by immediate centrifugation at 1,000–3,000 rpm for 5 min.
8. Repeat for a total of two washes, discarding supernatant between washes.
9. Resuspend the pellet in 500 μL of cold staining buffer and transfer samples to either round bottom test tubes or a 96-well round bottom microtiter plate.
10. [OPTIONAL] Add a viability dye to each sample to exclude dead cells from analysis.
11. Acquire data on a flow cytometer. A representative data set is displayed in Fig. 3.
12. Use forward scatter (indicator of size) and side scatter (indicator of surface roughness, granularity, and internal complexity) pulse width, height, and area to discriminate single cells from cell doublets and cell debris.
13. Collect 20,000–100,000 single-cell events for the gated population.

4.2 Experimental Considerations

1. Single-cell suspensions at a density of 10^6–10^7 cells/mL are ideal.
2. All QD labeling steps should be done with minimal exposure to light.
3. Always include a rigorous set of controls and use a negative control to set voltages and gates. Recommended controls include unstained cells, NTT-expressing cells labeled with QD only, non-expressing cells labeled with biotinylated NTT-specific ligand and QDs, and NTT-expressing cells pre-incubated with a high-affinity inhibitor to block the transporter–ligand interaction.
4. Include BSA and/or FBS to block non-specific interactions of QDs with the cell surface.
5. Determine the number of single-cell events that must be collected to ensure that your sampling is statistically significant (10^4–10^5 single viable cells per final scatter gate per sample is usually sufficient).
6. For multicolor experiments, always include single-color controls to determine fluorescence compensation settings to address the spectral overlap of fluorophores used.

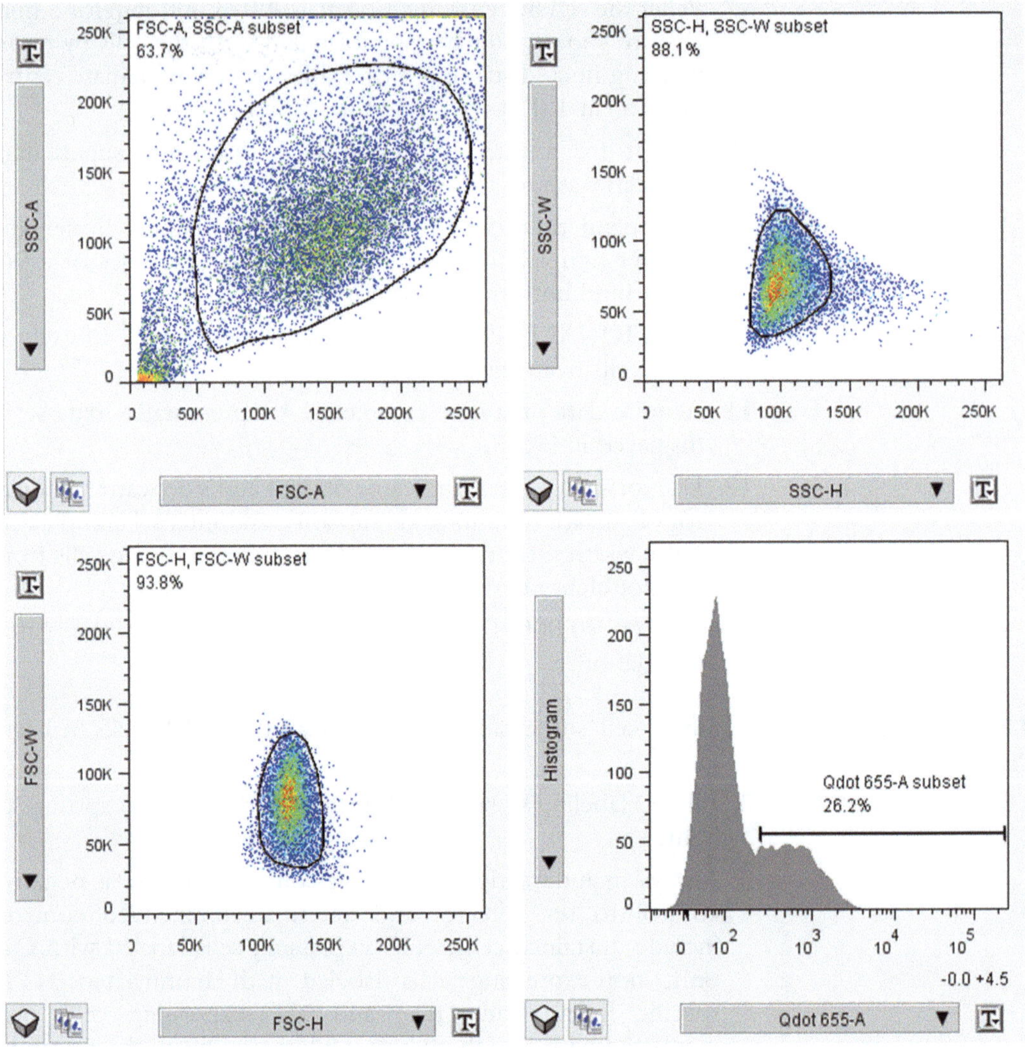

Fig. 3 Flow cytometric detection QD-labeled cells transiently expressing NTT of interest. Forward scatter (FSC) and side scatter (SSC) pulse area, height, and width are used to discriminate single-cell events from cell doublets and cell debris. The histogram of the gated population shows a fluorescence intensity distribution of single cells. The characteristic bimodal pattern indicates the presence of a dim population (non-transfected cells, hence low QD fluorescence intensity) and a bright fraction (transfected cells, higher QD fluorescence intensity)

5 Single Particle Tracking of Neurotransmitter Transporters

In contrast to ensemble approaches discussed in Sects. 3 and 4, SPT permits real-time observation of NTTs in the plasma membrane at a subdiffraction-limited spatial resolution and imaging rates at $\geq$10 Hz. As surface NTT trafficking appears to be an important

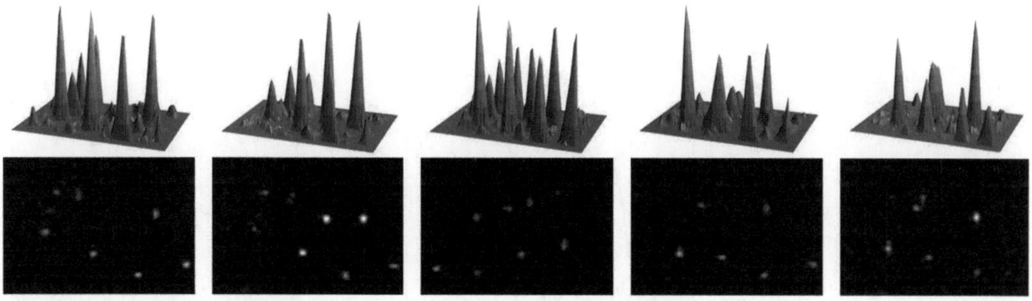

Fig. 4 Typical video frame sequence of single QDs bound to NTT molecules. *Bottom*: Raw image sequence of diffusing QDs. *Top*: Corresponding surface intensity plots used for QD localization

post-translational regulatory mechanism, SPT analysis can provide invaluable information about NTT lateral diffusion dynamics, membrane compartmentalization, and dynamic interactions with its binding partners. QDs are bright, robust point-like emitters characterized by a narrow, well-defined point-spread function (PSF) and therefore a probe choice in SPT studies (Fig. 4) [11]. QDs act as point emitters (Fig. 4) that can be modeled with the point-spread function (PSF) for use in SPT. The point emitters must be separated farther than the diffraction-limited regime and thus must be in sufficiently low concentration. The centroid positions can then be localized with subpixel accuracy by fitting the intensity distribution to a 2D Gaussian function:

$$I_{xy} = A_0 + A \times e^{\frac{(x-x_0)^2+(y-y_0)^2}{w^2}}$$

where I_{xy} is the pixel intensity, A is signal amplitude, A_0 is local background, x_0 and y_0 are the local maximum coordinates of the Gaussian fit, and w is the width of the curve [12]. The accuracy of the fit depends on the signal-to-noise ratio (SNR):

$$\text{SNR} = \frac{I_0}{\sqrt{\sigma_{\text{bg}}^2 + \sigma_{I_0}^2}}$$

where I_0 is the signal intensity, σ_{bg} is the variance of the background intensity, and σ_I is the variance of the signal intensity [12]. QDs significantly increase the SNR compared to common organic fluorophores, providing another advantage as single molecule imaging probes. Once precise localization data (x, y) is determined, QD positions in successive frames must be linked to construct trajectories. We previously described in detail basic analysis using ImageJ [12]. Once trajectories are constructed, net displacement and velocities can be determined, as well as mean square displacement (MSD) (Fig. 5):

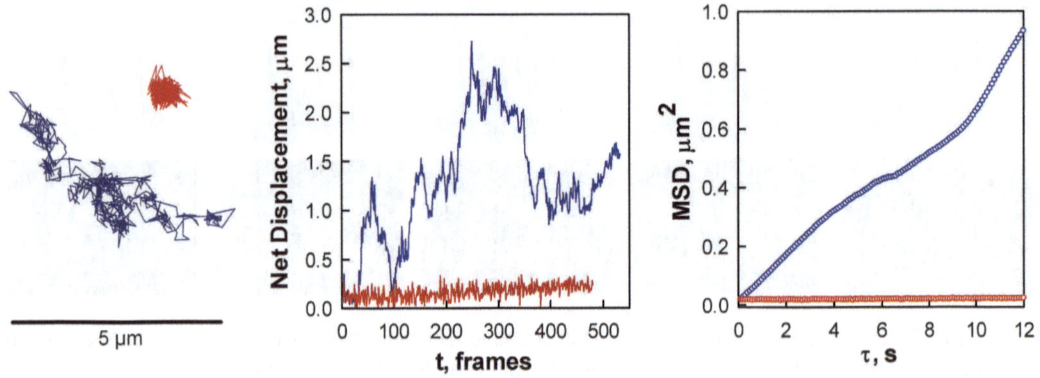

Fig. 5 Basic analysis of example trajectories shows plots of displacement from the starting position and MSD for a confined (*red*) and freely diffusing (*blue*) particle

$$\text{MSD}\,(n\Delta t) = (N - n)^{-1} \sum_{i=1}^{N-n} \left[(x_{i+n} - x_i)^2 + (y_{i+n} - y_i)^2 \right]$$

where $n\Delta t$ is the time interval in which the MSD is calculated, N is the total number of frames, and x_i and y_i are positions of the particle over time. Our recent group efforts focused on visualizing and analyzing NTT membrane dynamics using antagonist-conjugated QDs. We first utilized a SERT antagonist, IDT318, to specifically label individual SERT proteins within the membranes of serotonergic RN46A cells. We observed two distinct populations, one that freely diffused across the membrane and one that displayed restricted mobility and was confined to membrane microdomains. Upon SERT stimulation, individual proteins remained confined within microdomains but displayed untethering from cytoskeletons that allowed for increased mobility through a p38 MAPK pathway [8]. Next we visualized the dynamics of wild-type and ADHD-associated mutant (R615C) DAT using a high-affinity biotinylated cocaine analog, IDT444, and streptavidin-conjugated QDs in living transfected HEK293 cells. From QD trajectory analysis, the R615C mutant showed increased lateral mobility in the membrane compared to the wild-type protein and lacked response to cholesterol depletion and amphetamine stimulation [13]. These findings demonstrate, for the first time, how a disease-associated mutation affects the surface dynamics of single DAT molecules, which opens up possibilities for future studies linking disease states with individual molecular behavior. This protocol outlines QD labeling in NTT-expressing adherent cell lines and provides a set of common troubleshooting tips.

5.1 Protocol

1. Culture NTT-expressing cells in glass-bottom dishes until they are 50–60% confluent.
2. Prepare a solution of 0.01–0.1 nM Sav-QDs in warm Fluorobrite™ DMEM media containing 2–5% BSA or 1–2% dialyzed FBS. Vortex for 5 s to break up aggregates.
3. Add 0.1–0.5 nM biotinylated NTT-specific ligand directly to culture media and incubate for 5 min.
4. Wash cells gently 3–5 times with phenol-red free DMEM or Fluorobrite™ DMEM.
5. Add QD solution to cells and incubate at 37°C for 5 min.
6. Wash cells gently 3–5 times with warm Fluorobrite™ DMEM.
7. Place the dish on a heated (37°C) microscope stage and acquire time-lapse images.

5.2 Experimental Considerations

1. A frame rate of 10 frames per second or faster is recommended using an appropriate filter for the type of QD. A 600/30 band-pass filter is used for 605 QDs.
2. Image in a low-background buffer, like Fluorobrite™ DMEM, to improve signal-to noise ratio.
3. Include a blocking reagent, such as BSA, dialyzed FBS, or casein, to decrease non-specific labeling of QDs.
4. Avoid QD internalization by imaging immediately after labeling.
5. Decrease QD incubation time if QDs are internalizing.
6. Determine ligand specificity using controls (see Sect. 3.2 for example control experiments).

References

1. Bruchez M, Moronne M, Gin P, Weiss S, Alivisatos AP (1998) Semiconductor nanocrystals as fluorescent biological labels. Science 281:2013–2016
2. Chan WCW, Nie S (1998) Quantum dot bioconjugates for ultrasensitive nonisotopic detection. Science 281:2016–2018
3. Dahan M et al (2003) Diffusion dynamics of glycine receptors revealed by single-quantum dot tracking. Science 302:442–445
4. Bouzigues C, Morel M, Triller A, Dahan M (2007) Asymmetric redistribution of GABA receptors during GABA gradient sensing by nerve growth cones analyzed by single quantum dot imaging. Proc Natl Acad Sci 104:11251–11256
5. Rosenthal SJ, Chang JC, Kovtun O, McBride JR, Tomlinson ID (2011) Biocompatible quantum dots for biological applications. Chem Biol 18:10–24
6. Frischknecht R et al (2009) Brain extracellular matrix affects AMPA receptor lateral mobility and short-term synaptic plasticity. Nat Neurosci 12:897–904
7. Chang JC et al (2011) A fluorescence displacement assay for antidepressant drug discovery based on ligand-conjugated quantum dots. J Am Chem Soc 133:17528–17531
8. Chang JC et al (2012) Single molecule analysis of serotonin transporter regulation using antagonist-conjugated quantum dots reveals restricted, p38 MAPK-dependent mobilization underlying uptake activation. J Neurosci 32:8919–8929
9. Kovtun O et al (2011) Visualization of the cocaine-sensitive dopamine transporter with ligand-conjugated quantum dots. ACS Chem Nerosci 2:370–378

10. Kovtun O, Ross EJ, Tomlinson ID, Rosenthal SJ (2012) A flow cytometry-based dopamine transporter binding assay using antagonist-conjugated quantum dots. Chem Commun 48:5428–5430
11. Chang JC, Rosenthal SJ (2013) A bright light to reveal mobility: single quantum dot tracking reveals membrane dynamics and cellular mechanisms. J Phys Chem Lett 4:2858–2866
12. Chang JC, Rosenthal SJ (2013) In: Sandra Rosenthal J, David Wright W (eds) NanoBiotechnology protocols. Humana Press, New York, pp 71–84
13. Kovtun O et al (2015) Single-quantum-dot tracking reveals altered membrane dynamics of an attention-deficit/hyperactivity-disorder-derived dopamine transporter coding variant. ACS Chem Nerosci 6:526–534

Generation of Synthetic Antibody Fragments to Detergent Solubilized Membrane Proteins

Serdar Uysal and Anthony Kossiakoff

Abstract

Structural determination of membrane proteins is extremely challenging due to the physical characteristics of membrane proteins themselves and the lack of adequate tools and technologies to perform the studies. Recent developments in micro-focus X-ray beams, novel detergents, protein thermo-stabilization, and protein engineering have been essential in expanding the pool of membrane proteins deposited in PDB. Despite these advances, crystallization of membrane proteins still remains the main bottleneck in obtaining high quality structures. Recently, the use of antibody and non-antibody scaffold binding partners as crystallization "chaperones" has emerged as a powerful method to obtaining well-diffracting crystals of membrane proteins. In this chapter, a protocol is provided to generate synthetic antibody fragments for use as crystallization chaperones for membrane proteins.

Keywords: Crystallization chaperones, Phage display, Synthetic antibodies

1 Introduction

Membrane proteins (MPs) represent one-third of the proteins found in eukaryotic genomes [1]. Their functions are governed by myriad types of stimuli including tension, light, and voltage [2–4]. In order to couple these diverse stimuli to function, membrane proteins have evolved to utilize dynamic transitions between different energy states, which are manifested in programmed conformational movements. The challenge for studying these conformational intermediates by crystallography is that many are transient in nature making these molecules highly dynamic and extremely sensitive to local environment.

MPs reside in cellular membranes and their native environment is comprised of diverse lipids providing necessary chemical composition for folding, stability, and function. However, to conduct any structural study, they need to be extracted from membrane and their unique amphipathic nature requires the use of detergents. It is appreciated that detergents can disrupt membrane protein lipid

Arun K. Shukla (ed.), *Chemical and Synthetic Approaches in Membrane Biology*, Springer Protocols Handbooks, (2017) 231–243, DOI 10.1007/8623_2016_11, © Springer Science+Business Media LLC 2017, Published online: 26 February 2017 interaction and cover hydrophobic surface of MP thus creating a protein detergent complex (PDC) [5]. Even though PDC is water soluble and further used for crystallization, in most cases, it is far from being an ideal medium for structural studies. Once the MP is extracted from lipid bilayer as a PDC, the lateral pressure applied by membrane is dissipated, increasing the already existing flexibility of MP. Consequently, the conformationally non-homogeneous species formed within PDC diminishes the probability of forming well-diffracting crystals. The problem is further exacerbated by the size and the shape of detergent micelles. Assembling around the hydrophobic face of an MP with an MW of as large as 100 kD [6], detergent micelles may impede formation of 3D crystal lattice by inhibiting lattice contacts needs to be provided by protein–protein interactions.

Virtually all MP structures deposited in PDB were determined from three-dimensional (3D) crystals formed either as type I or type II [7]. In meso methods such as lipidic cubic phase and bicelle are known to form type I crystals whereby proteins assemble in two-dimensional planar sheets via hydrophobic faces which then 2D planar sheets stack on top of each other using polar interactions to form the third dimension [8, 9]. In contrast, water soluble PDCs are observed to form type II crystals; hydrophilic portions of protein mediate the necessary contacts for 3D lattice formation. Detergent micelles masking the hydrophobic surface of the protein are not involved in ordered crystal lattice formation resulting in crystals with large water content.

To circumvent the problem, especially in case of generating type II crystals, application of both antibody fragments and non-antibody scaffolds emerged as a promising method. The potential of using antibody fragments to crystallize MPs has been established. For example, application of an Fv and a Fab fragment enabled structure determinations of bacterial cytochrome c and potassium channel KcsA, respectively [10, 11]. In each case, antibody fragments expanded the hydrophilic surface dramatically and provided all the crystal contacts within the lattice demonstrating the validity of the antibody-assisted MP crystallography. Early on MP structures solved through the application of antibody fragments mainly relied on chaperones generated via animal immunization [12, 13]. While successful, generation of antibody fragments by hybridoma technology is costly and time consuming. In addition, producing antibodies to MPs is even harder due to difficulties associated with maintaining the stability of detergent solubilized MP in serum for extended period of time after injection.

In vitro display technologies have emerged as an alternative method in creating binding partners to MPs (Fig. 1). Phage, yeast, ribosome, bacterial in vitro display systems all share a common principle; coupling genotype to phenotype (binding protein) [14–17]. Fabs, Camelid single chain antibodies also known as

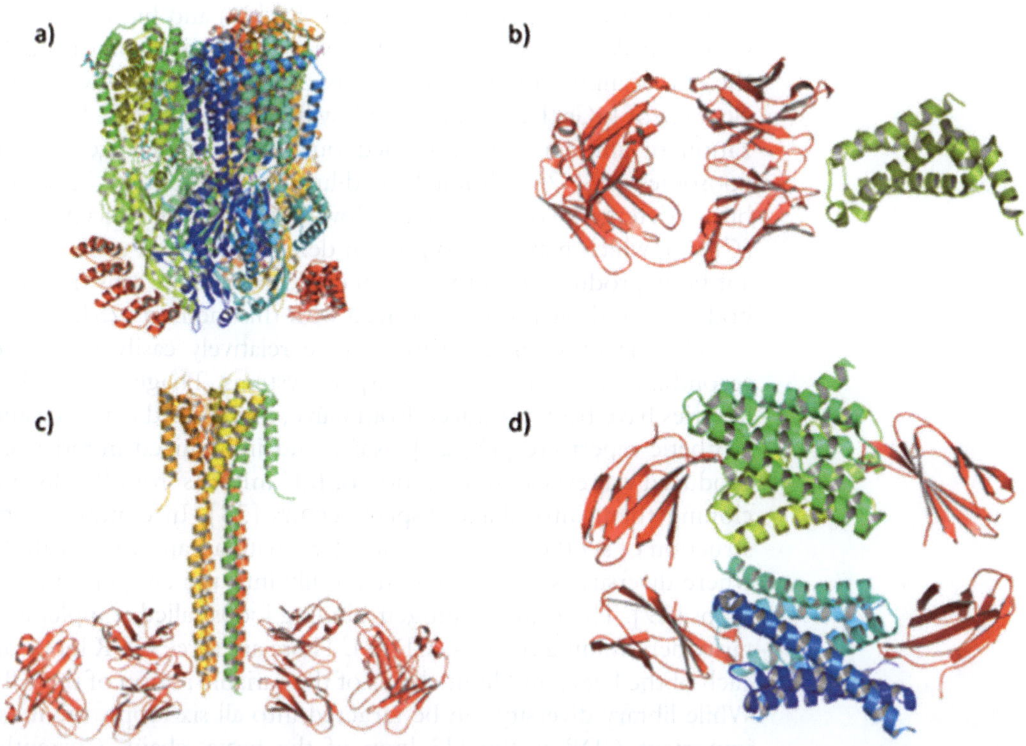

Fig. 1 Representatives of membrane protein structures with antibody and non-antibody scaffolds generated via in vitro display systems (PDB IDs: (**a**) 4DX5, (**b**) 4G7V, (**c**) 3EFF, (**d**) 5FXB). Crystallization chaperones are colored *red*. All binding proteins are produced through library sorting in MPB with proper detergent. Restricted amino acid diversity library was used to generate Fabs for structures (**b**) and (**c**)

nanobodies (VHH), and scFvs are among the antibody scaffolds that have utilized display methods to create binding partners for biological applications [18–20]. These methods also provide for expanding the possible chaperone types to include non-antibody scaffolds. For instance, designed ankyrin repeat proteins (DARPins), Fibronectin type III domains (FN3) monobodies, anticalins, and affibodies are successfully employed as synthetic binding scaffolds to generate affinity reagents to various targets [21–24]. What makes display methods so powerful is that diversity is introduced at the DNA level that encodes binding proteins to be displayed and tailored in vitro selection schemes can be highly controlled allowing generation of binding molecules with desired properties for a given target. Importantly, contrary to animal immunization, ability to control both selection conditions and presentation of target molecules, display systems provide a unique opportunity to generate binding proteins to specific epitopes and conformations.

While each scaffold type has its strengths and weaknesses, arguably, the Fab domain has been the most widely used scaffold in facilitating the crystallization of MPs. This success can be attributed to its relatively large size (~50 kD) and high structural stability making it a prime candidate as crystallization chaperone. In the case of animal immunization, Fabs can be recovered after papain cleavage of IgGs that are produced by hybridoma B-cells. Immunization of antigen can be carried out either in detergent or in liposomes [25, 26]. Immediate dilution of detergent in serum brings detergent concentration below critical micelle concentration (CMC), which may lead to protein denaturation and unsuccessful antibody production but yet, even conformationally sensitive antibodies were shown to be produced with this method [25].

On the other hand, Fabs can be relatively easily generated recombinantly using in vitro display systems. Phage display Fab libraries have been produced from naïve, immunized (biased), and synthetic repertoire [18, 27]. Naïve and immunized libraries are produced via reverse transcription of IgG mRNAs from B-cells and cloning them into phage display vectors [28]. In contrast, construction of synthetic Fab libraries starts with a framework scaffold where diversity is introduced chemically into the antigen-binding loops [29]. There are six antigen-binding loops called complementarity determining regions (CDRs). There are three CDR loops on each of the heavy and light chains of the variable region of the Fab. While library diversity can be targeted into all six loops, the most important CDR is the H3 loop of the heavy chain. Currently, synthetic library repertoires can easily exceed 10^{10} variants creating great potential to generate binders to variety of antigens including MPs and ribonucleic acids [30, 31]. Together with designed in vitro selection schemes for particular desired applications, generating engineered Fab fragments at nanomolar affinity with great specificity from synthetic libraries is routine nowadays. Additionally, through the sub-libraries one can affinity mature already generated antibodies to the sub-nanomolar range, thereby exceeding antibody affinities that of natural immune system [32].

Any important attribute of display approaches is that it is possible to design the selection step to preferentially bind to a particular conformational state or region on the target protein's surface. This provides a way to isolate and study particular properties of the membrane protein. For instance, synthetic Fabs enabled structure determinations of closed and open conformations of full-length KcsA and voltage sensor domain of Ciano intestinalis (Ci-VSD) [30, 33, 34]. In each instance, Fabs were generated against detergent solubilized MPs under selection conditions suited for MP stability. For the full-length structure of KcsA, Fabs could be specifically selected for the cytoplasmic domain of the channel known to be highly flexible and removed at earlier structural studies [35]. Selecting binders specific for the cytoplasmic domain helped to stabilize the full-length channel leading to its structure determination.

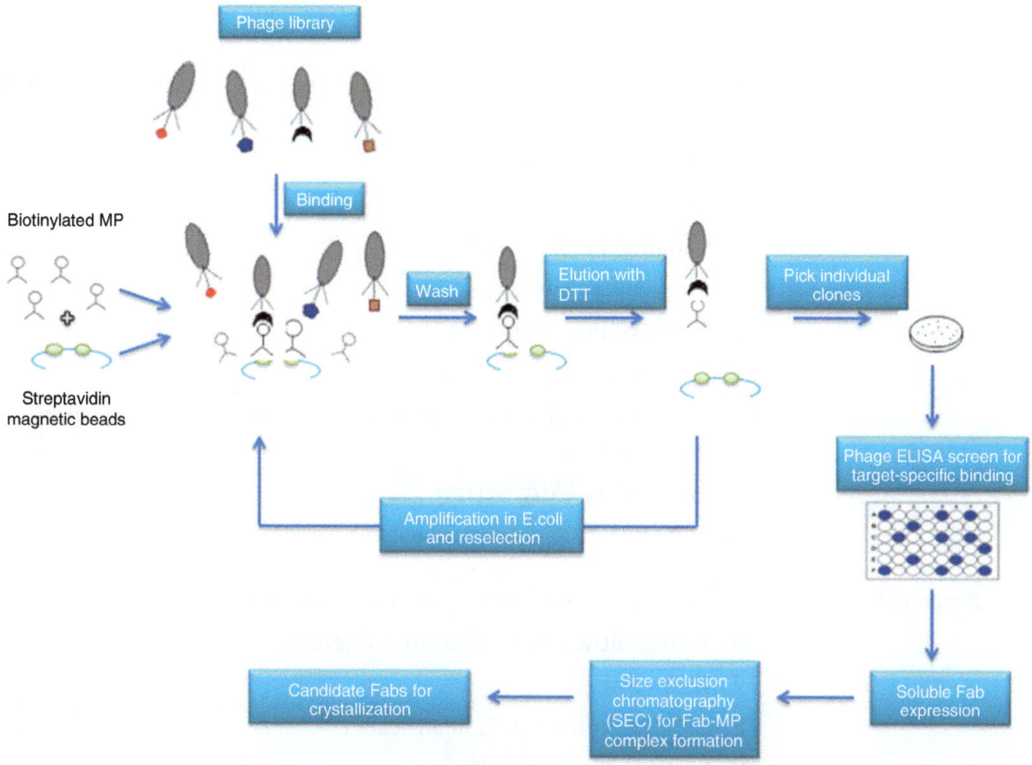

Fig. 2 Flowchart of library sorting with phage display and characterization of binding Fabs

Within this chapter, an outline will be given on how to generate synthetic antibody fragments to detergent solubilized MPs, which can be used as crystallization chaperones (Fig. 2). The phage display library that was used in these experiments is built upon Herceptin scaffold. An unconventional and revolutionary approach was taken by the authors in constructing the libraries whereby the CDRs were randomized by limited number of amino acids. Restricting chemical diversity to limited number of amino acids which are highly populated in antibody antigen interfaces, authors were able to randomize many more positions than conventional synthetic phage libraries. The details of the library construction are beyond the scope of this chapter but there are excellent literature and reviews on this issue [36–39]. However, the principles and methods presented below are not limited to any particular library and can be used with minor modifications with other scaffold types.

2 Materials

2.1 Phage Selection

1. Phage display library
2. KingFisher™ Magnetic Particle Processors (Thermo Scientific)
3. EZ-Link NHS-SS-PEG4-Biotin (Thermo Scientific)

4. Streptavidin magnetic beads: Streptavidin MagneSphere Paramagnetic Particles (Promega)
5. MagneSphere® Technology Magnetic Separation Stand (Promega)
6. XL-1 Blue *E. coli*
7. 2xYT
8. Kanamycin (Biobasic)
9. Ampicillin (Biobasic)
10. Tetracycline (Biobasic)
11. M13 KO7 helper phage (NEB)
12. Reservoir strips and combs for Kingfisher Magnetic particle processor
13. 10% DDM (Affymetrix)
14. SDS-PAGE and proper electrophoresis system

2.2 Phage ELISA

15. HRP/anti-M13 antibody conjugate (GE Healthcare)
16. 1-step ultra TMB substrate (Thermo)
17. BSA (Fisher)
18. PD10 columns (GE healthcare)
19. 96-well Maxisorp plates (Nunc)
20. 96-well deep well plates (Greiner)
21. Breathable and non-breathable sealing film (Corning)
22. Streptavidin (NEB, 1 mg/ml)
23. 1 M H_3PO_4

3 Methods

3.1 Biotinylation of MP

In order to carry out phage selection, the MP needs to be displayed on a surface. We use streptavidin magnetic beads and biotinylate target protein for immobilization purposes. EZ-Link NHS-SS-PEG4-Biotin reagent is chosen since it contains a cleavable disulfide bond within the pegulated arm to release protein with bound phages off the beads at the end of selection. This reduces the possibility of enriching streptavidin binding phages that could be detrimental to selection. Chemical biotinylation via amine groups requires buffer composition to be free from primary amines (Tris-based buffers are not recommended). The manufacturer's biotinylation reaction is robust enough to be carried out in a variety of detergents where MPs are solubilized. In some cases, heavy precipitation can occur due to either aggressive surface modification or modification of "structural lysines" by the biotin moiety which necessitates to optimize biotinylation reaction. Excess biotin can be removed using PD10 columns at the end of the reaction. Note, however, that chemical biotinylation can affect the MP stability and it is best to use the smallest amount possible.

3.2 Testing Biotinylated MP Pull Down

To determine the efficiency of biotinylation and amount of beads required for phage selection, a pull down assay needs to be performed.

1. Wash 750 μl of Streptavidin MagneSphere Paramagnetic Particles twice with membrane protein buffer (MPB). Aliquot 50, 150, 450 μl of magnetic beads to 1.5 ml eppendorf tubes. Using magnetic stand, remove buffer and suspend beads in 50 μl MPB.
2. Add 2 μg of biotinylated protein to each tube including one with no beads (50 μl buffer only) as a control. Incubate for 15 min on a rotator.
3. Recover 50 μl of protein solution using magnetic stand.
4. Run SDS-PAGE of the collected 50 μl samples. With increasing amount of magnetic beads, gradual decrease of protein amount within the samples needs to be observed as an indication of capture efficiency. Around 50% of pull down is sufficient at this step.
5. Biotinylated protein will be kept at 4°C for further phage selection experiments.

3.3 Phage Selection: First Round

The first round of library sorting will be carried out manually to recover as many binders as possible at this stage. Recovered phages will be propagated to be used as an input phage pool in successive rounds. The stringency for the selection will be determined by defining protein concentration for phage binding. In the first round, 200 nM of MP may be used for immobilization on the surface of magnetic beads. Protein concentration will decrease throughout phage selection to increase stringency of selection in order to enrich high affinity binders.

1. Start growing XL-1 Blue *E. coli* cells in 2 ml of 2xYT supplemented with 5 μg/ml tetracycline the day before phage selection at 37°C and 230 rpm over night (O/N).
2. Next day, inoculate 200 μl O/N grown XL-1 blue cells into 4 ml of 2xYT and grow 1–2 h at 37°C and 230 rpm until OD reaches 0.5–0.8 in order to collect log phase cells for phage infection.
3. Calculate the amount of magnetic beads needs to be used to immobilize 200 nM of MP based on pull down experiment. Wash the beads twice with 1 ml MPB.
4. Resuspend magnetic beads in 500 μl of MPB and add MP to have the final protein concentration to be 200 nM in 1 ml volume. Incubate beads and MP for 15 min to capture protein.

5. In parallel to protein capture, start to prepare phage solution.
6. Dilute 1 ml of phage library (10^{11}–10^{12}/ml particles) in 15 ml PBS and add 4 ml of PEG/NaCl (20% PEG 8000, 2.5 M NaCl). Mix by inversion and let the mixture sit on ice for 20 min.
7. Spin down phage solution in 50 ml conical tubes at 9,000 rpm for 15 min. Discard supernatant and white pellet should be visible. Resuspend pellet in 1 ml of MPB-0.1%BSA solution. Fabs on the surface of phage are prone to proteolytic degradation; therefore phage solution needs to be kept on ice until further use.
8. From step 2, wash beads twice with MPB to remove excess protein that is not captured. Resuspend beads with 1 ml phage solution (step 7) and rotate for 15 min at RT for phage binding.
9. Wash beads twice with MPB and resuspend beads in 300 μl of PBS.
10. 150 μl of phage solution is mixed with tenfold excess (1.5 ml) of log phase XL-1 blue *E. coli* cells and place in a shaker at 37°C for 20 min at 230 rpm for phage infection.
11. Dispense 150 μl of M13KO7 (10^{10} pfu/ml) helper phage into XL-1 blue *E. coli* cells from step 10 and shake the cells at 37°C for 40 min at 230 rpm for helper phage infection.
12. Inoculate 1.9 ml of phage and helper phage infected XL-1 *E. coli* cells into 35 ml 2xYT supplemented with 100 μg/ml ampicillin and 25 μg/ml kanamycin.
13. Shake flask 20–22 h at 37°C and 230 rpm for phage propagation.
14. Spin down culture at 8,000 rpm for 10 min. Transfer supernatant to 50 ml conical tube.
15. Add 1/5th of the O/N culture volume of PEG/NaCl (20% PEG 8000, 2.5 M NaCl) into phage solution. Stand it on ice for 20 min.
16. Spin down solution at 9,000 rpm for 15 min. Resuspend phage pellet in 3 ml of MPB-0.1%BSA. Keep on ice until next round of phage selection.

3.4 Phage Selection: Second, Third and Fourth Rounds

Phage selection for the following three rounds will be performed semiautomatically using Kingfisher Magnetic particle processor. Kingfisher reservoir strip can be prepared according to the following protocol:

(a) Well A: 100 nM of biotinylated membrane protein and 50 μl of phage particles from first round (Sect. 3.3, step #14), 50 μl of MPB

(b) Well B: 40 μl of magnetic beads (washed three times) and 60 μl of MPB

(c) Well C–G: 100 μl of MPB

(d) Well H: 100 μl of 100 mM DTT solution

1. Place the comb and load the plate with the strip into KingFisher. Perform library sorting according to the protocol:
   (a) Mix vigorously well A for 15 min
   (b) Collect beads from well B, transfer to well A, and mix vigorously for 15 min
   (c) Collect beads from well A, transfer to well C, and mix vigorously for 40 s
   (d) Transfer beads into wells D–G and mix vigorously for 40 s.
   (e) Collect beads from well G, transfer to well H, and mix vigorously for 10 min
   (f) Collect beads from well H and transfer back to well G and release the beads.
2. Transfer 50 μl of phage solution from well H into 500 μl of log phase XL-1 Blue *E. coli* cells. Incubate in a shaker at 37°C, 230 rpm for 20 min.
3. Add 50 μl M13KO7 (10^{10} pfu/ml) into phage infected *E. coli* cells and incubate in a shaker at 37°C, 230 rpm for 40 min.
4. Inoculate 600 μl of phage and helper phage infected XL-1 *E. coli* cells into 30 ml 2xYT supplemented with 100 μg/ml ampicillin and 25 μg/ml kanamycin.
5. Follow steps 11–14 from Sect. 3.3
6. Input protein concentration for phage selection will be 100 nM and 10 nM for the third and the fourth round, respectively.
7. Perform phage tittering for third and fourth rounds of selection. Make up to five tenfold serial dilutions of eluted phages in a 96-well plate. Mix 10 μl of phages with 30 μl of log phase XL-1 blue cells. Incubate for 20 min in a shaker. Plate 40 μl of cells on LB-agar plates with 100 μg/ml ampicillin. Grow plates O/N at 37°C. Enrichment of more than fivefold is a good indication of a successful library sorting.

3.5 Phage ELISA

In order to determine individual binders, phage ELISA experiments need to be performed. Individual colonies can be picked from plates used for phage tittering from Sect. 3.4.

1. Dispense 350 μl of 2xYT supplemented with 100 μg/ml ampicillin and 10^9 pfu/ml M13KO7 into 96-well deep well plate. Pick 24 colonies to inoculate individual wells. Seal the plate with breathable seal for proper aeration. Place plate in a shaker at 37°C, 230 rpm for 20–22 h.
2. Prepare 6 ml of 2 μg/ml of streptavidin in 50 mM sodium carbonate pH 8.5 coating buffer. Dispense 100 μl of streptavidin solution into 48 well in a 96-well maxisorp plate (24 wells as negative control) and place the plate O/N at 4°C.
3. Spin down 96-well deep well plate at 4,000 rpm for 10 min at 4°C. In parallel, dispense 125 μl of twice the concentration of MPB-0.1%BSA solution into 48 wells in a 96-well plate (MBP-BSA solution will be diluted with phage solution). Transfer 125 μl of phage solution from deep well plate into MPB-BSA solution plate and mix well with pipetting.
4. Discard coating solution and wash once with PBS. Block the plate with 2% BSA for 2 h at room temperature. Remove BSA solution and wash twice with MPB. Dispense 100 μl of 5–10 μg/ml biotinylated MP in MPB into 24 wells. Transfer 100 μl of MPB into 24 control wells. Rotate at RT, 150 rpm for 20 min.
5. Discard solution and wash three times with MPB. For each individual phage, transfer 100 μl of individual phage solution from step 3 into each MP and control well. Rotate at RT, 150 rpm for 1 h.
6. Discard phage solution and wash four times with MPB.
7. Prepare 5 ml of HRP-conjugated anti M13 antibody in MPB. Dispense 100 μl into each well. Rotate at RT, 150 rpm for 30 min.
8. Remove HRP-conjugated anti M13 antibody solution and wash 4 times with MPB.
9. Add 100 μl of turbo TMB solution to each well and develop for 4–7 min. Blue signal will appear for binding phages. Quench the reaction with 100 μl of 1M H_3PO_4. Signal will turn yellow, and read plate at 450 nm with a plate reader.
10. Pick colonies having more than threefold ELISA signal difference between protein and streptavidin only coated control wells and grow them in 5 ml of LB supplemented with 100 μg/ml ampicillin O/N at 37°C.
11. Prepare miniprep DNA for sequencing. Determine individual binders based on DNA sequences.

4 Notes

1. It is imperative to confirm individual binders further using size exclusion chromatography (SEC). Detailed protocols for creation of expression plasmid and bacterial Fab expression can be found elsewhere [40, 41]. Tandem affinity purification (TAP) can be carried out first with protein A, then with IMAC (We prefer to express Fabs with 4His at carboxy terminus of the heavy chain). TAP will yield Fabs with purity more than 90%. Purified Fabs are incubated and complexed with MP and run on SEC using FPLC. Generally, an obvious shift in the chromatograms is observed for Fab-MP complex with respect to MP alone. In some cases, due to micelle size and detergent displacement by antibody binding, a clear shift may not be seen. Nonetheless, running fractions from FPLC with SDS-PAGE will reveal complex formation.
2. Chemical biotinylation can be harsh for some proteins. It is important to optimize reaction with modifying biotin-protein ratio by decreasing biotin amount. Adding biotinylation reagent into reaction solution with incremental amounts may also help to resolve precipitation problem. If precipitation persists, in vitro or in vivo Avitag BirA biotinylation system can be used instead [41]. This system requires expression of MP with Avitag though.
3. It is important to include detergent during library sorting to eliminate "sticky" phage binders. Tween 20 (0.05%) is generally the choice of detergent for library sorting for soluble proteins. To maintain the stability throughout phage selection, the detergent used to stabilize membrane protein should replace Tween 20. So far, common detergents (CHAPS, DDM, DM, LMNG) used for MP solubilization in literature were found to be not problematic in eliminating "sticky" binders. For crystallization purposes, it is preferred to use detergent concentration close to CMC to avoid phase separation. In contrast, it may be required to increase detergent concentration to increase efficiency of washing steps in library sorting for successful phage select.

References

1. Wallin E, Heijne GV (1998) Genome-wide analysis of integral membrane proteins from eubacterial, archaean, and eukaryotic organisms. Protein Sci 7:1029–1038. doi:10.1002/pro.5560070420
2. Coste B, Mathur J, Schmidt M et al (2010) Piezol and Piezo2 are essential components of distinct mechanically activated cation channels. Science 330:55–60. doi:10.1126/science.1193270
3. Hubbard R, Kropf A (1958) The action of light on rhodopsin. Proc Natl Acad Sci U S A. doi:10.1126/science.1193270
4. Catterall AW (1995) Structure and function of voltage-gated ion channels. Annu Rev Biochem 64:493–531. doi:10.1146/annurev.bi.64.070195.002425
5. Privé GG (2007) Detergents for the stabilization and crystallization of membrane proteins. Methods. doi:10.1146/annurev.bi.55.070186.004513

6. Seddon AM, Curnow P, Booth PJ (2004) Membrane proteins, lipids and detergents: not just a soap opera. Biochim Biophys Acta 1666:105–117. doi:10.1016/j.bbamem.2004.04.011
7. Michel H (1983) Crystallization of membrane proteins. Trends Biochem Sci 8:56–59. doi:10.1016/0968-0004(83)90390-0
8. Caffrey M, Li D, Dukkipati A (2012) Membrane protein structure determination using crystallography and lipidic mesophases: recent advances and successes. Biochemistry 51:6266–6288. doi:10.1021/bi300010w
9. Faham S, Bowie JU (2002) Bicelle crystallization: a new method for crystallizing membrane proteins yields a monomeric bacteriorhodopsin structure. J Mol Biol 316:1–6. doi:10.1006/jmbi.2001.5295
10. Iwata S, Ostermeier C, Ludwig B, Michel H (1995) Structure at 2.8 Å resolution of cytochrome c oxidase from Paracoccus denitrificans. Nature 376:660–669
11. Zhou Y, Morais-Cabral JH, Kaufman A, MacKinnon R (2001) Chemistry of ion coordination and hydration revealed by a K+ channel–Fab complex at 2.0 Å resolution. Nature 414:43–48. doi:10.1038/35102009
12. Hino T, Arakawa T, Iwanari H et al (2012) G-protein-coupled receptor inactivation by an allosteric inverse-agonist antibody. Nature 482:237–240. doi:10.1038/nature10750
13. Fang Y, Jayaram H, Shane T et al (2009) Structure of a prokaryotic virtual proton pump at 3.2 Å resolution. Nature 460:1040–1043. doi:10.1038/nature08201
14. Smith GP (1985) Filamentous fusion phage: novel expression vectors that display cloned antigens on the virion surface. Science 228:1315–1317. doi:10.1126/science.4001944
15. Boder ET, Wittruo KD (1997) Yeast surface display for screening combinatorial polypeptide libraries. Nat Biotechnol 15:553–557. doi:10.1038/nbt0697-553
16. Hanes J, Plückthun A (1997) In vitro selection and evolution of functional proteins by using ribosome display. Proc Natl Acad Sci U S A. doi:10.1038/nbt.1791
17. Jostock T, Dübel S (2005) Screening of molecular repertoires by microbial surface display. Comb Chem High Throughput Screen 8:127–133. doi:10.2174/1386207053258479
18. de Haard HJ, van Neer N, Reurs A et al (1999) A large non-immunized human Fab fragment phage library that permits rapid isolation and kinetic analysis of high affinity antibodies. J Biol Chem 274:18218–18230
19. Vaughan TJ, Williams AJ, Pritchard K et al (1996) Human antibodies with sub-nanomolar affinities isolated from a large non-immunized phage display library. Nat Biotechnol 14:309–314. doi:10.1038/nbt0396-309
20. Arbabi Ghahroudi M, Desmyter A, Wyns L et al (1998) Selection and identification of single domain antibody fragments from camel heavy-chain antibodies. FEBS Lett 414:521–526. doi:10.1016/S0014-5793(97)01062-4
21. Binz HK, Amstutz P, Kohl A et al (2004) High-affinity binders selected from designed ankyrin repeat protein libraries. Nat Biotechnol 22:575–582. doi:10.1038/nbt962
22. Koide A, Bailey CW, Huang X, Koide S (1998) The fibronectin type III domain as a scaffold for novel binding proteins. J Mol Biol 284:1141–1151. doi:10.1006/jmbi.1998.2238
23. Schönfeld D, Matschiner G, Chatwell L et al (2009) An engineered lipocalin specific for CTLA-4 reveals a combining site with structural and conformational features similar to antibodies. Proc Natl Acad Sci U S A 106:8198–8203. doi:10.1073/pnas.0813399106
24. Cyranka-Czaja A, Otlewski J (2012) A novel, stable, helical scaffold as an alternative binder—construction of phage display libraries. Acta Biochim Pol 59(3):383–390
25. Krishnamurthy H, Gouaux E (2012) X-ray structures of LeuT in substrate-free outward-open and apo inward-open states. Nature 481:469–474. doi:10.1038/nature10737
26. Hino T, Iwata S, Murata T (2013) Generation of functional antibodies for mammalian membrane protein crystallography. Curr Opin Struct Biol 23:563–568. doi:10.1016/j.sbi.2013.04.007
27. Fellouse FA, Wiesmann C, Sidhu SS (2004) Synthetic antibodies from a four-amino-acid code: a dominant role for tyrosine in antigen recognition. Proc Natl Acad Sci U S A 101:12467–12472. doi:10.1073/pnas.0401786101
28. Ponsel D, Neugebauer J, Ladetzki-Baehs K, Tissot K (2011) High affinity, developability and functional size: the holy grail of combinatorial antibody library generation. Molecules 16:3675–3700. doi:10.3390/molecules16053675
29. Lee CV, Liang W-C, Dennis MS et al (2004) High-affinity human antibodies from phage-displayed synthetic Fab libraries with a single framework scaffold. J Mol Biol 340:1073–1093. doi:10.1016/j.jmb.2004.05.051

30. Uysal S, Vásquez V, Tereshko V et al (2009) Crystal structure of full-length KcsA in its closed conformation. Proc Natl Acad Sci U S A 106:6644–6649. doi:10.1073/pnas.0810663106
31. Ye J-D, Tereshko V, Frederiksen JK et al (2008) Synthetic antibodies for specific recognition and crystallization of structured RNA. Proc Natl Acad Sci U S A 105:82–87. doi:10.1073/pnas.0709082105
32. Razai A, Garcia-Rodriguez C, Lou J et al (2005) Molecular evolution of antibody affinity for sensitive detection of botulinum neurotoxin type A. J Mol Biol 351:158–169. doi:10.1016/j.jmb.2005.06.003
33. Uysal S, Cuello LG, Cortes DM et al (2011) Mechanism of activation gating in the full-length KcsA K+ channel. Proc Natl Acad Sci U S A 108:11896–11899. doi:10.1073/pnas.1105112108
34. Li Q, Wanderling S, Paduch M et al (2014) Structural mechanism of voltage-dependent gating in an isolated voltage-sensing domain. Nat Struct Mol Biol 21:244–252. doi:10.1038/nsmb.2768
35. Doyle DA, Cabral JM, Pfuetzner RA et al (1998) The structure of the potassium channel: molecular basis of K+ conduction and selectivity. Science 280:69–77. doi:10.1126/science.280.5360.69
36. Fellouse FA, Esaki K, Birtalan S et al (2007) High-throughput generation of synthetic antibodies from highly functional minimalist phage-displayed libraries. J Mol Biol 373:924–940. doi:10.1016/j.jmb.2007.08.005
37. Fellouse FA, Barthelemy PA, Kelley RF, Sidhu SS (2006) Tyrosine plays a dominant functional role in the paratope of a synthetic antibody derived from a four amino acid code. J Mol Biol 357:100–114. doi:10.1016/j.jmb.2005.11.092
38. Birtalan S, Zhang Y, Fellouse FA et al (2008) The intrinsic contributions of tyrosine, serine, glycine and arginine to the affinity and specificity of antibodies. J Mol Biol 377:1518–1528. doi:10.1016/j.jmb.2008.01.093
39. Adams JJ, Nelson B, Sidhu SS (2014) Recombinant genetic libraries and human monoclonal antibodies. Methods Mol Biol 1060:149–170. doi:10.1007/978-1-62703-586-6_9
40. Paduch M, Koide A, Uysal S et al (2013) Generating conformation-specific synthetic antibodies to trap proteins in selected functional states. Methods 60:3–14. doi:10.1016/j.ymeth.2012.12.010
41. Zhong N, Loppnau P, Seitova A et al (2015) Optimizing production of antigens and Fabs in the context of generating recombinant antibodies to human proteins. PLoS One 10:e0139695. doi:10.1371/journal.pone.0139695

Pulse-Chase Covalent Labeling Technique for Monitoring GPCR Endocytosis

Hidetoshi Kumagai and Yuichi Ikeda

Abstract

G protein-coupled receptors represent one of the largest families of cell-surface receptors. They play important roles in various physiological systems in response to a range of ligands with diverse physicochemical properties. Ligand stimulation triggers G protein-coupled receptor (GPCR) endocytosis, which controls cellular desensitization and resensitization processes by regulating the number of GPCRs on the plasma membrane. Notably, it has recently been shown that endosome-localized GPCRs contribute significantly to both G protein-dependent and G protein-independent signaling in the acute and subacute phase. Studying individual GPCR behaviors during endocytosis is thus becoming increasingly more important in GPCR research. Here we describe a simple and versatile protocol for monitoring GPCR endocytosis utilizing a pulse-chase covalent labeling technique based on the HaloTag technology.

Keywords Covalent bond formation, Endocytosis, GPCR, HaloTag technology, Pulse-chase labeling, Trafficking

1 Introduction

The superfamily of G protein-coupled receptors (GPCRs), one of the largest receptor families expressed on the cell surface, plays significant roles in various physiological systems in response to a range of ligands with diverse physicochemical properties [1]. Ligand binding initiates conformational alterations of the receptor and subsequently activates heterotrimeric G proteins. The receptor in its active conformation is then phosphorylated by GPCR kinases (GRKs), resulting in the recruitment of β-arrestins, ubiquitously expressed adapter proteins, from the cytosol to the phosphorylated receptor. β-Arrestin recruitment shuts off the canonical G protein-dependent signaling through steric hindrance at the plasmamembrane and simultaneously promotes receptor endocytosis by facilitating association of receptors with the structural components of clathrin-coated pits (CCPs). By internalizing CCPs, GPCR-β-arrestin complex formation leads to the activation

Arun K. Shukla (ed.), *Chemical and Synthetic Approaches in Membrane Biology*, Springer Protocols Handbooks, (2017) 245–252, DOI 10.1007/8623_2016_2, © Springer Science+Business Media LLC 2016, Published online: 25 September 2016

of G protein-independent signaling, such as MAP kinase activation, which is influenced by the endocytic behavior of individual GPCRs [2]. Notably, it has been shown that endosome-localized GPCRs contribute significantly to G protein-dependent signaling in the subacute phase by directly activating G proteins, and that β-arrestins accumulated in CCPs, which have been primed into their active forms at the plasma membrane, can trigger G protein-independent signaling even in the absence of GPCR-β-arrestin complex formation in CCPs [3–6]. These conceptually novel findings demonstrate that GPCR endocytosis does not simply control cellular desensitization and resensitization processes by regulating the number of functional GPCRs on the cell surface, but also affects both G protein-dependent and G protein-independent signaling in the acute and subacute phase, providing another layer of complexity to the conventional paradigm of GPCR signaling and trafficking. Hence studying individual GPCR behaviors during endocytosis is increasingly more important in GPCR research.

A catalytically inactive derivative of a bacterial hydrolase named HaloTag was recently developed [7, 8]. The HaloTag is able to form a covalent bond with synthetic HaloTag ligands possessing a reactive chloroalkane linker. Utilizing this HaloTag as an artificially designed tag, a chimeric GPCR, genetically engineered to have the HaloTag at its N-terminal end (HT-GPCR), can be irreversibly and covalently labeled with various kinds of HaloTag ligands conjugated with functional reporters, including fluorophores or a high-affinity handle (such as biotin) (Fig. 1). Significantly, a HaloTag PEG ligand, which cannot enter the cell because of its hydrophilic ethylene glycol moiety, allows selective labeling of HT-GPCRs on the cell surface at any given moment. This nonradioactive pulse-chase covalent labeling thus offers a simple and versatile assay for monitoring GPCR endocytosis (*see* **Note 1**) [9].

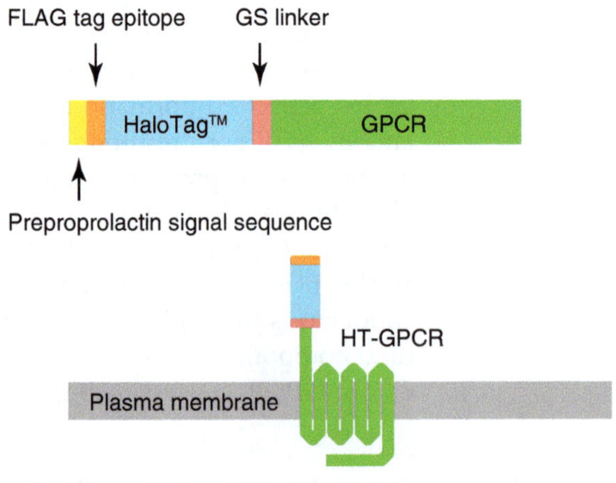

Fig. 1 A schematic of HT-GPCR [9]

2 Materials

2.1 Plasmid

1. HT-GPCR-expressing plasmid: The bovine prolactin signal sequence (MDSKGSSQKGSRLLLLLVVSNLLLCQGVVS), followed by a FLAG epitope tag (DYKDDDDK), the HaloTag sequence without the first methionine, the GS linker (G4S x3), and the GPCR sequence without the first methionine were fused in-frame and subcloned into a mammalian expression vector (Fig. 1).

2.2 Cell Culture Preparation

1. CHO-K1 cells (American Type Culture Collection) stably expressing the HT-GPCR of interest
2. CHO-K1 growth medium: Dulbecco's modified Eagle medium (DMEM) supplemented with 5% fetal calf serum (FCS) and 1% nonessential amino acids
3. Phosphate-buffered saline (PBS), Ca^{2+-} and Mg^{2+}-free
4. 0.05% trypsin-EDTA
5. 10 cm tissue culture-treated dishes
6. Tissue culture-treated 96-well plates

2.3 HaloTag GPCR Endocytosis Assay

1. HaloTag PEG-Biotin Ligand (Promega)
2. GPCR ligands
3. Streptavidin-horseradish peroxidase (SA-HRP, Pierce)
4. HRP substrates (1-Step Ultra TMB-ELISA, Pierce)
5. 4% paraformaldehyde (PFA)
6. 2N H_2SO_4
7. 0.3% Triton X-100
8. Streptavidin (SA, Pierce)
9. 0.3% H_2O_2/PBS
10. Victor V plate reader (PerkinElmer)

3 Methods

3.1 Pulse-Chase Covalent Labeling of HT-GPCRs on the Cell Surface

CHO-K1 cells, stably expressing the HT-GPCR (CHO-HT-GPCR cells), are maintained in 10 cm tissue culture-treated dishes and trypsinized and plated in tissue culture-treated 96-well plates at a cell density of 2–4 × 10^4 cells/well. After 24 h, HaloTag PEG-Biotin Ligand is added to the wells to label HT-GPCRs for 15 min at 37°C (HaloTag PEG-Biotin Ligand, final conc. 5 μM) (*see* **Note 2**). Following 15 min incubation, CHO-HT-GPCR cells are washed three times in fresh CHO-K1 growth medium to remove unreacted HaloTag PEG-Biotin Ligands (*see* **Note 3**).

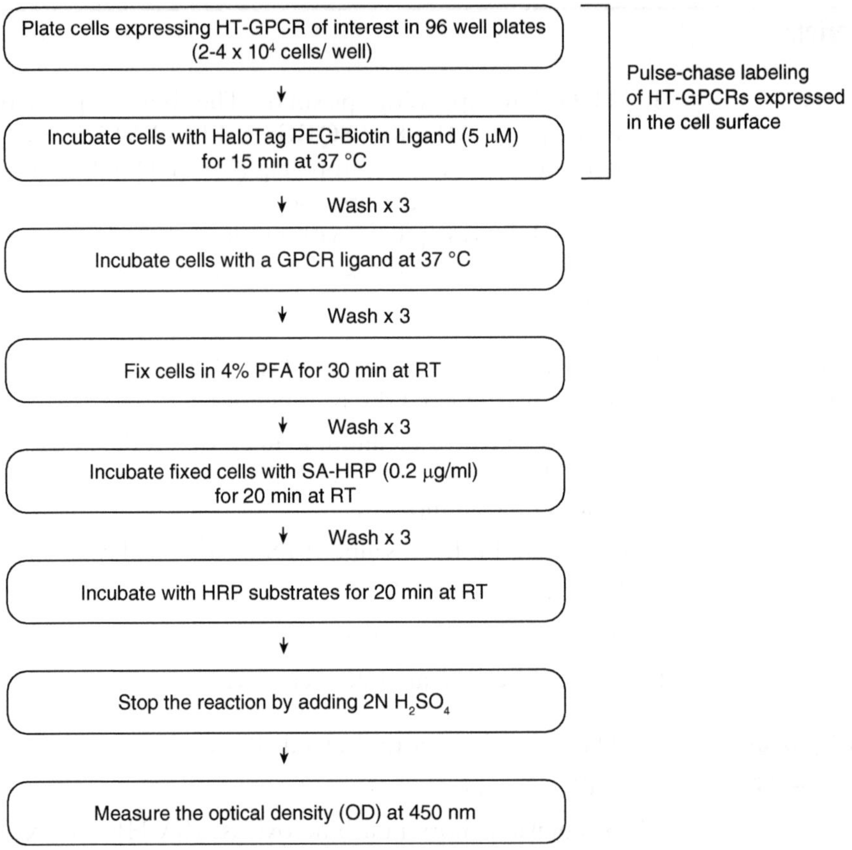

Fig. 2 A brief overview of the protocol for measuring the relative amount of HT-GPCRs remaining on the cell surface [9]

3.2 Quantitative Measurement of HT-GPCRs on the Cell Surface (Fig. 2)

After pulse-chase labeling performed by following the protocol described above, CHO-HT-GPCR cells are treated with various concentrations of GPCR agonists for 1 h (*see* **Note 4**). Following 1 h stimulation with the agonists, CHO-HT-GPCR cells are washed three times in PBS and immediately fixed in 4% PFA for 30 min at room temperature (RT). After washing in PBS, fixed cells are treated with SA-HRP (0.2 μg/ml) for 20 min at RT. After eliminating unreacted SA-HRP by washing three times in PBS, cells are incubated with 50 μl of HRP substrates (1-Step Ultra TMB-ELISA) for 20 min at RT. The reaction is suspended by directly adding 50 μl of 2N H_2SO_4 to the wells, and the optical density (OD) at 450 nm is then measured with a plate reader.

Following this standard protocol, we measured the relative amount of the labeled HT-hBRS3s on the cell surface after HT-hBRS3s were stimulated with various concentrations of the surrogate hBRS3 ligand ([D-Phe^{6}, β-Ala^{11}, Phe^{13}, Nle^{14}]-Bombesin (6–14); Phoenix Pharmaceuticals, Inc.). The number of labeled HT-hBRS3s on the cell surface decreased in response to the surrogate ligand in a dose-dependent manner (Fig. 3).

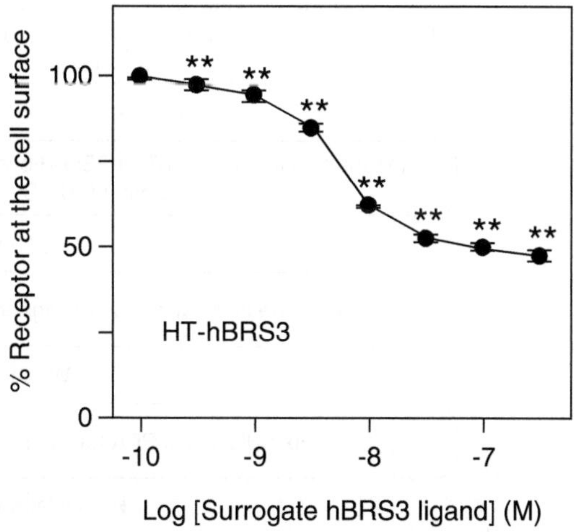

Fig. 3 Agonist-induced GPCR internalization. The number of HT-hBRS3s on the cell surface decreased in response to the surrogate hBRS3 ligand in a dose-dependent manner [9]

3.3 Quantitative Measurement of Internalized HT-GPCRs (Fig. 4)

After pulse-chase labeling, CHO-HT-GPCR cells are stimulated with various concentrations of GPCR agonists for 1 h (*see* **Note 4**). Following 1 h incubation with the agonists, CHO-HT-GPCR cells are washed three times in PBS and immediately fixed in 4% PFA for 30 min at RT. After washing in PBS, fixed cells are incubated with SA (5 μg/ml) for 20 min at RT (*see* **Note 5**). After removal of unreacted SA by washing three times in PBS, cells are incubated with 0.3% Triton X-100 for 10 min at RT in order to permeabilize the cell membrane (*see* **Note 6**). Following three washes in PBS, cells are treated with 0.3% H_2O_2 /PBS for 30 min at RT to block endogenous peroxidase activities. Cells are then incubated with SA-HRP (0.2 μg/ml) for 20 min at RT. After eliminating unreacted SA-HRP by PBS washing, cells are incubated with 50 μl of HRP substrates (1-Step Ultra TMB-ELISA) for 20 min at RT. The reaction is suspended by directly adding 50 μl of 2N H_2SO_4 to the wells, and the optical density (OD) at 450 nm is then measured with a plate reader.

Following this protocol, we measured the relative amount of the internalized HT-hOX2Rs after HT-hOX2Rs were stimulated with various concentrations of orexin-A (OXA) peptides. The number of the internalized HT-hOX2Rs increased in response to OXA in a dose-dependent manner (Fig. 5).

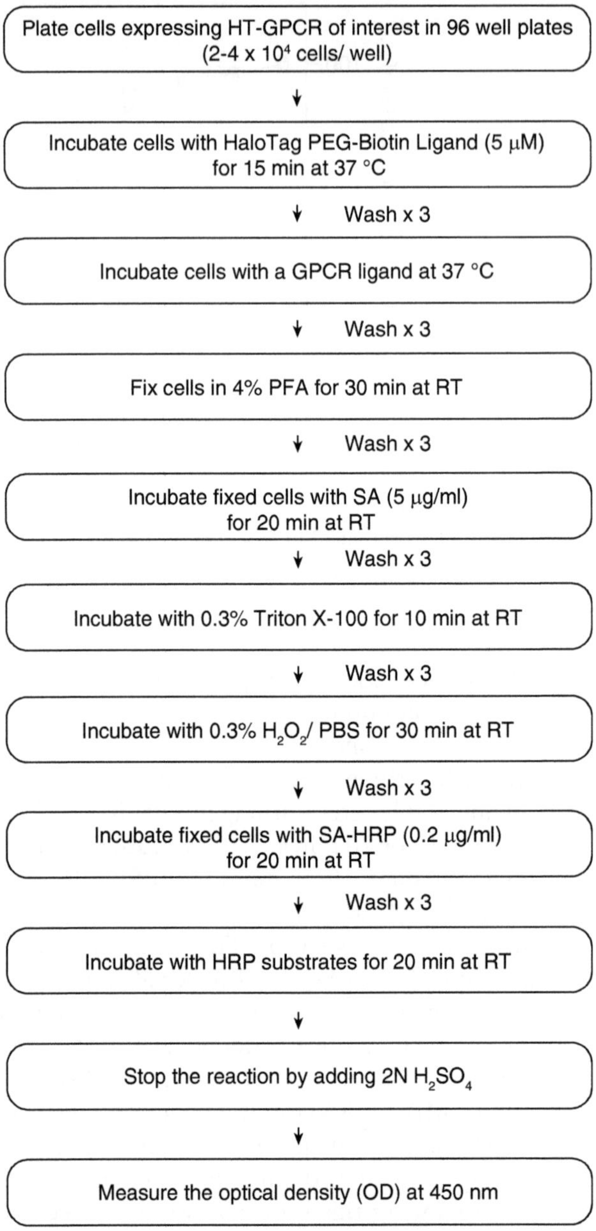

Fig. 4 A brief overview of the protocol for measuring the relative amount of internalized HT-GPCRs [9]

4 Notes

1. In general, the assays described in this chapter can be applied to any GPCR of interest. However, it is possible that attaching the HaloTag to the N-terminal end of GPCRs interferes with ligand-receptor interactions because some GPCRs are known

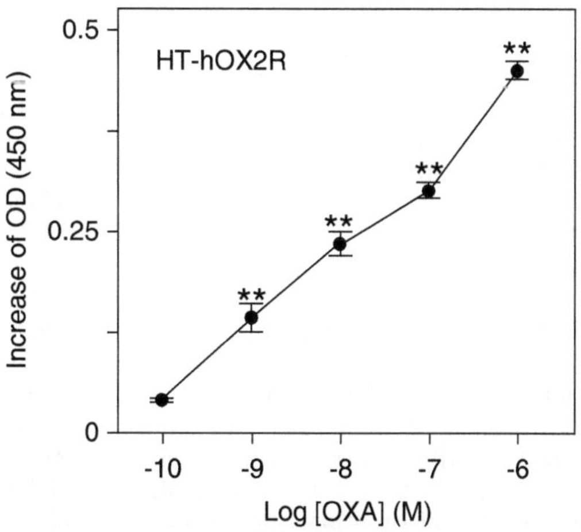

Fig. 5 Direct measurement of internalized HT-GPCRs. The number of internalized HT-hOX2Rs increased in response to OXA in a dose-dependent manner [9]

to recognize the ligands through their N-terminal extracellular domain (Fig. 1).

2. Since its hydrophilic ethylene glycol moiety prevents the HaloTag PEG-Biotin Ligand from entering the cell, HT-GPCRs on the cell surface are selectively labeled by the HaloTag PEG-Biotin Ligand.
3. Stable covalent labeling of HT-GPCRs with HaloTag ligands allows extensive washing to minimize the background and also ensures that labeled HT-GPCRs can be tracked until they are degraded in the lysosome.
4. Individual GPCRs have incredibly diverse basal turnover rates. We noted that some receptors exhibit less than a 2 h elimination half-life even without agonist stimulation. We therefore examined agonist-induced GPCR internalization by stimulating the receptor for an hour in our standard protocol. However, incubation time with the ligands should be adjusted depending on the GPCR of interest and the purpose of the experiment.
5. SA treatment blocks all the free biotin sites that have been attached to the pulse-chase labeled HT-GPCRs remaining on the cell surface.
6. Permeabilization of the cell membrane makes the internalized HT-GPCRs, which have been pulse-chase labeled with the HaloTag PEG-Biotin Ligand, accessible to the SA-HRP.
7. An important caveat is that endocytic behaviors of individual GPCRs may be dependent on the cell type in which the assay is performed. Although we used CHO cells in our protocol, more relevant cell types may have to be chosen depending on the purpose(s) of the assay.

References

1. Vassilatis DK, Hohmann JG, Zeng H, Li F, Ranchalis JE, Mortrud MT, Brown A, Rodriguez SS, Weller JR, Wright AC, Bergmann JE, Gaitanaris GA (2003) The G protein-coupled receptor repertoires of human and mouse. Proc Natl Acad Sci U S A 100(8):4903–4908. doi:10.1073/pnas.0230374100
2. Rajagopal S, Rajagopal K, Lefkowitz RJ (2010) Teaching old receptors new tricks: biasing seven-transmembrane receptors. Nat Rev Drug Discov 9(5):373–386. doi:10.1038/nrd3024
3. Irannejad R, Tomshine JC, Tomshine JR, Chevalier M, Mahoney JP, Steyaert J, Rasmussen SGF, Sunahara RK, El-Samad H, Huang B, von Zastrow M (2013) Conformational biosensors reveal GPCR signalling from endosomes. Nature 495(7442):534–538. doi:10.1038/nature12000
4. Eichel K, Jullie D, von Zastrow M (2016) beta-Arrestin drives MAP kinase signalling from clathrin-coated structures after GPCR dissociation. Nat Cell Biol 18(3):303–310. doi:10.1038/ncb3307
5. Ranjan R, Gupta P, Shukla AK (2016) GPCR Signaling: beta-arrestins Kiss and Remember. Curr Biol 26(7):R285–R288. doi:10.1016/j.cub.2016.02.056
6. Nuber S, Zabel U, Lorenz K, Nuber A, Milligan G, Tobin AB, Lohse MJ, Hoffmann C (2016) beta-Arrestin biosensors reveal a rapid, receptor-dependent activation/deactivation cycle. Nature 531(7596):661–664. doi:10.1038/nature17198
7. Los GV, Wood K (2007) The HaloTag: a novel technology for cell imaging and protein analysis. Methods Mol Biol 356:195–208
8. Los GV, Encell LP, McDougall MG, Hartzell DD, Karassina N, Zimprich C, Wood MG, Learish R, Ohana RF, Urh M, Simpson D, Mendez J, Zimmerman K, Otto P, Vidugiris G, Zhu J, Darzins A, Klaubert DH, Bulleit RF, Wood KV (2008) HaloTag: a novel protein labeling technology for cell imaging and protein analysis. ACS Chem Biol 3(6):373–382. doi:10.1021/cb800025k
9. Kumagai H, Ikeda Y, Motozawa Y, Fujishiro M, Okamura T, Fujio K, Okazaki H, Nomura S, Takeda N, Harada M, Toko H, Takimoto E, Akazawa H, Morita H, Suzuki J, Yamazaki T, Yamamoto K, Komuro I, Yanagisawa M (2015) Quantitative measurement of GPCR endocytosis via pulse-chase covalent labeling. PLoS One 10(5), e0129394. doi:10.1371/journal.pone.0129394

Sample Preparation of Rhodopsins in the *E. coli* Membrane for In Situ Magic Angle Spinning Solid-State Nuclear Magnetic Resonance Studies

Meaghan E. Ward, Vladimir Ladizhansky, and Leonid S. Brown

Abstract

Determination of the structure and dynamics of membrane proteins in complex, native cellular environments is one of the primary targets of structural biology. Here, we present a protocol for the preparation of recombinant membrane proteins in the native *E. coli* membrane environment for solid-state NMR (SSNMR) studies. This protocol has been developed on *Anabaena* sensory rhodopsin (ASR), a seven-transmembrane α-helical light receptor, but should be easily transferable to similar recombinant membrane protein systems. In order for SSNMR studies to be possible on such complex systems, it is desirable to remove as much background signal as possible. This is achieved both through physically separating segments of the membrane containing ASR and through isotopic labeling strategies which strategically limit isotopic incorporation into background proteins. Through the implementation of these methods and 3D SSNMR spectroscopy, we find that it is possible to resolve and characterize 40% of the previously assigned residues of ASR in the *E. coli* membrane environment.

Keywords Chemical shift, *E. coli* membrane, Isotope labeling, Multidimensional spectroscopy, Over-expression, Rhodopsin, Solid-state NMR

Abbreviations

ASR	*Anabaena* Sensory Rhodopsin
EM	*E. coli* membrane
IM	Inner membrane
MAS	Magic angle spinning
NA	Natural abundance
NIC	Non-induced cells
NMR	Nuclear magnetic resonance
OM	Outer membrane
rbUCN/rbUN	Reduced-background UCN/UN labeled
SSNMR	Solid-state NMR
UCN	Uniformly ^{13}C,^{15}N-labeled
UN	Uniformly ^{15}N-labeled

Arun K. Shukla (ed.), *Chemical and Synthetic Approaches in Membrane Biology*, Springer Protocols Handbooks, (2017) 253–267, DOI 10.1007/8623_2016_5, © Springer Science+Business Media LLC 2016, Published online: 25 September 2016

1 Introduction

Membrane proteins are a diverse class of biomolecules which account for a third of the proteome of living cells. They play many critical roles in all types of cells, being responsible for signal transduction, transport of various substances across lipid membranes, enzymatic activities, etc. Studying the structure and dynamics of membrane proteins in a cellular environment in which they are subjected to a plethora of interactions with the natively present cellular components, e.g., other proteins, nucleic acids, lipids, cofactors, etc., is the ultimate goal of structural biology. Chemical heterogeneity of such systems often translates into structural heterogeneity, and, with a few exceptions [1, 2], such systems are not readily amenable to high-resolution methods of structural biology that require long-range order.

Nuclear magnetic resonance (NMR)-based methods, however, do not require long-range order and can be used to characterize many biological systems, including globular proteins in native-like solubilized states, lipid-associated proteins, proteins in an aggregated state, or large macromolecular complexes, which are not amenable to other high-resolution structural methods. In particular, solid-state NMR (SSNMR) methodologies have matured over the past two decades to a point where three-dimensional structures can be reliably determined, and protein interactions and dynamics can be studied at atomic resolution [3–9]. In addition, both solution and solid-state NMR have been successfully applied to study molecules in cellular environments, either in the cytosol or in cellular membranes, respectively [10–21].

Below we present a protocol for the preparation of isotopically labeled recombinant membrane proteins in the *E. coli* membrane. The procedures described produce samples which are suitable for studies of the effects of the cellular membrane environment on protein structure and dynamics using magic angle spinning (MAS) solid-state NMR. These methods were optimized on the C-terminally truncated 6-His-tagged version of *Anabaena* sensory rhodopsin (ASR) from *Anabaena* sp. PCC 7120 [22]. ASR is a seven-helical transmembrane light sensor which is believed to be involved in gene regulation in cyanobacteria. It can be conveniently overexpressed and isotopically labeled in *E. coli* cells (estimated crude yields are ~10–20 mg per liter of culture), and solid-state NMR studies on purified ASR reconstituted into proteoliposomes (PL-ASR) have yielded extensive spectroscopic assignments [23, 24] as well as a high-resolution structure [25]. Though this protocol will be described in the context of its application to ASR, we believe that it should be easily transferable to similar recombinant systems. The application of this protocol to other rhodopsins is a particularly attractive option, as these proteins contain retinal molecules which absorb light and give the proteins a visible color, greatly simplifying purification procedures.

To initiate solid-state NMR studies of membrane proteins in the complex environment of native cell membranes, a few challenges must be addressed. Firstly, in order to suppress NMR signals from background proteins and to simplify the spectra and data analysis, isotopic ^{13}C and ^{15}N labels supplied through the growth media need to be incorporated primarily into the recombinant protein. Secondly, as the solid-state NMR sample volume is limited, it is beneficial to remove nonessential membrane components (i.e., parts of the membranes that do not interact with ASR). This not only further simplifies the spectra but also allows for an increased amount of the recombinant protein in the sample.

The main steps taken to address these challenges and prepare ASR in the *E. coli* membrane for SSNMR studies (we denote this sample as EM-ASR) are shown in Fig. 1. In order to preferentially incorporate isotopic labels into ASR, cells are initially grown to the exponential phase in unlabeled media and are later resuspended in a lower volume of isotopically enriched media for the expression of the recombinant protein (samples made from cells grown in this fashion are denoted as rbUN or rbUCN) [26]. In this way, the majority of cellular components are produced in unlabeled media and therefore do not contain isotopic labels, while ASR is produced in isotopically labeled media and therefore contains a higher proportion of isotope labels. The effectiveness of this process could potentially be increased by combining the above method with the use of a mutant *E. coli* BL21Star(DE3) strain which lacks the two major OM proteins OmpF and OmpA [11], the repression of endogenous protein expression through the addition of rifampicin [27] and/or lower incubation temperatures during expression [12].

Next, as ASR partitions primarily in the inner membrane (IM) of *E. coli* cells, the background proteins present in the outer membrane (OM) are removed through the application of a sucrose gradient [28]. Background proteins in the IM fraction are then further removed through taking advantage of the presence of a His-tag by applying this fraction to a two-phase separation system [29, 30] which uses Ni^{2+}–NTA–agarose resin to bind the His-tag of ASR and partition IM fragments containing ASR to a denser phase, while those which do not contain ASR partition to the less dense phase and can therefore be removed.

Through these steps ASR NMR signals were sufficiently isolated from the background signals, and basic three-dimensional SSNMR spectroscopy could be recorded to resolve individual ASR resonances. However, the low signal-to-noise ratio and spectral overlap precluded obtaining de novo chemical shift assignments. Therefore, interpretation of the NMR experiments required two additional reference samples: (1) ASR reconstituted in synthetic

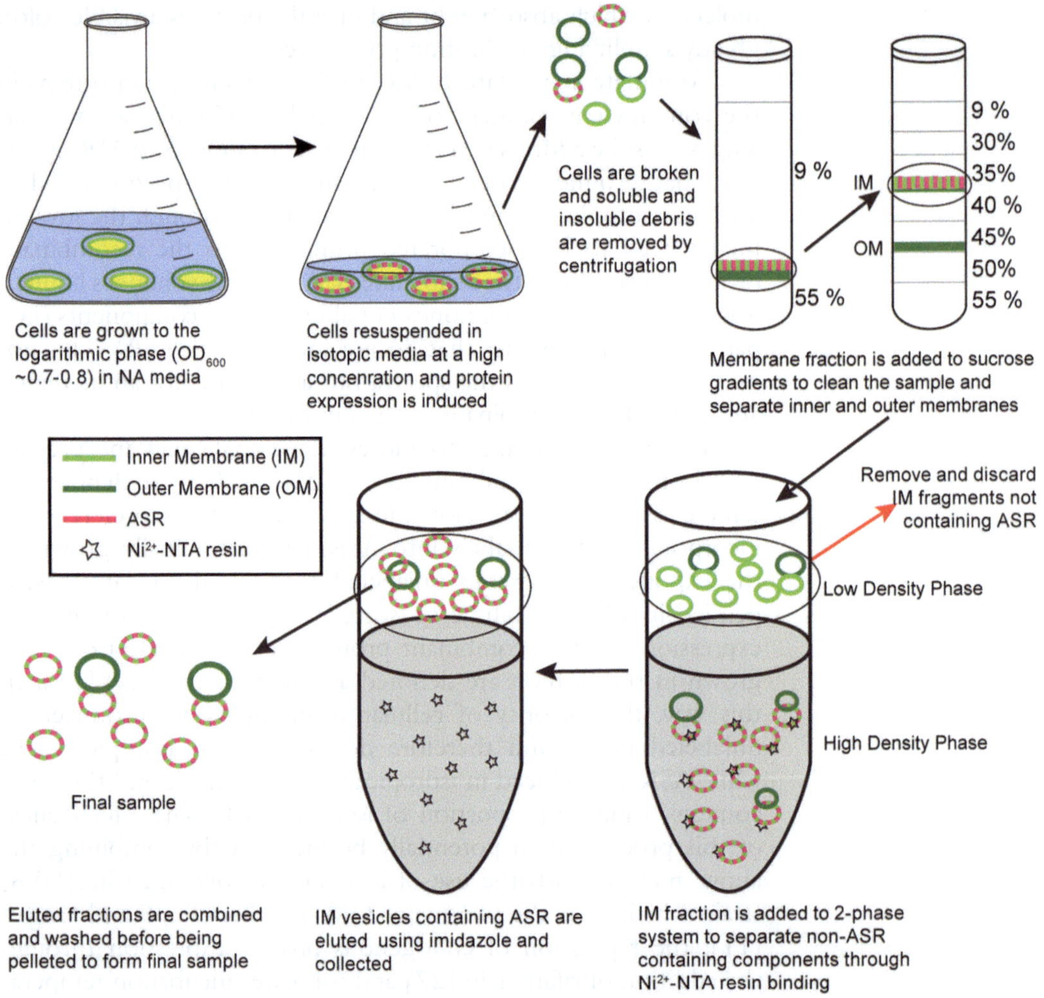

Fig. 1 Summary of isotopic labeling and sample preparation strategies. To reduce NMR signal from the background proteins, cells are grown to the exponential phase in NA media and then resuspended at a high concentration in labeled media before protein expression is induced (denoted as rbUN or rbUCN labeling). The inner and outer membranes are then separated through a sucrose gradient. IM vesicles not containing ASR are then separated using a two-phase system which partitions ASR-containing vesicles to the heavy phase through resin binding

lipids [23, 24], on which spectroscopic assignment experiments have previously been carried out (PL-ASR), and (2) an additional negative control sample which contains all cell membrane components but excludes ASR (EM-NIC). The latter sample was obtained by performing all steps of the isolation procedure described below on cells which were not induced to express recombinant ASR. In this way, NMR spectra obtained on all three samples can be compared to confidently identify spectroscopic signals which arise from ASR. Through these methods and the implementation of 3D SSNMR spectroscopy, we were able to confidently identify and analyze 40% of the previously assigned residues in PL-ASR, in the more complex environment of the *E. coli* membrane [31].

2 Materials

2.1 Protein Expression with Reduced Isotopic Labeling of Background Proteins

1. *E. coli* host strain transformed with vector containing a 6×His-tagged rhodopsin sequence (*see* **Note 1**).
2. LB medium: 10 g/L NaCl, 10 g/L tryptone, 5 g/L yeast extract. Autoclave. Add filtered 100 mg/L ampicillin and 50 mg/L chloramphenicol (*see* **Note 2**).
3. M9 minimal medium: 6 g/L Na_2HPO_4, 3 g/L KH_2PO_4, 1 g/L NH_4Cl or $^{15}NH_4Cl$, and 0.5 g/L NaCl. Autoclave. Add filtered 1 mM $MgSO_4$, 0.1 M $CaCl_2$, 500 μl trace metals (see below), 4 g/L glucose or $^{13}C_6$ glucose, 10 ml/L 100× BME Vitamins solution, 100 mg/L ampicillin, and 50 mg/L chloramphenicol.
4. Trace metals: 100 mM $Fe_2(SO_4)_3 \cdot n(H_2O)$, 180 mM $FeSO_4 \cdot 7(H_2O)$, 38 mM $ZnCl_2$, 32 mM $CuSO_4$, 32 mM H_3BO_3, 33 mM $MnSO_4 \cdot 4(H_2O)$, and 29 mM $Na_2MoO_4 \cdot 2(H_2O)$ (filter sterilized).
5. M9 salt solution: 6 g/L Na_2HPO_4, 3 g/L KH_2PO_4, 0.5 g/L NaCl. Autoclave.
6. Retinal stock solution in ethanol (~10–20 mM).
7. 1 M IPTG (filter sterilized).
8. 150 mM NaCl.

2.2 Lysis of Cells and Preparation of Membrane Fraction

1. Lysis buffer: 0.15 M NaCl (pH 7.2), 0.05 M Tris, 1 mM $MgCl_2$, 2 μg/ml DNase 1, and 0.2 mg/ml lysozyme

2.3 Sucrose Gradient to Separate Inner and Outer Membranes

1. Ultracentrifuge with swinging bucket rotor (i.e., Thermo Scientific Sorvall WX Ultra Series, SureSpin 630)
2. Sucrose gradient buffer: 50 mM TEA (pH 7.5), 1 mM EDTA, and 1 mM DTT, with 0, 9, 30, 35, 40, 45, 50, and 55% sucrose
3. 10 mM Tris (pH 8)

2.4 Two-Phase System and Final Sample Preparation

1. Two-phase system: 10 mM Tris (pH 8), 0.3 M NaCl, 6.45% (w/w) PEG 3350, 8.45% (w/w) dextran, 320 μl of Ni^{2+}–NTA–agarose resin slurry per milliliter of system (*see* **Note 3**)
2. Washing buffer: 10 mM Tris (pH 8), 0.3 M NaCl, 6.45% (w/w) PEG 3350
3. Elution buffer: 5 mM Tris (pH 8), 0.5 M imidazole
4. NMR buffer: 10 mM NaCl, 24 mM CHES (pH 9)

3 Methods

3.1 Protein Expression with Reduced Isotopic Labeling of Background Protein

1. Inoculate 2 ml of LB media (*see* **Note 2**) with transformed *E. coli* cells. Shake at 37°C and 240 rpm for ~15 h.
2. Once the culture has reached the maximal stationary OD_{600}, use the 2 ml culture to inoculate a 25 ml culture. The starting OD_{600} must be at least 0.1. Shake at 37°C, 240 rpm for ~15 h.
3. Once the culture has reached the maximal stationary OD_{600}, use this 25 ml to inoculate 1 L of LB media. The starting OD_{600} must be at least 0.1. Shake at 240 rpm and 37°C until the OD_{600} reaches ~0.7–0.8.
4. Collect the cells through gentle centrifugation (15 min; 3,500 × *g*; 4°C) in autoclaved tubes. Discard supernatant under aseptic conditions.
5. Under aseptic conditions, resuspend pellet in ~50 ml of M9 salt solution. Likely **Step 4** will be performed with multiple tubes, so at this point, combine the pellets from all tubes into one.
6. Collect cells through gentle centrifugation (15 min; 3,500 × *g*; 4°C). Discard supernatant under aseptic conditions.
7. Under aseptic conditions, resuspend pellet in 250 ml of M9 minimal media labeled with appropriate isotopes (*see* **Note 4**).
8. Allow cells to shake in M9 media at 240 rpm at 30°C for 1 h.
9. Induce the cells with IPTG at a concentration of 1 mM, and add retinal to a final concentration of 7.5 μM. Increase the rate of shaking to 275 rpm.
10. After ~4 h collect the cells by centrifugation (15 min; 3,500 × *g*; 4°C) (*see* **Note 5**).
11. Discard the supernatant. Resuspend the cell pellet in ~50 ml of 150 mM NaCl and combine fractions.
12. Collect cells by centrifugation (15 min; 3,500 × *g*; 4°C).

3.2 Lysis of Cells and Preparation of Membrane Fraction

1. Resuspend the cell pellet in 10 ml of lysis buffer and mix gently at room temperature for ~3 h. Freeze.
2. Thaw cell pellet and sonicate cells to break cell membranes (*see* **Note 6**).
3. Centrifuge the cell lysate at low speed (15 min; 3,500 × *g*; 4°C) to check for unbroken cells and to remove insoluble cell debris. If there is recombinant protein evident in the pellet, then resuspend the pellet and repeat sonication. If there is

none, collect the supernatant, and repeat low-speed centrifugation until there is no or minimal pellet.

4. Centrifuge the membrane fraction in order to collect the membrane pellet and separate soluble debris (50 min; 37,000 × *g*; 4°C).
5. Discard supernatant. Resuspend pellet in ~10 ml of sucrose gradient buffer (0% sucrose).

3.3 Sucrose Gradient to Separate Inner and Outer Membranes

1. Apply the membrane fraction to the top of a two-step sucrose gradient composed of discrete 9 and 55% (from top to bottom) sucrose layers.
2. Centrifuge the gradients in a swinging bucket rotor (2.5 h; 210,000 × *g*; 4°C).
3. Collect the membrane fraction from the top of the 55% sucrose layer.
4. Dilute the collected membrane fraction by at least 3× with sucrose gradient buffer (0% sucrose) and apply to the top of a six-step sucrose gradient composed of discrete 30, 35, 40, 45, 50, and 55% (from top to bottom) sucrose fractions.
5. Centrifuge gradients in a swinging bucket rotor (15 h; 210,000 × *g*; 4°C).
6. Collect the inner membrane fraction from the top of the 40% sucrose layer (*see* **Note** 7).
7. Dilute the inner membrane fraction by at least 3× with 10 mM Tris, and centrifuge to collect the pellet (30 min; 30,000 × *g*; 4°C).
8. To ensure excess sucrose has been removed from the sample, resuspend pellet in 10 mM Tris, and recollect the pellet through centrifugation (30 min; 30,000 × *g*; 4°C).

3.4 Two-Phase System and Final Sample Preparation

1. Incubate the inner membrane fraction with the two-phase system for ~30 min with gentle mixing (*see* **Note 8**).
2. Separate the phases by low-speed centrifugation (5 min; 1,600 × *g*; 4°C).
3. Collect the top phase and measure an absorption spectrum to determine the protein content (OD_{280}). The phases should be clearly visible due to the color of the protein. Discard the top phase.
4. Replace the top phase with an equal volume of washing buffer.
5. Repeat **Steps 2–4** until the absorption spectrum shows a lack of protein in the top phase.
6. Elute the membranes from the affinity beads by replacing the top phase with an equal volume of elution buffer and incubating with gentle mixing.

7. Use low-speed centrifugation (2 min; 1,000 × *g*; 4°C) to sediment the affinity beads, and collect the membranes from the top phase. Repeat **Steps 6–7** as necessary (*see* **Note 9**).
8. Pellet the membranes through gentle centrifugation (30 min; 30,000 × *g*; 4°C).
9. Resuspend pellet in ~10 ml of NMR buffer.
10. Repeat **Steps 8–9** once.
11. Repeat **Step 8**.
12. Resuspend membrane pellet using 1 ml NMR buffer.
13. Ultracentrifuge sample (3 h; 900,000 × *g*; 4°C).
14. Remove supernatant and replace with 20 μl of NMR buffer placed directly on the cell pellet.
15. Centrifuge the sample (3 h; 900,000 × *g*; 4°C). Remove as much liquid as possible from the membrane pellet.
16. Pack sample into an NMR rotor.

3.5 Creation of Reference Sample Not Containing ASR

Though we have found the above protocol to be effective in reducing the level of isotopic labeling in background proteins, these signals are not completely eliminated, and we still see a significant level of background proteins in our samples (Fig. 2b). Distinguishing cross-peaks of ASR from those originating from the background cannot be done with confidence in regions of high spectral overlap. In order to facilitate assignments, a control sample must be produced in which all steps of culture growth and inner membrane isolation are identical to those performed above, but

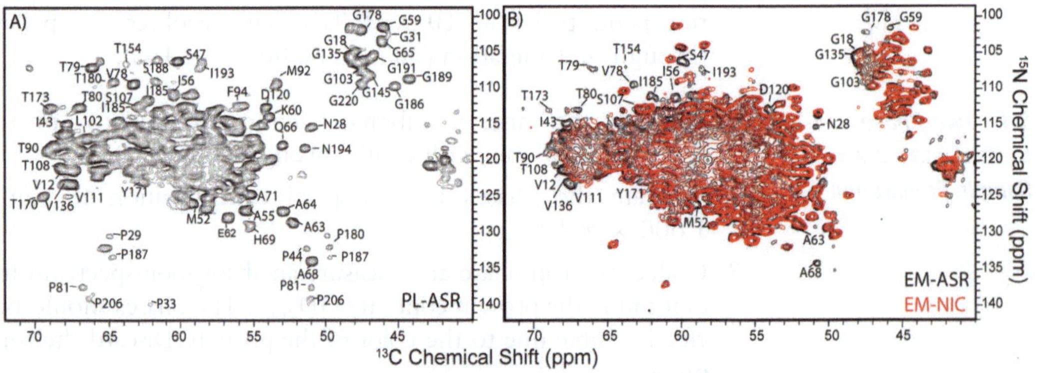

Fig. 2 NCA SSNMR spectra of (**a**) UCN PL-ASR, (**b**) rbUCN EM-ASR (*black*), and rbUCN EM-NIC (*red*). All peaks are labeled according to their assignments in PL-ASR. Overlap of the two EM spectra in (**b**) reveals many peaks in the EM-ASR sample which are not present in the EM-NIC sample, indicating that these peaks originate from ASR. Comparisons of the chemical shifts of these peaks to the PL-ASR sample in (**a**) allow for amino acid assignments to be transferred between the two samples. Reprinted with modifications from Biophysical Journal, 108, Ward ME et al., In Situ Structural Studies of *Anabaena* Sensory Rhodopsin in the *E. coli* Membrane, 1683–1696 (2015), with permission from Elsevier

without the induction of ASR expression (i.e., **Step 9** in Sect. 3.1 is omitted). We refer to this sample as EM-NIC (non-induced cells). Such a negative control sample contains only the background proteins but no ASR signals, and a direct comparison of the spectra collected on the two samples can be used to confirm the ASR cross-peak assignments, as demonstrated in Fig. 2.

3.6 Monitoring Protein Separation

1. Complete the entire protocol using natural abundance (NA) M9 minimal media at **Step 7** in Sect. 3.1. This will allow you to determine the optimal length of induction, as described in **Note 5**. The effectiveness of the removal of background proteins in this sample can be monitored through a number of methods, as shown in Fig. 3. SDS/PAGE can be used to monitor the removal of background proteins. While many stains, such as Sypro Ruby Protein stain (Fig. 3a), display all proteins in the sample, more selective stains, such as InVision His_6-tag stain (Fig. 3b), can be used to selectively stain proteins, like recombinant ASR, which contain a His_6-tag. Alternatively, a Western blot could provide the same information. If

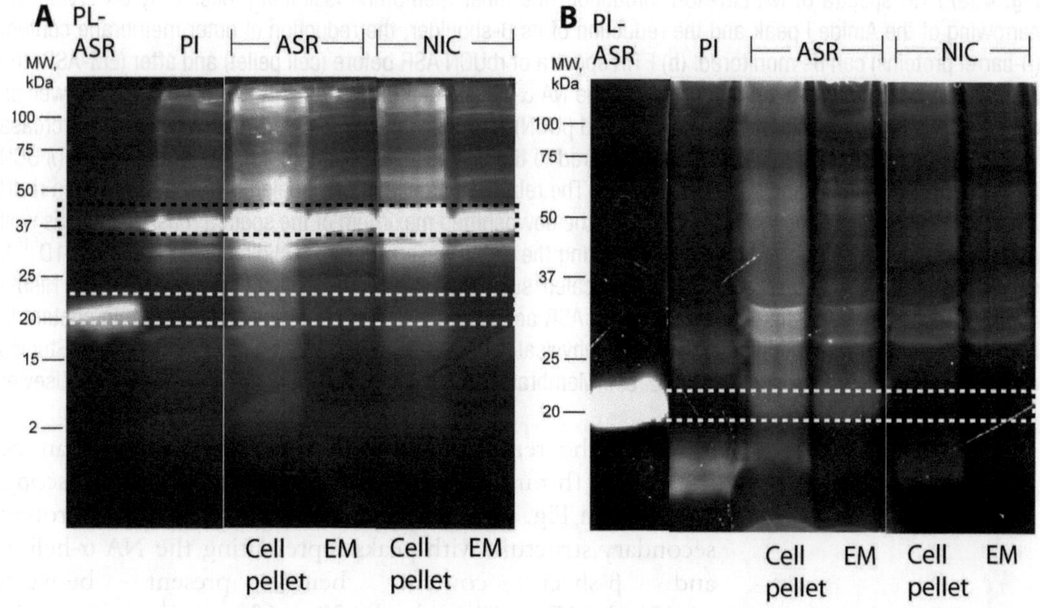

Fig. 3 (**a**) Sypro Ruby Protein and (**b**) InVision His_6-tag-stained SDS/PAGE analysis of *E. coli* membranes from BL21-Codon-plus-RIL cells containing the plasmid for ASR which have been induced (ASR) and not induced (NIC). Samples were taken pre-induction (PI) as well as before (cell pellet) and after EM purification. Membranes were boiled in SDS before electrophoresis. The positions of PL-ASR and a high abundance 37 kDa protein, which we have identified to be an outer membrane protein, are indicated with *white and black-dashed boxes*, respectively. Reprinted from Biophysical Journal, 108, Ward ME et al., In Situ Structural Studies of *Anabaena* Sensory Rhodopsin in the *E. coli* Membrane, 1683–1696 (2015), with permission from Elsevier

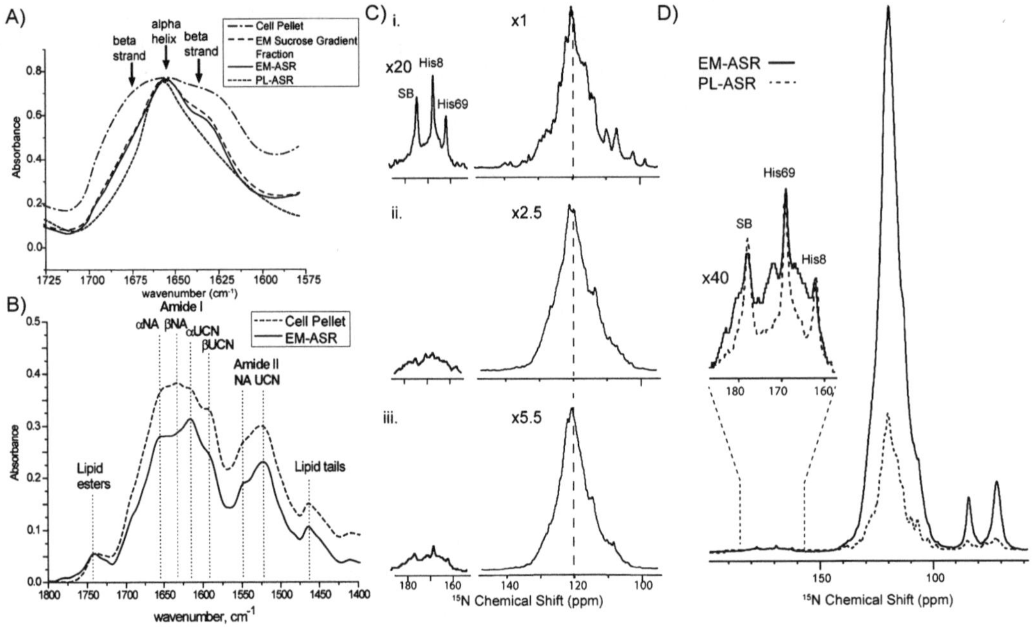

Fig. 4 (**a**) FTIR spectra of NA EM-ASR throughout the inner membrane isolation process. By observing the narrowing of the Amide I peak and the reduction of its β-shoulder, the reduction of outer membrane content (β-barrel proteins) can be monitored. (**b**) FTIR spectra of rbUCN ASR before (cell pellet) and after (EM-ASR) the IM isolation process. Peaks corresponding to the NA α-helical and β-strand (αNA and βNA) peaks as well as their isotopically labeled counterparts (αUCN and βUCN) are indicated with *dashed lines*. A noticeable increase in the UCN α-helical peak in EM-ASR as compared to the cell pellet is observed. (**c**) 1D ^{15}N spectra of (*i*) UCN PL-ASR, (*ii*) UN EM-ASR, and (*iii*) rbUN EM-ASR. The relative increase in UN-labeled α-helical proteins in rbUN EM-ASR is evident from the relative increase in the downshifted maximum of the spectra (*dashed line*), as well as the appearance of resolved peaks representing the Schiff base (SB) and histidine side chains. (**d**) 1D ^{15}N spectra of UCN PL-ASR and rbUN EM-ASR scaled such that the three resolved peaks (SB, His8, His69) approximately match in intensity, showing that ASR accounts for only ~1/4 of the labeled protein content in EM-ASR. Reprinted with modifications from Biophysical Journal, 108, Ward ME et al., In Situ Structural Studies of *Anabaena* Sensory Rhodopsin in the *E. coli* Membrane, 1683–1696 (2015), with permission from Elsevier

available, the removal of outer membrane proteins can be monitored through Fourier transform infrared spectroscopy, as shown in Fig. 4a. The Amide I band is sensitive to protein secondary structure with peaks representing the NA α-helical and β-sheet content being present between ~1,658–1,647 cm^{-1} and ~1,638–1,632 cm^{-1}, respectively. Narrowing of the Amide I peak around the NA α-helical frequencies indicates the removal of outer membrane β-barrel proteins.

2. Create a sample using ^{15}N M9 minimal media (UN) at **Step 7** in Sect. 3.1 (rbUN labeling) and sample where cells are grown without the media swap performed in Sect. 3.1 (UN labeling). Comparisons of the 1D ^{15}N spectra of these samples should indicate preferential incorporation of the ^{15}N labels into α-

helical proteins in the rbUN sample, as shown in Fig. 4c. In addition, comparisons between resolved resonances in the PL sample and the EM sample should allow for the quantification of the amount of recombinant protein in the EM sample. In the case of ASR, three resolved resonances are visible in the 160–180 ppm region and correspond to the Schiff base (SB) and side-chain histidine resonances, as seen in Fig. 4c. By scaling the spectra with respect to these resonances, the relative amount of UN ASR in the EM-ASR sample can be estimated, as seen in Fig. 4d.

3. Create a sample grown on ^{13}C-^{15}N M9 minimal media (UCN). The relative content of UCN-labeled α-helical proteins in the EM-ASR and EM-NIC samples can be further analyzed using FTIR through observation of the Amide I band, which is dominated by CO stretching and is therefore sensitive to ^{13}C labeling, and the Amide II band, which is sensitive to both ^{13}C and ^{15}N labeling. Isotopic labeling causes a downshift of the vibrational frequency of the amino acids containing the heavier atoms by ~20–50 cm^{-1}, and therefore the peaks representing the ^{13}C and/or ^{15}N-labeled proteins are downshifted from the frequencies observed in unlabeled proteins. As seen in Fig. 4b, even in the cell pellet spectrum the Amide II band is clearly split into two peaks representing isotopically labeled and natural abundance proteins. The Amide II peak representing isotopically labeled proteins shows a relative increase in the EM-ASR spectrum, indicating the removal of unlabeled proteins. Similarly, Amide I peaks representing the NA α-helical and β-sheet content are present between ~1,658–1,647 cm^{-1} and ~1,638–1,632 cm^{-1}, respectively, with their isotopically labeled counterparts being significantly downshifted by ~40 cm^{-1} from these wavelengths. Although all four peaks have similar intensities in the cell pellet spectrum, the increase in the relative content of isotopically labeled α-helical proteins is evident in the EM-ASR spectrum.

4 Notes

1. Before attempting this protocol, it is important to ensure that the *E. coli* strain used correctly expresses the recombinant protein in M9 minimal medium and incorporates it into the membrane. As the preparation of a purified, lipid-reconstituted protein sample (i.e., PL-ASR) is necessary for data analysis and for determining the resin binding affinity of the protein (*see* **Note 8**), it is recommended that this procedure be developed and implemented before proceeding to create a sample of the recombinant protein in the native *E. coli* membrane environment.

2. The LB medium used to initially grow the cells could be replaced with NA M9 medium. This may lead to a reduction in isotopic labeling of background proteins. Other methods for reducing the isotopic labeling of background proteins have been developed by the Baldus group and include the use of a mutant *E. coli* BL21Star(DE3) strain which lacks the two major OM proteins OmpF and OmpA [11] and the repression of endogenous protein expression through the addition of rifampicin [27] at the time of induction. Additionally, lowering the incubation temperature during protein expression has also been shown to result in reduced isotopic incorporation in background proteins [12].
3. As is recommended by the manufacturer, we find that a high concentration of NaCl is necessary to bind ASR to the Ni^{2+}–NTA–agarose resin. In cases when binding the affinity resin is possible without the presence of a high salt concentration, it is beneficial to remove the NaCl from the system as it reduces lipid aggregation.
4. It is advisable to test this protocol using NA M9 medium and then subsequently ^{15}N M9 medium and monitor the effectiveness of the selective isotopic incorporation into the recombinant protein, as well as the removal of background proteins, as described in Sect. 3.6 before proceeding to ^{13}C- and ^{15}N-labeled media.
5. The length of protein expression time should be optimized on a case-to-case basis. While longer expression periods are often necessary to increase the concentration of the recombinant membrane protein and to ensure uniform posttranslational modifications, they also lead to increased isotopic labeling of the background proteins and in some cases protein degradation and cell death. Optimization of the length of expression can be done by inducing a cell culture grown in NA media and taking 1 ml samples every hour or half hour after the induced culture begins to show protein color. After spinning down the media, the cell pellets can be visually compared for color intensity. If there is no noticeable increase in color intensity between subsequent tests, then continued incubation likely will not lead to an increase in recombinant protein content. Incubation can be continued if it is necessary to ensure that the resulting protein is uniformly modified; however, this could lead to an increase in isotopic labeling of background proteins or protein degradation and cell death which will be indicated by a brown color in the pellet.
6. It is important to ensure that the temperature of the cell lysate does not rise too high during sonication. Therefore, it is advisable to keep the cell lysate on ice during sonication, to sonicate in short pulses (~30 s) followed by an equal rest period for a

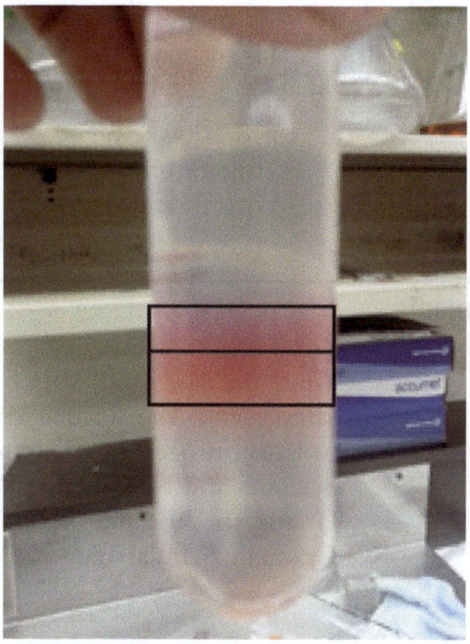

Fig. 5 Example of the smearing of the inner membrane fraction observed during the six-phase sucrose gradient. The inner membrane fraction sits on top of the 40% sucrose layer and is separated into a light and a heavy fraction which are distinguishable based on the differing color. FTIR analysis of the two fractions indicates that the lighter fraction contains significantly less β-barrel proteins than the heavier fraction, and therefore the two fractions were kept separate, with only the lighter fraction being used in the final NMR sample

maximum pulse time of ~150 s before allowing the sample to rest for approximately the total sonication cycle time (~300 s). The duration and power of sonication necessarily will vary between *E. coli* strains and equipment used.

7. During this step we saw that the IM fraction was quite smeared throughout the 35% sucrose layer (Fig. 5) with the color clearly changing from the bottom to the top. Careful collection of these two layers and FTIR analysis reveals that the β content of the lighter layer is significantly lower than that of the heavier layer. Therefore, if similar observations are made, it is recommended to separately collect the fractions and analyze their content before continuing the protocol.

8. To determine the volume of two-phase system to use, you can first determine the binding affinity for your protein to the resin by solubilizing and quantifying your protein and testing resin binding. Based on this result and an estimation of your yield, the amount of resin necessary to bind your protein can be estimated. For ASR, we used approximately 10 ml of resin per liter of culture. In addition, it is often easier to resuspend the IM fraction in the appropriate volume of 10 mM Tris before

adding the remaining components of the two-phase system to this suspension.

9. This step can be difficult as the membranes will often sit on top of the lower phase. Often it is easiest to put the tube in the refrigerator and wait for the resin to fall out without centrifugation.

Acknowledgments

We thank Ms. Emily Ritz for providing us with the UCN PL-ASR sample and Ms. Rachel Munro for her preliminary work on the isolation procedure. This research was supported by the Natural Sciences and Engineering Research Council of Canada (Discovery Grants to V.L. and L.S.B.). MEW is a recipient of the Ontario Graduate Scholarship.

References

1. Grigorieff N, Ceska TA, Downing KH et al (1996) Electron-crystallographic refinement of the structure of bacteriorhodopsin. J Mol Biol 259:393–421. doi:10.1006/jmbi.1996.0328
2. Unger VM, Kumar NM, Gilula NB, Yeager M (1997) Projection structure of a gap junction membrane channel at 7 A resolution. Nat Struct Biol 4:39–43
3. Tang M, Comellas G, Rienstra CM (2013) Advanced solid-state NMR approaches for structure determination of membrane proteins and amyloid fibrils. Acc Chem Res 46:2080–2088. doi:10.1021/ar4000168
4. Lewandowski JR (2013) Advances in solid-state relaxation methodology for probing site-specific protein dynamics. Acc Chem Res 46:2018–2027. doi:10.1021/ar300334g
5. Murray DT, Das N, Cross TA (2013) Solid state NMR strategy for characterizing native membrane protein structures. Acc Chem Res 46:2172–2181. doi:10.1021/ar3003442
6. Opella SJ (2013) Structure determination of membrane proteins by nuclear magnetic resonance spectroscopy. Annu Rev Anal Chem 6:305–328. doi:10.1146/annurev-anchem-062012-092631
7. Ladizhansky V (2014) Recent advances in magic-angle spinning solid-state NMR of proteins. Isr J Chem 54:86–103. doi:10.1002/ijch.201300096
8. Wang S, Ladizhansky V (2014) Recent advances in magic angle spinning solid state NMR of membrane proteins. Prog Nucl Magn Reson Spectrosc 82:1–26. doi:10.1016/j.pnmrs.2014.07.001
9. Polenova T, Gupta R, Goldbourt A (2015) Magic angle spinning NMR spectroscopy: a versatile technique for structural and dynamic analysis of solid-phase systems. Anal Chem 87:5458–5469. doi:10.1021/ac504288u
10. Fu R, Wang X, Li C et al (2011) In situ structural characterization of a recombinant protein in native *Escherichia coli* membranes with solid-state magic-angle-spinning NMR. J Am Chem Soc 133:12370–12373. doi:10.1021/ja204062v
11. Renault M, Tommassen-van Boxtel R, Bos MP et al (2012) Cellular solid-state nuclear magnetic resonance spectroscopy. Proc Natl Acad Sci U S A 109:4863–4868. doi:10.1073/pnas.1116478109
12. Miao Y, Qin H, Fu R et al (2012) M2 proton channel structural validation from full-length protein samples in synthetic bilayers and *E. coli* membranes. Angew Chem Int Ed Engl 51:8383–8386. doi:10.1002/anie.201204666.M2
13. Renault M, Pawsey S, Bos MP et al (2012) Solid-state NMR spectroscopy on cellular preparations enhanced by dynamic nuclear polarization. Angew Chem Int Ed Engl 51:2998–3001. doi:10.1002/anie.201105984
14. Jacso T, Franks WT, Rose H et al (2012) Characterization of membrane proteins in isolated native cellular membranes by dynamic nuclear polarization solid-state NMR spectroscopy without purification and reconstitution.

Angew Chem Int Ed Engl 51:432–435. doi:10.1002/anie.201104987

15. Yamamoto K, Caporini MA, Im S-C et al (2015) Cellular solid-state NMR investigation of a membrane protein using dynamic nuclear polarization. Biochim Biophys Acta 1848:342–349. doi:10.1016/j.bbamem.2014.07.008
16. Sakakibara D, Sasaki A, Ikeya T et al (2009) Protein structure determination in living cells by in-cell NMR spectroscopy. Nature 458:102–105. doi:10.1038/nature07814
17. Leis A, Rockel B, Andrees L, Baumeister W (2009) Visualizing cells at the nanoscale. Trends Biochem Sci 34:60–70. doi:10.1016/j.tibs.2008.10.011
18. Matwiyoff NA, Needham TE (1972) Carbon-13 NMR spectroscopy of red blood cell suspensions. Biochem Biophys Res Commun 49:1158–1164. doi:10.1007/s13398-014-0173-7.2
19. Selenko P, Frueh DP, Elsaesser SJ et al (2008) In situ observation of protein phosphorylation by high-resolution NMR spectroscopy. Nat Struct Mol Biol 15:321–329. doi:10.1038/nsmb.1395
20. Serber Z, Keatinge-Clay AT, Ledwidge R et al (2001) High-resolution macromolecular NMR spectroscopy inside living cells. J Am Chem Soc 123:2446–2447. doi:10.1021/ja0057528
21. Kaplan M, Cukkemane A, van Zundert GCP et al (2015) Probing a cell-embedded megadalton protein complex by DNP-supported solid-state NMR. Nat Methods 12:5–9. doi:10.1038/nmeth.3406
22. Jung KH, Trivedi VD, Spudich JL (2003) Demonstration of a sensory rhodopsin in eubacteria. Mol Microbiol 47:1513–1522. doi:10.1046/j.1365-2958.2003.03395.x
23. Shi L, Kawamura I, Jung K-H et al (2011) Conformation of a seven-helical transmembrane photosensor in the lipid environment. Angew Chem Int Ed 50:1302–1305. doi:10.1002/anie.201004422
24. Wang S, Shi L, Okitsu T et al (2013) Solid-state NMR ^{13}C and 15N resonance assignments of a seven-transmembrane helical protein *Anabaena* Sensory Rhodopsin. Biomol NMR Assign 7:253–256. doi:10.1007/s12104-012-9421-y
25. Wang S, Munro RA, Shi L et al (2013) Solid-state NMR spectroscopy structure determination of a lipid-embedded heptahelical membrane protein. Nat Methods 10:1007–1012. doi:10.1038/nmeth.2635
26. Marley J, Lu M, Bracken C (2001) A method for efficient isotopic labeling of recombinant proteins. J Biomol NMR 20:71–75
27. Baker LA, Daniëls M, van der Cruijsen EAW et al (2015) Efficient cellular solid-state NMR of membrane proteins by targeted protein labeling. J Biomol NMR 62:199–208. doi:10.1007/s10858-015-9936-5
28. Wagner S, Baars L, Ytterberg AJ et al (2007) Consequences of membrane protein overexpression in *Escherichia coli*. Mol Cell Proteomics 6:1527–1550. doi:10.1074/mcp.M600431-MCP200
29. Barinaga-Rementeria Ramírez I, Abedinpour P, Jergil B (2004) Purification of caveolae by affinity two-phase partitioning using biotinylated antibodies and NeutrAvidin-dextran. Anal Biochem 331:17–26. doi:10.1016/j.ab.2004.04.044
30. Everberg H, Clough J, Henderson P et al (2006) Isolation of *Escherichia coli* inner membranes by metal affinity two-phase partitioning. J Chromatogr A 1118:244–252. doi:10.1016/j.chroma.2006.03.123
31. Ward ME, Wang S, Munro R et al (2015) In situ structural studies of *Anabaena* sensory rhodopsin in the *E. coli* membrane. Biophys J 108:1683–1696. doi:10.1016/j.bpj.2015.02.018

INDEX

Arun K. Shukla (ed.), *Chemical and Synthetic Approaches in Membrane Biology*, Springer Protocols Handbooks, DOI 10.1007/978-1-4939-6836-7, © Springer Science+Business Media LLC 2017

R

S

T

MIX
Papier aus verantwortungsvollen Quellen
Paper from responsible sources
FSC® C105338

If you have any concerns about our products,
you can contact us on
ProductSafety@springernature.com

In case Publisher is established outside the EU,
the EU authorized representative is:
Springer Nature Customer Service Center GmbH
Europaplatz 3, 69115 Heidelberg, Germany

Printed by Libri Plureos GmbH
in Hamburg, Germany